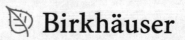

Annals of the International Society of Dynamic Games

Volume 15

More information about this series at http://www.springer.com/series/4919

Joseph Apaloo • Bruno Viscolani

Editors

Advances in Dynamic and Mean Field Games

Theory, Applications, and Numerical Methods

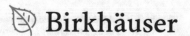 **Birkhäuser**

Editors
Joseph Apaloo
Department of Mathematics, Statistics
and Computer Science
St. Francis Xavier University
Antigonish, NS, Canada

Bruno Viscolani
Department of Mathematics
University of Padova
Padova, Italy

ISSN 2474-0179   ISSN 2474-0187   (electronic)
Annals of the International Society of Dynamic Games
ISBN 978-3-030-09975-6   ISBN 978-3-319-70619-1   (eBook)
https://doi.org/10.1007/978-3-319-70619-1

Mathematics Subject Classification: 91A23, 91A25, 91A15, 91A22, 91A24, 49N90

Printed on acid-free paper

This book is published under the trade name Birkhäuser, www.birkhauser-science.com
The registered company is Springer International Publishing AG
The registered company address is: Gewerbestrasse 11, 6330 Cham, Switzerland

Preface

This fifteenth volume of the *Annals of the International Society of Dynamic Games* (ISDG) is composed of some of the papers presented at the International Symposium on Dynamic Games and Applications which took place in Urbino, Italy, in July 2016.

The plenary lectures of the symposium were given by Roberto Cellini, Catania, Italy, and by Hanna Kokko, Zurich, Switzerland, both of them contributors of this volume. One tutorial lecture was given by Pierre Cardaliaguet, who is also a contributor of this volume. The 2016 Isaacs Awards laureates, for "outstanding contribution to the theory and applications of dynamic games," were Martino Bardi, Padova, Italy, and Ross Cressman, Waterloo, Ontario, Canada.

The symposium was cosponsored by the Department of Economics, Society and Politics (DESP), University of Urbino Carlo Bo, Italy; the Group for Research in Decision Analysis (GERAD) and the Chair in Game Theory and Management, HEC Montréal, Canada; and the Italian Association for Mathematics Applied to Social and Economic Sciences (AMASES).

This contributed book covers a variety of topics from theory to applications of dynamic games. We want to provide a unitary exhibition of the main themes treated in the symposium and that are, in fact, objects of active research today.

All chapters submitted for consideration for publication in the Annals were peer-reviewed according to the standards of international journals in game theory and applications. Each chapter was reviewed by two experts who were chosen by an associate editor, as in the tradition of the ISDG Annals. The volume contains 16 chapters which are organized into 4 parts.

Part I is devoted to *mean field games (MFGs) and applications* and contains six chapters. In the chapter "Non-memoryless Pedestrian Flow in a Crowded Environment with Target Sets," *Fabio Bagagiolo and Raffaele Pesenti* propose an interesting model of masses of tourists visiting multiple hotspots in a historic city. Venice is considered as a case study. Using an approach based on mean field game theory, they examine the possibility of directing these masses through different routes to minimize congestion.

In the chapter "Limit Game Models for Climate Change Negotiations," *Olivier Bahn, Alain Haurie, and Roland Malhamé* propose a mean field game approach to the problem of limiting the carbon emissions between groups of countries to keep the climate as stable as possible. They present first a deterministic game where m coalitions of countries share a global emission budget, and they then consider a stochastic game framework. Both approaches take into account the limit as the number of players increases. Thus, the mean field game theory is used to characterize the stochastic approach.

In the chapter "A Segregation Problem in Multi-Population Mean Field Games," *Pierre Cardaliaguet, Alessio Porretta, and Daniela Tonon* study a two-population mean field game in which the coupling between populations becomes increasingly singular. In the case of a quadratic Hamiltonian, they show that the limit system corresponds to a partition of the space into two components in which the players have to solve an optimal control problem with state constraints and mean field interactions.

In the chapter "Evolutionary Game of Coalition Building Under External Pressure," *Alekos Cecchin and Vassili N. Kolokoltsov* address coalition-building Markovian models for N players and their convergence as N approaches infinity. The paper begins by describing the fragmentation-coagulation model. Next, the authors discuss the convergence of the value function of a controlled problem for N players to a mean field-like control problem. They also obtain the almost optimality of the trajectories of the mean field control problem for the problem with N players.

In the chapter "The Execution Problem in Finance with Major and Minor Traders: A Mean Field Game Formulation," *Dena Firoozi and Peter E. Caines* extend the theory of partially observed mean field games to indefinite linear-quadratic-Gaussian (LQG) problems and apply it to the interesting problem of optimal execution in finance. An institutional investor aims to liquidate the amount of shares and has partial observations of its own state. Two populations of high-frequency traders wish to liquidate or acquire numbers of shares within a specific time, and each one of them has partial observations of its own state and of the institutional investor's state. Each agent aims to maximize its own wealth and to avoid the occurrence of large execution prices, large rates of trading, and large trading accelerations. The authors establish the existence of ε-Nash equilibria together with the individual agents' trading strategies yielding them.

In the chapter "Mean Field Limits Through Local Interactions," *Suhail M. Shah and Vivek S. Borkar* consider stochastic processes on directed and connected graphs with nearest neighbor affine interactions at a faster time scale than the underlying dynamics of the systems at the nodes. In the limit, as the time scale separation diverges and as the network grows to an infinite graph, mean field dynamics are established for the entire, infinite system. A key feature of the model is the fixed finite-degree connections at each node, which makes the model intrinsically different from the standard mean field game (MFG) model. In the latter, the node degree is implicitly infinite in the limit since each node receives state information from all the others.

Part II is devoted to *dynamic games and applications* and contains four chapters. In the chapter "Differential Games in Health-Care Markets: Models of Quality Competition with Fixed Prices," *Kurt R. Brekke, Roberto Cellini, Luigi Siciliani, and Odd Rune Straume* propose a review of the models investigating quality competition in health-care markets under price regulation, taking a differential game approach. The different scenarios, which are connected in a unified modeling framework, deal with specific characteristics of health-care markets, including social welfare issues. The authors derive and compare both open-loop and feedback equilibrium strategies and address potential policy implications from these analyses.

In the chapter "Open-Loop Nash Equilibria for Dynamic Games Involving Volterra Integral Equations," *Dean A. Carlson* establishes conditions which ensure the existence of an open-loop equilibrium in a class of finite horizon differential games with the state equation being a Volterra integral equation with infinite delay. The result is obtained by a fixed point argument using normalized equilibria and relies on convexity and seminormality conditions. The author provides an applied example considering a competitive economic model originally appearing in the works of C. F. Roos in 1925.

In the chapter "A Discrete Model of Conformance Quality and Advertising in Supply Chains," *Pietro De Giovanni and Fabio Tramontana* analyze a game in which a retailer adjusts their advertising efforts according to the type of defects that consumers experience, whereas the type of defects depends on a manufacturer's conformance quality investments. If neither of the players has a complete knowledge of the required information to compute their optimal decisions, then the firms adopt the heuristic of adjusting their advertising and quality according to the respective marginal profits. The paper concludes by analyzing the long-run behavior and stability of the solutions.

In the chapter "Sexual Reproduction as Bet-Hedging," *Xiang-Yi Li, Jussi Lehtonen, and Hanna Kokko* give a clear exposition of the evolutionary adaptiveness of sexual reproduction as a strategy which is robust to changes in the environment. In environments that change over time, it is optimal to maximize the geometric mean of the number of offspring, rather than the arithmetic mean, and this is often achieved by a strategy being quite successful in a range of environments, rather than very successful in a particular type of environment.

Part III is devoted to *stochastic and pursuit-evasion games* and contains four chapters. In the chapter "On Exact Construction of Solvability Set for Differential Games with Simple Motion and Non-convex Terminal Set," *Liudmila Kamneva and Valerii Patsko* consider two-player zero-sum differential games with simple motion in the plane and suggest a way of constructing the solvability set exactly in the case of non-convex polygonal terminal set and polygonal constraints of the players' controls. In fact, an explicit formula describing the solvability set is well known from B. N. Pshenichnyy and M. I. Sagaydak in the early 1970s, but only for the case of a convex terminal set.

In the chapter "Effects of Players' Random Participation to the Stability in LQ Games," *Ioannis Kordonis and George P. Papavasilopoulos* consider a linear-quadratic (LQ) game with randomly arriving players, staying in the game for a random period of time. Motivations for such a model are interaction patterns which appear among customers of a bank, in competitive markets and intergenerational cooperation and competition. They characterize the Nash equilibria of the game by a set of coupled Riccati-type equations and prove the existence of an equilibrium. They then consider the game in the limit as the number of players becomes large, assuming a partially Kantian behavior, and discuss the system stability.

In the chapter "Interval Computing of the Viability Kernel with Application to Robotic Collision Avoidance," *Stéphane Le Ménec* considers the problem of an approximate computation of a guaranteed capture zone and evasion zone in the context of pursuit-evasion differential games. Concepts from viability theory are used to rewrite the considered game dynamics as a differential inclusion, where the flows represent the different possible strategies of the players. Tools from interval analysis are used to approximate the viability kernel and capture basin. An interval contractor operator is used to provide viability kernel and capture basin algorithms. As an application, the author considers collision avoidance between two noncooperative ground mobile robots.

In the chapter "On Linear-Quadratic Gaussian Dynamic Games," *Meir Pachter* deals with linear-quadratic-Gaussian-dynamic games (LQ-G-DG). After recalling the complete solution of the perfect information LQ zero-sum dynamic game using Riccati equations, he considers the partial information case, where players' information is modeled through probability density functions. The goal is to obtain a Nash equilibrium in delayed commitment strategies such that the optimal solution is time consistent/subgame perfect.

Part IV is devoted to *computational methods for dynamic games* and contains two chapters. In the chapter "Viability Approach to Aircraft Control in Wind Shear Conditions," *Nikolai Botkin, Johannes Diepolder, Varvara Turova, Matthias Bittner, and Florian Holzapfel* address the analysis of aircraft control capabilities in the presence of wind shears to determine numerically the safe area of an aircraft. Their approach is a combination of Krassovski methods and viability theory.

In the chapter "Modeling Autoregulation of Cerebral Blood Flow Using Viability Approach," *Varvara Turova, Nikolai Botkin, Ana Alves-Pinto, Tobias Blumenstein, Esther Rieger-Fackeldey, and Renée Lampe* set up a game-theoretical model to analyze the self-regulation of cerebral blood flow. Sudden changes in partial pressure of oxygen and carbon dioxide in the arterial blood are considered as disturbances, and medicines dilating or restricting blood vessels are considered as controls. The authors use a viability approach for this game framework and propose a numerical approximation of the viability kernel.

Antigonish, NS, Canada   Joseph Apaloo
Padova, Italy   Bruno Viscolani
August 2017

Acknowledgments

We want to thank the associate editors and all the referees who made it possible to obtain this volume with their competent advice. We received important help from the editorial staff at Birkhäuser and especially Christopher Tominich, and we extend our thanks for their assistance throughout the editing process. We thank the Urbino symposium organizing committee: Gian Italo Bischi, Domenico De Giovanni, Laura Gardini, and Fabio Lamantia. We would moreover like to thank GERAD for its continuing financial and administrative support to ISDG activities and all the contributors to this volume of the Annals.

Contents

Part I Mean Field Games and Applications

**Non-memoryless Pedestrian Flow in a Crowded Environment
with Target Sets** 3
Fabio Bagagiolo and Raffaele Pesenti

Limit Game Models for Climate Change Negotiations 27
Olivier Bahn, Alain Haurie, and Roland Malhamé

A Segregation Problem in Multi-Population Mean Field Games 49
Pierre Cardaliaguet, Alessio Porretta, and Daniela Tonon

Evolutionary Game of Coalition Building Under External Pressure 71
Alekos Cecchin and Vassili N. Kolokoltsov

**The Execution Problem in Finance with Major and Minor Traders:
A Mean Field Game Formulation** 107
Dena Firoozi and Peter E. Caines

Mean Field Limits Through Local Interactions 131
Suhail M. Shah and Vivek S. Borkar

Part II Dynamic Games and Applications

**Differential Games in Health-Care Markets: Models of Quality
Competition with Fixed Prices** 145
Kurt R. Brekke, Roberto Cellini, Luigi Siciliani, and Odd Rune Straume

**Open-Loop Nash Equilibria for Dynamic Games Involving Volterra
Integral Equations** 169
Dean A. Carlson

**A Discrete Model of Conformance Quality and Advertising
in Supply Chains** 199
Pietro De Giovanni and Fabio Tramontana

Sexual Reproduction as Bet-Hedging 217
Xiang-Yi Li, Jussi Lehtonen, and Hanna Kokko

Part III   Stochastic and Pursuit-Evasion Games

**On Exact Construction of Solvability Set for Differential Games
with Simple Motion and Non-convex Terminal Set** 237
Liudmila Kamneva and Valerii Patsko

Effects of Players' Random Participation to the Stability in LQ Games .. 259
Ioannis Kordonis and George P. Papavassilopoulos

**Interval Computing of the Viability Kernel with Application
to Robotic Collision Avoidance** 279
Stéphane Le Ménec

On Linear-Quadratic Gaussian Dynamic Games 301
Meir Pachter

Part IV   Computational Methods for Dynamic Games

Viability Approach to Aircraft Control in Wind Shear Conditions 325
Nikolai Botkin, Johannes Diepolder, Varvara Turova, Matthias Bittner,
and Florian Holzapfel

**Modeling Autoregulation of Cerebral Blood Flow
Using Viability Approach** 345
Varvara Turova, Nikolai Botkin, Ana Alves-Pinto,
Tobias Blumenstein, Esther Rieger-Fackeldey, and Renée Lampe

Associate Editors

Contributors

Ana Alves-Pinto Orthopedic Department of the Clinic 'rechts der Isar' of the Technical University of Munich, Munich, Germany

Fabio Bagagiolo Department of Mathematics, Università di Trento, Povo-Trento, Italy

Olivier Bahn GERAD and Department of Decision Sciences, HEC Montréal, Montreal, QC, Canada

Matthias Bittner Institute of Flight System Dynamics, Technische Universität München, Garching near Munich, Germany

Tobias Blumenstein Orthopedic Department of the Clinic 'rechts der Isar' of the Technical University of Munich, Munich, Germany

Vivek S. Borkar Department of Electrical Engineering, Indian Institute of Technology Bombay, Powai, Mumbai, India

Nikolai Botkin Department of Mathematics, Technische Universität München, Garching near Munich, Germany

Kurt R. Brekke Department of Economics, Norwegian School of Economics, Bergen, Norway

Peter E. Caines The Centre for Intelligent Machines (CIM) and the Department of Electrical and Computer Engineering (ECE), McGill University, Montreal, QC, Canada

Pierre Cardaliaguet Université Paris-Dauphine, PSL Research University, CNRS, Ceremade, Paris, France

Dean A. Carlson Mathematical Reviews, American Mathematical Society, Ann Arbor, MI, USA

Alekos Cecchin Department of Mathematics, University of Padua, Padova, Italy

Roberto Cellini Department of Economics, University of Catania, Corso Italia, Catania, Italy

Pietro De Giovanni Department of Operations Management, ESSEC Business School, Paris, France

Johannes Diepolder Institute of Flight System Dynamics, Technische Universität München, Garching near Munich, Germany

Dena Firoozi The Centre for Intelligent Machines (CIM) and the Department of Electrical and Computer Engineering (ECE), McGill University, Montreal, QC, Canada

Alain Haurie ORDECSYS, Chêne-Bougeries, Switzerland; University of Geneva, Geneva, Switzerland; GERAD, HEC Montréal, Montreal, QC, Canada

Florian Holzapfel Institute of Flight System Dynamics, Technische Universität München, Garching near Munich, Germany

Liudmila Kamneva Institute of Mathematics and Mechanics, Ekaterinburg, Russia

Hanna Kokko Department of Evolutionary Biology and Environmental Studies, University of Zurich, Zurich, Switzerland

Vassili N. Kolokoltsov Department of Statistics, University of Warwick, Coventry, UK

Ioannis Kordonis CentraleSupélec, Cesson-Sévigné, France

Renée Lampe Orthopedic Department of the Clinic 'rechts der Isar' of the Technical University of Munich, Munich, Germany

Stéphane Le Ménec Airbus/MBDA, Le Plessis-Robinson, France

Jussi Lehtonen Evolution and Ecology Research Centre, School of Biological, Earth and Environmental Sciences, University of New South Wales, Sydney, NSW, Australia

Xiang-Yi Li Department of Evolutionary Biology and Environmental Studies, University of Zurich, Zurich, Switzerland

Roland Malhamé GERAD and Department of Electrical Engineering, POLY Montréal, Montreal, QC, Canada

Meir Pachter Department of Electrical and Computer Engineering, Air Force Institute of Technology, Wright-Patterson A.F.B., OH, USA

George P. Papavassilopoulos School of Electrical and Computer Engineering, National Technical University of Athens, Zografou, Athens, Greece

Valerii Patsko Institute of Mathematics and Mechanics, Ekaterinburg, Russia

Raffaele Pesenti Department of Management, Università Ca' Foscari Venezia, Venezia, Italy

Alessio Porretta Dipartimento di Matematica, Università di Roma "Tor Vergata" Roma, Italy

Esther Rieger-Fackeldey Frauenklinik und Poliklinik of the Clinic 'rechts der Isar' of the Technical University of Munich, Munich, Germany

Suhail M. Shah Department of Electrical Engineering, Indian Institute of Technology Bombay, Powai, Mumbai, India

Luigi Siciliani Department of Economics and Related Studies, University of York, Heslington, York, UK

Odd Rune Straume Department of Economics/NIPE, EEG, University of Minho, Braga, Portugal; Department of Economics, University of Bergen, Bergen, Norway

Daniela Tonon Université Paris-Dauphine, PSL Research University, CNRS, Ceremade, Paris, France

Fabio Tramontana Department of Mathematical Sciences, Mathematical Finance and Econometrics, Catholic University of Milan, Milano, Italy

Varvara Turova Department of Mathematics, Technische Universität München, Garching near Munich, Germany

Part I
Mean Field Games and Applications

Non-memoryless Pedestrian Flow in a Crowded Environment with Target Sets

Fabio Bagagiolo and Raffaele Pesenti

Abstract This work deals with the problem of managing the excursionist flow in historical cities. Venice is considered as a case study. There, in high season, thousands of excursionists arrive by train in the morning, spend the day visiting different sites, reach again the train station in late afternoon, and leave. With the idea of avoiding congestion by directing excursionists along different routes, a mean field model is introduced. Network/switching is used to describe the excursionists costs as a function of their position taking into consideration whether they have already visited a site or not, i.e., allowing excursionists to have memory of the past when making decisions. In particular, we analyze the model in the framework of Hamilton-Jacobi/transport equations, as it is standard in mean field games theory. Finally, to provide a starting datum for iterative solution algorithms, a second model is introduced in the framework of mathematical programming. For this second approach, some numerical experiments are presented.

1 Introduction

1.1 Motivation and Aim of the Work

This work is motivated by the authors' observation of excursionists' movement in the city of Venice, Italy, during high season. In the morning, many tens of thousands of excursionists arrive by train at the historical center of Venice. Most of them walk to Saint Mark's Square and other attractions following very few routes. That

F. Bagagiolo (✉)
Department of Mathematics, Università di Trento, Via Sommarive 14, 38123 Povo-Trento, Italy
e-mail: fabio.bagagiolo@unitn.it

R. Pesenti
Department of Management, Università Ca' Foscari Venezia, Fondamenta San Giobbe 873, Cannaregio, 30121 Venezia, Italy
e-mail: pesenti@unive.it

© Springer International Publishing AG 2017
J. Apaloo, B. Viscolani (eds.), *Advances in Dynamic and Mean Field Games*,
Annals of the International Society of Dynamic Games 15,
https://doi.org/10.1007/978-3-319-70619-1_1

3

movement in the morning, and the opposite in late afternoon, causes congestion along the narrow alleys of the city, at the entrances of the historical monuments and museums, as well as at restaurants and at other tourist facilities.

Keeping this example in mind, we introduce a mean field game that describes the excursionists' movement through the pedestrian zone of a historical city center. Excursionists minimize a cost penalizing congestion along the chosen path, as well as their miss of city attractions. In fact, this type of analysis may support city administrators in deciding which real-time information is best to be provided to the public, and in which places, in order to manage efficiently the excursionists' flow through the city.

1.2 Methodology and Literature Analysis

We undertake the analysis of the problem described above with two coordinate approaches. Firstly, we frame it into the recent mean field game theory and study the properties of the system. Secondly, to provide an initial solution for a possible iterative fixed-point solution algorithm, we introduce an associated model in the framework of mathematical programming. Nonetheless this procedure is not straightforward. The fact that excursionists/agents have more than one potential target to reach, namely, the different attractions of the city (with possible different degrees of discomfort associated to failure of visit), represents one of the main criticalities that we need to overcome. More specifically, except for the initial time, at all subsequent instants, different excursionists may occupy the same position but with a different status, depending on which attractions they have already visited: for example (in the case of two attractions P_1 and P_2), some agents have already visited P_1 only, some P_2 only, some both, and some none.

It is readily understood that information about agents' past history need to be encoded into our model. Otherwise the usual framing into mean field games (when one writes a unique system of Hamilton-Jacobi-Bellman and transport/conservation equations) proves insufficient to describe the problem, as the presence of more than one target leads to failure of the dynamic programming principle (briefly, DPP): the terminal part of an optimal trajectory is not necessarily optimal, as it does not enclose information about which targets have already been touched.

This problem with more than one target is reminiscent of the traveling salesman problem, for which several versions of suitable dynamic programming algorithms are intensively studied via optimization techniques. In that framework, we cite a brief contribution by R. Bellman [7], where a possible dynamic programming approach to the problem is proposed. There, some external variables (the labels of targets still to be reached) are inserted to recover DPP, and calculation of the numbers of subproblems is performed, but no Hamilton-Jacobi equation is written down. Indeed, to the best of the authors' knowledge, the first attempt to deduce a possible Hamilton-Jacobi equation for an optimal control, with several targets, is performed in Bagagiolo-Benetton [4] where a hysteresis feature is used to model the evolution of the external memory variables. Differently, in this paper we use some switching binary variables w_i, one per each target. Specifically, a variable w_i assumes value 1 if the corresponding target P_i has not been visited and 0 otherwise. Then, the state variable of our model is a vector $(\theta, w_1, \ldots, w_n)$ subject to a hybrid evolution (continuous/discrete), where θ is the one-dimensional continuous component of the state that describes the excursionist's position along his/her path.

The main results of the paper are the proof of existence, under suitable hypotheses, of a generalized (via convexification) mean field equilibrium for such switching mean field game.

Regarding the first approach in this work, the state evolution described by our model can be seen as a delayed-relay type, which is a typical example of thermostatic hysteresis. We recall that, in this context, optimal control problems via DPP and Hamilton-Jacobi-Bellman equations in the viscosity sense are studied in Bagagiolo [2] (see also Bagagiolo-Danieli [5]). In the framework of mean field games, a typical example of that situation is the ON/OFF status of a population of heating devices. The members of this population change in time their status, depending on some other parameters, and hence the model experiences a continuous transfer between the two subpopulations ON and OFF. The mean field game approach to this kind of situation is still to be investigated in depth (a first attempt of inserting a thermostatic behavior in a mean field game situation can be found in Bagagiolo-Bauso [3]), and the model studied here can be also seen as a possible step in that direction.

We recall that the mean field game theory goes back to the seminal work by Lasry-Lions [18] (see also Huang-Caines-Malhamé [15]). In the present work, we particularly refer to the unpublished note by Cardaliaguet [10]. Some applications of a mathematical model to the study of dynamics on networks and/or pedestrian movement can be found, for example, in Camilli-Carlini-Marchi [8], in Cristiani-Priuli-Tosin [12], and in Camilli-De Maio-Tosin [9]. We also cite Bardi-Cirant [1] where a mean field game model with several populations is studied, but, differently from our model, the agents cannot change status during their evolution. For a comprehensive account to the viscosity solutions theory for Hamilton-Jacobi equations, we refer to Bardi-Capuzzo Dolcetta [6].

The second approach described in this work is of mathematical programming type. It casts the memory variables w_i inside the possible paths of a graph describing the trajectories of the excursionists, changing somehow the nature of the variables from dynamic to static. The technique can be framed within the dynamic traffic assignment (DTA) literature, which describes the movement of vehicles or pedestrian in a network, given an origin destination demand that changes with times (see, e.g., [20] for a review, [21] for a taxonomy, [17] for primers).

DTA considers two solution approaches: equilibrium and nonequilibrium approaches. In the first case, Wardrop's user equilibrium principle defines the equilibrium conditions: "In a model network with many possible routes for each O-D pair, all used routes have equal and lowest generalized cost. No user may lower their generalized cost by unilaterally changing to a different route." In case of DTA, the condition must hold at each departure time t and refers to the experienced generalized cost. However, agents that travel along the same route with different departure times may experience different generalized costs. More generally, in the equilibrium approach agents are assumed to have a complete knowledge on the state of the network in terms of generalized costs, as they are supposed to be able to assess the costs of alternative routes.

In the nonequilibrium approach, agents are assumed to make their decisions and implement route adjustments on the basis of local information that become available during their movement in the network.

Another branch of literature someway related to the problem of interest is that of pedestrian flow modeling, where time evolution of pedestrian flows is described within 1-manifold (e.g., a street), 2-manifolds (e.g., a surface), and possibly in presence of obstacles, corners, or converging flows (see, e.g., [13, 14, 16] and in particular [22] for pedestrian evacuation). However, a more detailed analysis of the field is beyond the scope of this paper.

The paper is organized as follows. The core section is the next, where the evolutive mean field game approach is introduced. Then, Section 3 presents the mathematical programming model that can be used to provide an initial solution to a iterative fixed-point algorithm. Section 4 displays a simple numerical example. Finally, in Section 5 the conclusions are drawn.

2   The Evolutive Mean Field Games Approach

Throughout this paper, we use the following simplifying assumptions. We consider two attractions only and that, at the initial time, there are no excursionists in the city center (i.e., the initial distribution is zero). The latter in particular will play a role in the study of the transport equation.

2.1   The Problem

Consider the following notation and hypotheses for our model:

A1)   We represent the state space by a circle with three marked points, the train station S, and the two attractions P_1, P_2; we parametrize the circle by $\theta \in [0, 2\pi]$ extended by periodicity. The station is in θ_s and P_i is in θ_i for $i \in \{1, 2\}$. We also suppose that $\theta_1 < \theta_s < \theta_2$ (the positive direction is counterclockwise). See Figure 1.

A2)   We use the pair $(w_1, w_2) \in \{(1, 1), (1, 0), (0, 1), (0, 0)\}$ to represent the extended (memory) state variables, where $w_i = 1$ if attraction P_i at θ_i has not yet been visited, 0 otherwise. See again Figure 1. The (memory) state is then (θ, w_1, w_2).

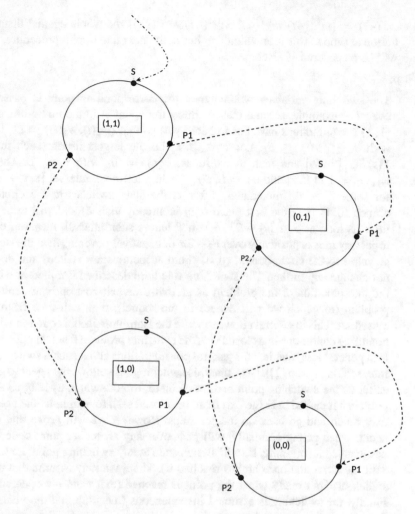

Fig. 1 Circle switching

We denote by:

A3) $T > 0$ the given finite horizon. This is a datum.

A4) $g : [0, T] \rightarrow [0, +\infty[$ a Lipschitz continuous function representing the incoming flow of agents at θ_s. This is a datum.

A5) $u(\cdot) : [0, T] \rightarrow \mathbb{R}$ measurable control of each agent.

A6) $\dot{\theta}(t) = u(t)$ controlled dynamics of each agent.

A7) $m_0^{w_1, w_2} : [0, 2\pi] \rightarrow [0, +\infty[$ the initial datum of the agents' distribution. We will always assume that $m_0^{w_1, w_2}(\theta) = 0$ for all θ and (w_1, w_2). And so, we will not display it in the subsequent transport equations.

A8) $\mathcal{M}(\cdot, t) = \left(m^{1,1}(\cdot, t), m^{0,1}(\cdot, t), m^{1,0}(\cdot, t), m^{0,0}(\cdot, t) \right)$ the whole agents' distribution at time t. This is an unknown, but in the next fixed-point procedure, it
will be considered as given.

Also:

A9)   The switching variables switch when the corresponding point is passed
over. For example, assume that at time t the agent is in the circle-branch
$(1, 1)$ (no attractions have been visited yet) so that $(w_1(t), w_2(t)) = (1, 1)$,
with $\theta_1 \leq \theta(t) \leq \theta_2$. Let $\tau \in [t, T]$ be the largest instant such that
$\theta(s) \in [\theta_1, \theta_2]$ and that, to fix ideas, $\theta(\tau) = \theta_1$ with $\tau < T$. Then
$w_1, w_2 = 1$ in $[t, \tau]$ and $w_1 = 0, w_2 = 1$ in a left-open interval $]\tau, \tau + \delta]$,
for some $\delta > 0$. Immediately after τ, the state switches to the circle-
branch $(0, 1)$ (i.e., the first place only is already visited) and will remain
there as long as $\theta \in [\theta_2 - 2\pi, \theta_2]$. If θ leaves such interval, meaning the
trajectory moves positively over $\theta_2 - 2\pi$ or negatively over θ_2, then the state
switches to the circle-branch $(0, 0)$ (both attractions are visited) and does
not change any further. The switching rule here described is coherent with
the interpretation of the problem as an exit-time/exit-cost optimal control
problem (to which we will adhere in the sequel), with exit-time from a
closed set. This in general allows to write the Hamilton-Jacobi equation with
boundary conditions in a closed set (the exit-time problem from an open set
is in general less "stable"; for general considerations about such a switching
rule, see Bagagiolo [2]). Note that this switching rule allows the agent to get
as far as the switching point and, once there, to not switch, as long as the
point is not passed over (no exit from the closed set); for example, the agent
may touch and go back on his/her steps. However we will prove that an
agent, when acting optimally, will pass over any switching point once he
has reached it (meaning, he will switch state at any switching point; he will
visit the attraction once he has reached it). Then, we may assume that the
switch occurs exactly when the point is reached (exit from the open set).
Finally, the switching is assumed instantaneous (meaning that the visit of
the attraction has a null time duration).

A10) For every $(w_1, w_2) \in \{(1, 1), (1, 0), (0, 1), (0, 0)\}$ and for every $t \in [0, T]$,
we indicate by $m^{w_1, w_2}(\cdot, t) : [0, 2\pi] \to [0, +\infty[$ the agents' distribution on
the circle-branch (w_1, w_2) at time t. This is an unknown. Formally, m^{w_1, w_2} is
a time-dependent measure on the circle-branch (w_1, w_2).

A11) By the final considerations of point A9), we can agree that the circle-branch
$(1,1)$ is identified with the interval $B_{1,1} := [\theta_1, \theta_2] \subseteq [0, 2\pi]$ (agents arrive
at the station $\theta_s \in]\theta_1, \theta_2[$). Similarly, the branch $(0,1)$ is identified with
$B_{0,1} := [\theta_2 - 2\pi, \theta_2]$ (agents enter that branch in position θ_1 and switch
out when they visit the attraction P_2). The branch $(1,0)$ is identified with
$B_{1,0} := [\theta_1, \theta_1 + 2\pi]$. The branch $(0,0)$ can be identified with $B_{0,0} = \mathbb{R}$ (we
do not switch away from that branch). See Figure 2.

A12) We define the whole (extended/memory) state space by $B = B_{1,1} \times \{1, 1\} \cup$
$B_{0,1} \times \{0, 1\} \cup B_{1,0} \times \{1, 0\} \cup B_{0,0} \times \{0, 0\}$. By conservation of mass principle,
$\mathcal{M}$ in point A8) must satisfy

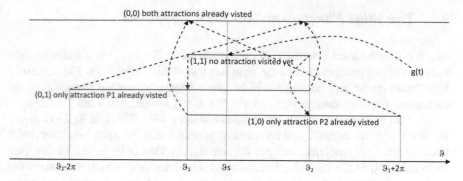

Fig. 2 Line switching

$$\int_B d\mathcal{M}(t) = \int_{B_{1,1}} dm^{1,1}(t) + \int_{B_{1,0}} dm^{1,0}(t) + \int_{B_{0,1}} dm^{0,1}(t) + \int_{B_{0,0}} dm^{0,0}(t) = \int_0^t g(s)ds$$

A13) We denote by $F^{(w_1,w_2)}(\mathcal{M}(\cdot,t))$ the running *distributional cost* for agents in the branch B_{w_1,w_2}. This is a datum and takes account of the agent's congestion cost along the trajectory, as well as of some costs specific to the running branch. Note that the distributional cost F depends on the state by means of the switching variables only (i.e., F does not depend on the value of θ inside the branch). This means that all the agents that at a certain time t are in the same branch B_{w_1,w_2} undergo the same cost F and, at the same time, that agents that at the same time t reside in different branches have in general different distributional costs. Such a hypothesis of independence of F on θ will play a role in the analysis of the transport equation.

A14) We assume that each agent, in his/her evolution ruled by means of the control $u(\cdot)$, faces a cost given by

$$J(\theta, w_1, w_2, t, \mathcal{M}, u(\cdot)) = \int_t^T \left(\frac{u(s)^2}{2} + F^{(w_1(s),w_2(s))}(\mathcal{M}(\cdot,s)) \right) ds +$$

$$+ c_1 w_1(T) + c_2 w_2(T) + c_3 (\theta(T) - \theta_s)^2 \quad (1)$$

where (θ, w_1, w_2) is the initial state at time t for the controlled switching evolution of the single agent and $\mathcal{M}$ is the evolution of the actual distribution of all the agents. The final costs penalize, with different given weights $c_i > 0$, failure of visiting the attractions, as well as the agent's distance from the station, at final time.

A15) The goal of each agent is to minimize the cost J, over the set of measurable controls $u(\cdot)$. The value function for a single agent $V : B \times [0, T] \to \mathbb{R}$ is then defined as

$$V(\theta, w_1, w_2, \mathcal{M}) = \inf_{u(\cdot)} J(\theta, w_1, w_2, t, \mathcal{M}, u(\cdot)). \quad (2)$$

2.2   The Value Function and the HJB Equations

Given a distributional evolution $\mathscr{M}$, the problem in Section 2 is a finite horizon optimal control problem, where the state has a switching evolution. The regularity hypotheses on $\mathscr{M}$ and on F^{w_1,w_2} will be given later. For now we assume that, for each given (w_1, w_2), the function $t \mapsto F^{w_1,w_2}(\mathscr{M}(t))$ is continuous and bounded.

We also assume that the distribution evolution $t \mapsto \mathscr{M}(\cdot, t)$ in $[0, T]$ is given, so that the finite horizon optimal control problem can be split into four finite horizon/exit-time problems, one per branch B_{w_1,w_2}. This is in particular due (see Bagagiolo [2]) to the very simple controlled dynamics A6), which guarantees the total controllability on the switching point, and to the fact that the switch occurs in a null time; see A9). Since we assume $\mathscr{M}$ given, we may drop the dependence on $\mathscr{M}$ of the value function, so that $V(\theta, w_1, w_2, t)$. In the first branch $(1,1)$, the problem is finite horizon and exit time. Given the initial time t, the initial state $(\theta, 1, 1)$, and a measurable control $u(\cdot)$, let τ be the first exit time from the closed interval $[\theta_1, \theta_2]$. Then by dynamic programming and total controllability ($\dot{\theta} = u$), arguing as in [2], the value function can be rewritten as

$$
\begin{aligned}
V(\theta, 1, 1, t) = \inf_{u(\cdot)} \Bigg\{ & \int_t^{\min(T,\tau)} \left(\frac{u^2(s)}{2} + F^{(1,1)}(\mathscr{M}) \right) ds \\
& + \chi_{T<\tau}(c_1 + c_2 + c_3(\theta(T) - \theta_s)^2) \\
& + \chi_{T>\tau, \theta(\tau)=\theta_1} V(\theta_1, 0, 1, \tau) \\
& + \chi_{T>\tau, \theta(\tau)=\theta_2} V(\theta_2, 1, 0, \tau) \\
& + \chi_{T=\tau, \theta(\tau)=\theta_1} \min(c_1 + c_2 + c_3(\theta_1 - \theta_s)^2, V(\theta_1, 0, 1, T)) \\
& + \chi_{T=\tau, \theta(T)=\theta_2} \min(c_1 + c_2 + c_3(\theta_2 - \theta_s)^2, V(\theta_2, 1, 0, T)) \Bigg\}
\end{aligned}
$$

where $\chi_{\{\cdot\}}$ equals 1 if $\{\cdot\}$ is true and 0 otherwise. In particular, the exit cost is equal to (a) the value function on the branch where the agent switches, if $\tau < T$, and (b) the minimum between that value and the final cost, if $\tau = T$. Note that (b) is the standard way of ensuring the continuity of the value function.

Similarly, in the branch $(0,1)$ the value function satisfies

$$
\begin{aligned}
V(\theta, 0, 1, t) = \inf_{u(\cdot)} \Bigg\{ & \int_t^{\min(T,\tau)} \left(\frac{u^2(s)}{2} + F^{(0,1)}(\mathscr{M}) \right) ds \\
& + \chi_{T<\tau}(c_2 + c_3(\theta(T) - \theta_s)^2) \\
& + \chi_{T>\tau, \theta(\tau)=\theta_2-2\pi,\theta_2} V(\theta_2, 0, 0, \tau) \\
& + \chi_{T=\tau, \theta(T)=\theta_2-2\pi,\theta_2} \min(c_2 + c_3(\theta_2 - \theta_s)^2, V(\theta_2, 0, 0, T)) \Bigg\}
\end{aligned}
$$

A symmetric representation holds for $V(\theta, 1, 0, t)$, while (note that on the branch $(0, 0)$ there is not exit time: we cannot switch away from that branch)

$$
V(\theta, 0, 0, t) = \inf_{u(\cdot)} \left\{ \int_t^T \left(\frac{u^2(s)}{2} + F^{(0,0)}(\mathscr{M}) \right) ds + c_3(\theta(T) - \theta_s)^2 \right\}.
$$

Note that when at $\tau = T$ we hit θ_i, switching is more convenient than not switching. Indeed, by definition $V(\theta_i, w_1, w_2, T)$ coincides with the final cost on the corresponding branch, while the value of $V(\theta, w_1, w_2, \tau)$, at a time τ immediately preceding T, has a higher final cost (higher of a quantity c_i). Hence we can rewrite the two formulas above as

$$
\begin{aligned}
V(\theta, 1, 1, t) = \inf_{u(\cdot)} & \left\{ \int_t^{\min(T,\tau)} \left(\frac{u^2(s)}{2} + F^{(1,1)}(\mathcal{M}) \right) ds \right. \\
& + \chi_{T<\tau}(c_1, +c_2 + c_3(\theta(T) - \theta_s)^2) \\
& + \chi_{T>\tau, \theta(\tau)=\theta_1} V(\theta_1, 0, 1, \tau) \\
& + \chi_{T>\tau, \theta(\tau)=\theta_2} V(\theta_2, 1, 0, \tau) \\
& + \chi_{T=\tau, \theta(T)=\theta_1}(c_2 + c_3(\theta_1 - \theta_s)^2) \\
& \left. + \chi_{T=\tau, \theta(T)=\theta_2}(c_1 + c_3(\theta_2 - \theta_s)^2) \right\}
\end{aligned}
$$

$$
\begin{aligned}
V(\theta, 0, 1, t) = \inf_{u(\cdot)} & \left\{ \int_t^{\min(T,\tau)} \left(\frac{u^2(s)}{2} + F^{(0,1)}(\mathcal{M}) \right) ds \right. \\
& + \chi_{T<\tau}(c_2 + c_3(\theta(T) - \theta_s)^2) \\
& + \chi_{T>\tau, \theta(\tau)=\theta_2-2\pi, \theta_2} V(\theta_2, 0, 0, \tau) \\
& \left. + \chi_{T=\tau, \theta(T)=\theta_2-2\pi, \theta_2}(c_3(\theta_2 - \theta_s)^2) \right\}
\end{aligned}
$$

Given the previous representation formulas, we have that in the branch $B_{1,1}$, the value function $(\theta, t) \mapsto V(\theta, 1, 1, t)$ is a continuous and bounded viscosity solution of the following HJB equation (subscript indicate derivatives)

$$
\begin{cases}
-V_t(\theta, 1, 1, t) + \frac{1}{2}|V_\theta(\theta, 1, 1, t)|^2 = F^{(1,1)}(\mathcal{M}(t)) & \text{in }]\theta_1, \theta_2[\times]0, T[\\
V(\theta_1, 1, 1, t) = V(\theta_1, 0, 1, t) & \text{in }]0, T] \\
V(\theta_2, 1, 1, t) = V(\theta_2, 1, 0, t) & \text{in }]0, T] \\
V(\theta, 1, 1, T) = c_1 + c_2 + c_3(\theta - \theta_s)^2 & \text{in }]\theta_1, \theta_2[
\end{cases}
\tag{3}
$$

A similar statement holds in $B_{0,1}$ for $(\theta, t) \mapsto V(\theta, 0, 1, t)$ with the HJB equation

$$
\begin{cases}
-V_t(\theta, 0, 1, t) + \frac{1}{2}|V_\theta(\theta, 0, 1, t)|^2 = F^{(0,1)}(\mathcal{M}(t)) & \text{in }]\theta_2 - 2\pi, \theta_2[\times]0, T[\\
V(\theta_2 - 2\pi, 0, 1, t) = V(\theta_2, 0, 0, t) & \text{in }]0, T] \\
V(\theta_2, 0, 1, t) = V(\theta_2, 0, 0, t) & \text{in }]0, T] \\
V(\theta, 0, 1, T) = c_2 + c_3(\theta - \theta_s)^2 & \text{in }]\theta_2 - 2\pi, \theta_2[
\end{cases}
\tag{4}
$$

in $B_{1,0}$, for $(\theta, t) \mapsto V(\theta, 1, 0, t)$ with the HJB equation

$$\begin{cases} -V_t(\theta,1,0,t) + \dfrac{1}{2}|V_\theta(\theta,1,0,t)|^2 = F^{(1,0)}(\mathcal{M}(t)) & \text{in }]\theta_1, \theta_1 + 2\pi[\times]0,T[\\ V(\theta_1,1,0,t) = V(\theta_1,0,0,t) & \text{in }]0,T] \\ V(\theta_1 + 2\pi,1,0,t) = V(\theta_1,0,0,t) & \text{in }]0,T] \\ V(\theta,1,0,T) = c_1 + c_3(\theta - \theta_s)^2 & \text{in }]\theta_1, \theta_1 + 2\pi[\end{cases}$$

(5)

and in $B_{0,0}$, $(\theta,t) \mapsto V(\theta,0,0,t)$ with the HJB equation

$$\begin{cases} -V_t(\theta,0,0,t) + \dfrac{1}{2}|V_\theta(\theta,0,0,t)|^2 = F^{(0,0)}(\mathcal{M}(t)) & \text{in } \mathbb{R}\times]0,T[\\ V(\theta,0,0,T) = c_3(\theta - \theta_s)^2 & \text{in } [0,2\pi] \end{cases}$$

(6)

Note that (6) is a finite horizon problem (with no exit/switching cost) and hence solvable, for a given $\mathcal{M}$. Then it has a unique viscosity solution $(\theta,t) \mapsto V(\theta,0,0,t)$ which is exactly the value function on the branch $B_{0,0}$. With (6) solved, the solution is used to specify the boundary conditions in θ_1 (respectively, θ_2) for the problem (5) (respectively, (4)). Then, those two problems also have a unique viscosity solution which is exactly the value function on the corresponding branch. Those solutions are used to specify the boundary conditions for (3). The final step is deriving $(\theta,t) \mapsto V(\theta,1,1,t)$ as the unique viscosity solution of (3). As a consequence of this process, the following theorem holds true.

Theorem 1. *Given a distributional evolution $t \mapsto \mathcal{M}(t)$, such that the functions $t \mapsto F^{w_1,w_2}(\mathcal{M}(t))$ are all continuous and bounded, then there exists a unique bounded continuous function $V : B \times [0,T] \to \mathbb{R}$ which is a viscosity solution of the four problems (3)–(6), that is, for every (w_1,w_2) the function $(\theta,t) \mapsto V(\theta,w_1,w_2,t)$ is a viscosity solution of the corresponding problem. Such unique solution V is the value function described in (2).*

2.3   Optimal Feedback and the Distribution Function

Due to the nature of the cost (1), in particular to the fact that the functions $F^{(w_1,w_2)}$ only depend on the whole distribution $\mathcal{M}$ and not also on the state θ (A13)), and due to the simple dynamics $\dot{\theta} = u$ (A6)), whence the distribution $\mathcal{M}$ is given, we then have the following facts:

B1)   No optimal trajectory θ^* can backtrack along its path, when inside a certain branch. Indeed, assume $\theta^*(t_1) = \theta^*(t_2)$ inside the same branch, with $t_1 < t_2$. The evolution of θ^* in $[t_1,t_2]$ does not affect the cost F, which is independent of θ, while $(\theta^*)'$ affects the running cost $u^2/2 = [(\theta^*)']^2/2$. Such cost is minimal when $(\theta^*)' = 0$, that is, when the trajectory stops at $\theta^*(t_1)$ in the time span $[t_1,t_2]$.

B2)   Taking this argument further, a moving optimal trajectory θ^* cannot hold still for a positive time, inside a certain branch. Indeed, suppose that θ^* moves from θ' to θ'', with $\theta'' > \theta'$, in a time interval $[t_1, t_2]$. By B1) the optimal control u^* cannot change sign, so that $u^*(t) \geq 0$ for every $t \in [t_1, t_2]$. Now, among all nonnegative controls u that lead from θ' to θ'' in the time interval $[t_1, t_2]$, the one which minimizes the integral of u^2 is the positive constant $u \equiv (\theta'' - \theta')/(t_2 - t_1) > 0$.

B3)   As a consequence of B1) and B2), once an optimal trajectory reaches a switching point, the switch must occur, as going back or stopping is not an optimal behavior.

B4)   From B1)–B3), we also deduce that, if an optimal trajectory has just reached the station θ_s at a time $\tau > 0$, then it must be $\tau = T$.

These facts will be used in the analysis of the optimal evolution of the density function $\mathcal{M}$.

We assumed (see (A7)) a null initial distribution m_0. Based on B1)–B4), and recalling that the evolution of the density depends on the paths of the agents, either the position θ is not occupied by anyone, or it is occupied by someone who is moving between the attractions or toward the station. In the first case, existence and uniqueness of an optimal feedback are irrelevant. In the second case, existence is a consequence of the fact that $\mathcal{M}$ is fixed, while uniqueness holds at every initial (θ, t) as a consequence of B1)–B4), except in a finite number of cases, which we list below. (For the next points, refer to Figure 2.)

C1) $(\theta_s, t, 1, 1)$. Here, we may have up to three optimal choices: to reach θ_1, to reach θ_2, and to remain all the time at θ_s, that is, the optimal controls may be one of them only, two of them, or even all of them.

C2) $(\theta_1, t, 0, 1)$, and t is exactly the instant at which we have reached the switching point θ_1 in the branch $B_{1,1}$. Here we may have up to four optimal choices: to reach $\theta_2 - 2\pi$, to reach θ_2, to move toward θ_s and (if reached) to stop there (if we reach it, then the arrival time is T), and to remain at θ_1 for all the remaining time.

C3) $(\theta_2, t, 1, 0)$, and t is exactly the instant at which we have reached the switching point θ_2 in the branch $B_{1,1}$. Here we may have up to four optimal choices: to reach θ_1, to reach $\theta_1 + 2\pi$, to move toward θ_s and (if reached) to stop there (if we reach it, then the arrival time is T), and to remain at θ_2 for all the remaining time.

C4) $(\theta_1, t, 0, 0)$, and t is exactly the instant at which we have reached the switching point $\theta_1 = \theta_1 + 2\pi$ in the branch $B_{1,0}$. Here we may have up to two optimal choices: to move toward θ_s and (if reached) to stop there (if we reach it, then the arrival time is T) and to remain at θ_1 for all the remaining time.

C5) $(\theta_2, t, 0, 0)$, and t is exactly the instant at which we have reached the switching point $\theta_2 = \theta_2 - 2\pi$ in the branch $B_{0,1}$. Here we may have up to two optimal choices: to move toward θ_s and (if reached) to stop there (if we reach it, then the arrival time is T) and to remain at θ_2 for all the remaining time.

Formally, for every branch B_{w_1,w_2}, for every t and θ, the optimal feedback u^* is given by

$$u^*(\theta, t, w_1, w_2) = -V_\theta(\theta, t, w_1, w_2), \tag{7}$$

and the relation holds at all points (θ, t, w_1, w_2) where V is differentiable with respect to θ, i.e., at points where the feedback optimal control is unique (see Cardaliaguet [10], Lemma 4.9). More in detail, C1)–C5) imply that this fact is true all points (θ, t, w_1, w_2), except for two points per branch. It is important to observe that, at those points, the branch is subject to an incoming flow (see next equations and the associated comments). We have the following four transport equations for the density m, one per every branch

$$
\begin{cases}
m_t^{1,1}(\theta, t) - [V_\theta(\theta, t, 1, 1)m^{1,1}(\theta, t)]_\theta = 0 \text{ in } B_{1,1} \times [0, T] \\
m^{1,1}(\theta_s, t) = g(t)
\end{cases}
\tag{8a}
$$

$$
\begin{cases}
m_t^{1,0}(\theta, t) - [V_\theta(\theta, t, 1, 0)m^{1,0}(\theta, t)]_\theta = 0 \text{ in } B_{1,0} \times [0, T] \\
m^{1,0}(\theta_2, t) = m^{1,1}(\theta_2, t)
\end{cases}
\tag{8b}
$$

$$
\begin{cases}
m_t^{0,1}(\theta, t) - [V_\theta(\theta, t, 0, 1)m^{0,1}(\theta, t)]_\theta = 0 \text{ in } B_{0,1} \times [0, T] \\
m^{0,1}(\theta_1, t) = m^{1,1}(\theta_1, t)
\end{cases}
\tag{8c}
$$

$$
\begin{cases}
m_t^{0,0}(\theta, t) - [V_\theta(\theta, t, 0, 0)m^{0,0}(\theta, t)]_\theta = 0 \text{ in } B_{0,0} \times [0, T] \\
m^{0,0}(\theta_1, t) = m^{1,0}(\theta_1, t) + m^{1,0}(\theta_1 + 2\pi, t) \\
m^{0,0}(\theta_2, t) = m^{0,1}(\theta_2, t) + m^{0,1}(\theta_2 - 2\pi, t)
\end{cases}
\tag{8d}
$$

Note that the switching nature of the problem is reflected in the boundary conditions in (8b)–(8d), which are given by the distribution in the previous branch before the switch. It has to be noted that the term "boundary conditions" is improper, as the aforementioned conditions hold at interior points (see Figure 2), and could rather be interpreted as incoming flows at those points. Usually this turns out to be troublesome in interpreting the equations, especially when the trajectories of the agents in the branch move back and forth through those points. However, this is not our case. Indeed, by B1)–B4), the optimal control moves the agents in the branch toward at most one such singularity. The direction depends on the time the branch is entered but, once chosen, does not change. In this way, for example, in (8a), for every t, we may interpret the relation $m^{1,1}(\theta_s, t) = g(t)$ exactly as a boundary condition for one of the two sides defined by θ_s in the $B_{1,1}$ and then possibly construct the solution by a stuitable adaptation of the method of characteristics (in the following: a characteristics-type solution). Moreover, for B1)–B4), we know that along an optimal trajectory, agents reaching a switching point do not reverse motion but switch to a new state. This means that, concerning the equations (8a)–(8c), switching points may be interpreted as sinks for the corresponding branch.

If then, given the optimal feedback control $-V_\theta$, we have a good solution concept for (8) implying uniqueness (recall that the initial distribution is null in every branch), then we may solve them in a forward-like manner: (8a) does not depend on the solutions of the other equations, and hence it can be solved separately, giving $m^{1,1}$ which enters as boundary conditions in (8b) and (8c). We then solve (8b) and (8c), getting $m^{1,0}$ and $m^{0,1}$, so that also (8d) is well posed and can be solved.

We now denote by $\mathscr{B}$ the space of Borel measures on B endowed by the weak-star topology (by the nature of the optimal feedback, and in view of the distribution of the agents, we may assume $B_{0,0}$ coinciding with $[\theta_1, \theta_2]$ (the optimal trajectories never go outside that interval) and hence assume both B and $\mathscr{B}$ to be compact). By $D \subset C([0,T], \mathscr{B})$ we denote the subset of time-dependent continuous measures $t \mapsto \mu(t) \in \mathscr{B}$ such that $\mu(0) \equiv 0$ and

$$\int_B d\mu(t) = \int_{B_{1,1}} d\mu(t,1,1) + \int_{B_{1,0}} d\mu(t,1,0) + \int_{B_{0,1}} d\mu(t,0,1) + \int_{B_{0,0}} d\mu(t,0,0)$$

$$= \int_0^t g(s)ds.$$

Note that D is convex and closed for the topology in $C([0,T], \mathscr{B})$ (uniformly in time, weak-star in $\mathscr{B}$).

The solution concept for (8) is a crucial point. In searching a characteristics-type solution, we need to make use of a rather strong assumption. Amending such assumption, or giving conditions to have it satisfied, will be the subject of a future work. However, note that the discretized model of the next session and the numerical examples seem to make it justified.

Assumption 1. There exists a positive integer $N > 0$ such that for any given distribution $\mathscr{M} \in D$, for any point $(\theta, \cdot, w_1, w_2)$ as the ones described in C1)–C5), the number of times such that, in the interval $[0,T]$, the set of optimal controls and its cardinality change, accordingly to C1)–C5), is not greater than N.

Given the optimal feedback control $u^*(= -V_\theta)$ satisfying Assumption 1, a solution of (8) is a function $\mathscr{M} = (m^{1,1}, m^{1,0}, m^{0,1}, m^{0,0})$, belonging to D such that for every (w_1, w_2), m^{w_1,w_2} solves the corresponding transport equation in the branch B_{w_1,w_2}, in the sense that it is a characteristics-type solution constructed on B_{w_1,w_2} transporting, via u^*, the values (concentration of agents) on the segments in the $(\theta - t)$ plane $\{\tilde{\theta}\} \times [0,T]$, where $(\tilde{\theta}, w_1, w_2)$ are the points in C1)–C5).

Some comments are due. (a) For each fixed t, $\mathscr{M}(\cdot, t)$ is a space distribution over the state space B (roughly speaking, a space-dependent function), and this is what must enter the cost functional (1). Formally the solution of (8) is a function which is both space and time dependent, and its boundary datum is a time-dependent function. The true quantity that must be inserted in (1) is then the space component in the disintegration of the measure $\mathscr{M}$, that is, what in Camilli-De Maio and Tosin [9] is called the conditional measure, or trace, with respect to t. In our case, this is the (time dependent) boundary datum somehow rescaled by the velocity of the flow. Note that the velocity is just the optimal control and that, whenever it is chosen at the starting point, it is not changed anymore (see B1–B4). Hence, in our case, being the optimal control u^* known when we solve (8), we can construct the right term to be inserted in (1). Finally note that all the time-dependent boundary conditions in (8) are given by the trace of $\mathscr{M}$ at some fixed-state point θ, but in this case, they exactly coincide with the function $\mathscr{M}(\theta, \cdot)$. b) The characteristics cannot cross each other. Indeed, if they are oriented in opposite directions, then it means that they are pointing toward the station θ_s from different sides, and then they may intersect only at the final time T. If instead they are oriented in the same direction, then they cannot intersect before the final time T because, otherwise, in the intersection point (which must be different from the ones of C1)–C5)) we will have two different optimal controls. Moreover the characteristics cannot rarefy, that is, when starting from an interval as in Assumption 1, they cannot form a blank zone between them. Indeed, this would bring a discontinuity in time of the (same direction pointing) optimal control u^* in a point as one of C1)–C5), which is not possible due to the boundary conditions in (3)–(6) which are continuous in time (they are given by V itself or by the final costs) and to the right hand side which is continuous in time (also see the optimal control representation given in B2)). Finally observe that some Dirac measures (delta functions) may arise at the station θ_s and in the other points of C1)–C5). The time-dependent coefficient of such Dirac masses is the accumulation of agents (the integral in time of the corresponding concentration) switching or arriving on those points, when possibly using the null optimal control. In any case we construct a solution belonging to D.

2.4   Mean Field Equilibrium

Let us now suppose that the cost functions F^{w_1,w_2} are all continuous as function from $\mathscr{B}$ to $[0, +\infty[$ and that they are bounded. So, for every $\mathscr{M} \in C([0, T]\mathscr{B})$, the function $t \mapsto F^{w_1,w_2}(\mathscr{M}(t))$ is continuous and bounded.

A pair of functions $(V, \mathscr{M}) \in C(B \times [0, T]) \times D$ is called a mean field equilibrium for our problem if the two functions, respectively, satisfy (branch by branch) the four Hamilton-Jacobi-Bellman equations (3)–(6) and the four transport equations (in the sense described above) (8).

We consider the multivalued map $\psi : D \to D$ such that to any $\mathscr{M} \in D$ associates the set $\psi(\mathscr{M}) \subseteq D$ of elements of D constructed inserting in the $\mathscr{M}$ Hamilton-Jacobi equations (3)–(6); getting the solution V, as well as the optimal feedback; and then solving the transport equations (8). Note that, due to the possible multiplicity of the optimal feedback, to any $\mathscr{M}$, the construction of the solution of the transport equations may give more than one result, and hence $\psi(\mathscr{M})$ is a set.

A mean field equilibrium is then identified by a function $\mathscr{M}$ (with the corresponding V) such that $\mathscr{M} \in \psi(\mathscr{M})$. However, in general, such a $\mathscr{M}$ does not exists, in particular due to the non-convexity of $\psi(\mathscr{M})$. For an optimal control/game point of view, this is linked to the fact that our agents may have more than one optimal choice in some particular situations (see C1)–C5)), which gives rise to different distribution evolutions. Indeed, at the instant and position where more than one optimal choice is admissible, the agents, formally, randomly split in a number of subgroups each one of them using one of the possible optimal controls. But this behavior is not captured by the mean field formulation for which the agents are homogeneous and indistinguishable. Roughly speaking, the agent must use mixed strategies, that is, they convexify.

We then introduce the notion of generalized mean field equilibrium. We say that a function $\mathcal{M} \in D$ is a generalized equilibrium if $\mathcal{M} \in \overline{co}\psi(\mathcal{M})$, where $\overline{co}$ stays for the closed convex envelope.

Theorem 2. *Given Assumption 1, and the hypotheses on F^{w_1,w_2} above, there exists at least one generalized mean field equilibrium.*

Proof. We used, as standard, a fixed-point procedure.

We first prove that ψ is closed as multivalued function. Let $\mathcal{M}_n$ be converging to $\mathcal{M}$ in D and $m_n \in \psi(\mathcal{M}_n)$ be converging to m in D, and then we have to prove that $m \in \psi(\mathcal{M}_n)$. Let V_n, V be the solutions of (3)–(6) corresponding to $\mathcal{M}_n$, $\mathcal{M}$, respectively. Hence by standard stability results for viscosity solutions, V_n uniformly converges to V. Moreover, possibly considering a subsequence, we may assume that the number of changes in Assumption 1 is the same for all m_n. The convergence of m_n to m, the convergence of V_n to V, the nature of the optimal feedback (see B1)–B4), C1)–C5)), and the construction of the solution of (8) imply that $m \in \psi(\mathcal{M})$.

Since there exists a constant $C > 0$ such that every possible optimal control satisfies $|u^*| \leq C$, then (see Cardaliaguet [10]) we get that, for every $\mathcal{M} \in D$, every element of $\psi(\mathcal{M})$, as element of $C([0,T], \mathcal{B})$ is Lipschitz continuous in time with the same Lipschitz constant C. Hence $\psi(\mathcal{M})$ is compact in $C([0,T], \mathcal{B})$.

We then consider the compact and convex set $D' \subseteq D$ of the time-Lipschitz continuous functions with constant C and the multivalued function $\overline{\psi} : D' \to D'$ defined as $\overline{\psi}(\mathcal{M}) = \overline{co}\psi(\mathcal{M})$. The function $\overline{\psi}$ is also closed and compact. Indeed, by Assumption 1 and C1)–C5), the number of elements of $\psi(\mathcal{M})$ is at most $192N^5$, that is, it is finite with uniform bound. Hence, by the closedness and compactness of $\psi(\mathcal{M})$, we get the same properties for $\overline{co}\psi(\mathcal{M})$.

By the Kakutani fixed-point theorem, we then get the existence of at least one fixed-point $\mathcal{M} \in \overline{co}\psi(\mathcal{M})$. □

3 An Initial Approximate Solution

3.1 Motivations

In the previous section, we assumed that an agent manages its dynamics by deciding the value of the control $u(t)$. In particular, we were interested in determining an optimal feedback control $u^*(\theta, w_1, w_2, t)$ function of the agent current state (θ, w_1, w_2) and of the actual optimal density $m^{(w_1,w_2)*}(\theta, t)$.

In this section, we introduce a methodology to numerically find a density function that approximates $\mathcal{M}^*(., t)$. This function can then be used as a starting solution, e.g., on the fixed-point procedure that determines iteratively the optimal solution of the model considered in the previous section. Specifically, we determine the optimal solution of a mathematical programming problem as similar as possible to the problem in Section 2. To this end, first, we discretize the time interval $[0, T]$ with a constant time step. Second, instead of determining the optimal strategy of

each agent and then taking the limit of the number of agents to infinite, we simplify
the problem by taking the limit to infinite before we perform the optimization (see,
e.g., [11] for an interesting comparison between the two approaches). In this way,
we can describe the behavior of all agents at the same time. In particular, we can
express it in terms of stochastic decisions made by a single representative agent.

In the problem in Section 2, the state of an agent represents the situation of
visiting an attraction or moving between two attractions, having already visited a
given set of attractions. Note that this definition of state implies that an agent may
enter or leave a state only when he/she reaches or leaves an attraction.

Within this context, hereinafter, we say that a *completion event* occurs when an
agent completes the visit to an attraction or reaches a new one, that is, when he exits
a state and is ready to enter a subsequent state. As an example, a completion event
is associated with each arc in Figure 1 and with a switching in Figure 2.

The introduction of completion events allows modeling agents' decisions in
terms of their movements on a network $\mathscr{G} = (\mathscr{V}, \mathscr{E})$ of states, where:

- the set of nodes $\mathscr{V}$ is the set of the possible completion events and of other two
  artificial events, *in* and *out*, occurring to agents, respectively, at the beginning and
  at the end of the day
- the set of oriented arcs $\mathscr{E}$ is the set of the states. The head of each arc $e \in \mathscr{E}$
  is the completion event of e, whereas the tail of e is the completion event of the
  previous state of the agent. Each arc $e \in \mathscr{E}$ is labeled by a value α_e that represents
  the minimum time that an agent must spend in state e.

3.2 Notation

Let:

- $A(v)$, $B(v)$: the set of arcs entering in, respectively, exiting from, the node $v \in \mathscr{V}$
- $\mathscr{T} = \{0, \dots, T\}$: discrete set of time instants in which the time interval $[0, T]$ is
  discretized. Time 0 is the beginning of the day and time T the end of the day
- $P_t = [p_{e,t} : e \in \mathscr{E}]$: probability vector at time $t \in \mathscr{T}$ whose generic entry $p_{e,t}$
  represents the fraction of agents that have just entered the state e at time t
- $Q_t = [q_{e,t} : e \in \mathscr{E}]$: probability vector at time $t \in \mathscr{T}$ whose generic entry $q_{e,t}$
  represents the fraction of agents that have just left state e at time t
- $R_t : \{r_{e,t}, e \in \mathscr{E}\}$: probability vector at time $t \in \mathscr{T}$ whose generic entry $r_{e,t}$
  represents the fraction of agents that remain in state e at time t
- $M_t : \{m_{e,t}, e \in \mathscr{E}\}$: probability vector at time $t \in \mathscr{T}$ whose generic entry $m_{e,t} =
  r_{e,t} + p_{e,t}$ represents the fraction of agents that are in state e at time t
- $c_t = [c_{e,t} : e \in \mathscr{I}]$: cost vector at time $t \in \mathscr{T}$ whose generic entry $c_{e,t}$ represents
  the cost paid for being in state e at time t; this cost is a function of M_t
- $d_t = [d_{v,t} : v \in \mathscr{V}]$: external flow vector at time $t \in \mathscr{T}$ whose generic entry $d_{v,t}$
  absolute value represents the fraction of agents entering or leaving the system
  at time t through node v. By hypothesis, $d_{in,0} = -1$, $d_{out,T} = 1$, and $d_{v,t} = 0$
  otherwise.

3.3   The Mathematical Programming Model

Consider the following mathematical programming problem:

$$J_0^* = min_{P,Q} \sum_{t \in \mathcal{T}} \sum_{e \in \mathcal{E}} c_{e,t}(M_t) m_t^e \tag{9a}$$

$$\sum_{e \in A(v)} q_{e,t-1} - \sum_{e \in B(v)} p_{e,t} = d_{v,t} \quad \forall v \in \mathcal{V}, t \in \mathcal{T} \tag{9b}$$

$$\sum_{\tau=0}^{t-\alpha_e} p_{e,\tau} \geq \sum_{\tau=0}^{t} q_{e,\tau} \quad \forall e \in \mathcal{E}, t \in \mathcal{T} \tag{9c}$$

$$r_{e,t+1} = r_{e,t} + p_{e,t} - q_{e,t} \quad \forall e \in \mathcal{E}, t \in \mathcal{T} \tag{9d}$$

$$m_t^e = r_{e,t} + p_{e,t} \quad \forall e \in \mathcal{E}, t \in \mathcal{T} \tag{9e}$$

$$p_{e,t}, q_{e,t} \geq 0 \quad \forall e \in \mathcal{E}, t \in \mathcal{T} \tag{9f}$$

$$r_{e,t} \geq 0 \quad \forall e \in \mathcal{E}, t \in \mathcal{T} \tag{9g}$$

$$r_{e,0} = 0 \quad \forall e \in \mathcal{E} \tag{9h}$$

Conditions (9b) are flow balancing conditions which impose that an agent, when leaving a state, immediately enters a new state. Conditions (9c) impose that an agent spends at least a time α_e in state e. Conditions (9d) are conservation conditions which impose that the fraction of agents that are in state e at time $t + 1$ is equal to the number of agents that were in the same state at time t plus the agents that have just entered state e minus the ones that have just left state e. Conditions (9e) define the variables m_t^e. Conditions (9f) and (9h) are the minimal conditions that allow interpreting the values of all problem variables as probabilities, as we show next.

Initially consider variables m_t^e, which can be seen as the probability that an agent is in state e for $t < \tau < t+1$. Indeed, Conditions (9f) and (9g) imply that $m_t^e \geq 0$, or all $e \in \mathcal{E}, t \in \mathcal{T}$. In addition, by hypothesis we have $\sum_{e \in \mathcal{E}} m_0^e = \sum_{e \in \mathcal{E}} (r_{e,0} + p_{e,0}) = 1$. Then, using an induction argument, we assume $\sum_{e \in \mathcal{E}} m_t^e = 1$ and we observe

$$\sum_{e \in \mathcal{E}} m_t^e = \sum_{e \in \mathcal{E}} r_{e,t} + p_{e,t} = \sum_{e \in \mathcal{E}} r_{e,t+1} + q_{e,t} = \sum_{e \in \mathcal{E}} r_{e,t+1} + p_{e,t+1} = \sum_{e \in \mathcal{E}} m_{t+1}^e = 1 \tag{10}$$

where the second equality is implied by Condition (9b) and the third equality by Condition (9d).

Consider now variables $p_{e,t}, q_{e,t}$. Conditions (9f) and (9g) together with (10) imply that $p_{e,t}, q_{e,t} \leq 1$ for all $e \in \mathcal{E}, t \in \mathcal{T}$ and then also the values of these variables can be interpreted as probabilities.

Problem (9) generalizes the one proposed in the seminal problem introduced in [19]. Specifically, Problem (9) presents:

- a more general objective function
- the additional Conditions (9e)
- a greater number of independent variables. In [19], an agent controls just the variables $p_{e,t}$, whereas the values of variables $q_{e,t}$ are fixed given m_t^e,

In addition, other constraints can be introduced in the formulation of Problem (9). As an example, bounds on the sum of the values of some subsets of variables m_t^e can be imposed to take account of the maximum number of visitors an attraction can host at each time.

We observe that Problem (9) is non-memoryless, as each state $e \in \mathscr{E}$ includes both information on what an agent is currently doing and on its past history. This fact can also be deduced by recognizing that Problem (9) can be seen as a min-cost flow problem, specifically a min path problem from the artificial event in to the artificial event out, on the space-time network $\mathscr{H} = ((\mathscr{V} \cup \mathscr{E}) \times \mathscr{T}, 2\mathscr{E} \times \mathscr{T})$. Network $\mathscr{H}$ presents a node for each possible state or completion event at each time instant in $\mathscr{T}$. Then, a flow along a path from $(in, 0)$ to (out, T) on $\mathscr{H}$ describes the states assumed by a given fraction of agents over times.

Medium-sized instances of Problem (9) can be easily solved by commercial software. A theoretical drawback of Problem (9) is that the size of its instances increases with the number of states possibly assumed by agents. In turn, such a number increases exponentially with the number of the considered attractions. However, in a real-world environment, it rarely occurs that a city includes more than a few dozens of attractions.

The optimal values of the variables m_t^e of Problem (9) can be used to obtain a first approximation of the density function $M^*(., t)$.

Compare Problem (9) objective function (9a) with the problem in Section 2 objective function (1). We can observe that (9) includes neither the final costs $c_1 w_1(T) + c_2 w_2(T) + c_3(\theta(T) - \theta_s)^2$ nor the control cost $\frac{u(s)^2}{2}$. Indeed, the former costs are not included as Conditions (9b) for $t = T$ impose that all the agents reach the final destination by T. The latter costs are not included, as Problem (9) does not explicitly present variables describing the intensity of the agents' controls. To partially overcome this last issue, we can change the structure of the cost of being in state e at time t so that it depends on the time length that an agent has already spent in this state. To this end, we have to modify the interpretation of variables r_t, objective function (9a), and conditions (9d) and (9e). Specifically, we should let variables $r_{e,t}$ represent the fraction of agents that remain in state e at time t in excess to the minimum time α_e, that is, we substitute (9d) with condition

$$r_{e,t+1} = r_{e,t} + p_{e,t-\alpha_e} - q_{e,t} \quad \forall e \in \mathscr{E}, t \in \mathscr{T} \tag{11}$$

and (9a) with the new objective function

$$J_0^* = \min_{P,Q} \sum_{t \in \mathscr{T}} \sum_{e \in \mathscr{E}} \left(\hat{c}_{e,t}(M_t) r_{e,t} + \sum_{\tau=0}^{\alpha_e - 1} \tilde{c}_{e,t-\tau}(M_t) p_{e,t-\tau} \right) \tag{12}$$

where $\hat{c}_t = [\hat{c}_{e,t} : e \in \mathscr{I}]$ and $\tilde{c}_t = [\tilde{c}_{e,t} : e \in \mathscr{I}]$ are appropriate cost vectors with possibly negative entries. Finally, we should rewrite Conditions (9e) as $m_t^e = r_{e,t} + \sum_{\tau=0}^{\alpha_e-1} p_{e,t-\tau}$.

4 A Simple Example

4.1 Example Description

Throughout this section, we will consider the toy example displayed in Figure 3. There excursionists (agents) are ready to arrive at the city, that is, they are in the node *in* of network $\mathscr{G}$, at time 8:00 am that we conventionally fix as $t = 0$. On the other hand, they have to leave the city, that is, they have to be in the node *out* of network $\mathscr{G}$, by time 08:00 pm that we conventionally fix as $t = T$. To allow our models to mimic the real distributions of the agents' actual arrivals at and departures from the train station, we assume that the excursionists may spend time at no cost on the two arcs of $\mathscr{G}$ that, respectively, join the node *in* with the train station and, vice versa, the train station with node *Out*.

The agents' main objective is visiting the major attraction P_1, but, if possible, they would also enjoy seeing the minor attraction P_2. Attraction P_1 is further away from the station than attraction P_2. Attraction P_1 is a closed place, e.g., a cathedral, whereas attraction P_2 is an open space, e.g., a square. Then, agents feel more uncomfortable in presence of a crowd, that is, they pay a greater congestion cost, at attraction P_1 than at attraction P_2.

4.2 Expected Behavior of the Excursionists

From a qualitative point of view, the expected behavior of the agents in the above example is the following. A preferred route should be $R_1 : S \to P_1 \to P_2 \to S$. An agent that follows this route visits first the major attraction and then the minor attraction and, finally, reaches the train station to leave the city. If this first route is

Fig. 3 Network of the possible tourist movements in the toy example

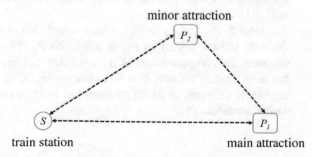

minor attraction

P_2

S

train station

P_1

main attraction

Fig. 4 Network $\mathscr{G}$ of the toy
example in Figure 3

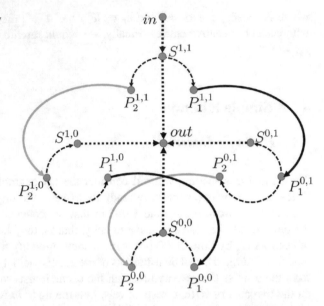

congested, then a second route $R_2 : S \to P_2 \to P_1 \to S$ could become attractive. An agent that follows this second route may hope to reach the major attraction after the time of its maximum congestion in order to better enjoy the place. Finally, if both the former routes are congested, an agent may resort to route $R_3 : S \to P_1 \to S$, or even, $R_4 : S \to P_2 \to S$, giving up visiting one of the attractions.

4.3 Model (9) Numerical Results

Figure 4 presents network $\mathscr{G}$ for the toy example considered. In $\mathscr{G}$:

- the solid arcs represent the states in which an agent is visiting an attraction; the black arcs are associated to attraction P_1 and the gray arcs to attraction P_2
- the dashed arcs represent the states in which an agent is moving from an attraction to the next or is moving from, or to, the train station
- the dotted arcs represent the states in which an agent is moving from, or to, the artificial events *in* and *out*
- the nodes S, P_1, and P_2 indicate that a completion event occurs at, respectively, the train station, attraction P_1, or attraction P_2. The node superscripts w_1, w_2 represent the components w_i of the state that indicate whether the attraction P_i has been already visited ($w_i = 0$) or not ($w_i = 1$). As an example the node $P_2^{0,1}$ represents the event in which an agent has reached attraction P_2 having already visited attraction P_1.

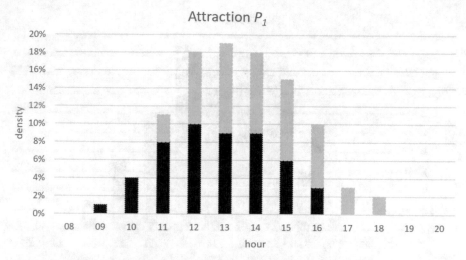

Fig. 5 Fraction of excursionists present in attraction P_1 at different hours of the day

In this example, we assume that the congestion (running) costs paid by an agent are proportional to the total density of agents present in each attraction or in the street joining the attractions that the agent is, respectively, visiting or traversing. In addition, other thing being equal, we assume that the congestion (running) cost paid by an agent visiting an attraction slightly increases over time (with the exception of lunch time), i.e., it is in general better for an agent to visit an attraction earlier than later. Finally, we assume that the costs decrease with the time that an agent takes visiting or reaching an attraction.

Figures 5 and 6 display the total density of the agents that are present, respectively, in attraction P_1 and P_2 at the different hours of the day. There, the black bars represent the agents that follow route $R_1 : S \rightarrow P_1 \rightarrow P_2 \rightarrow S$, and the gray bars represent the agents that follow route $R_2 : S \rightarrow P_2 \rightarrow P_1 \rightarrow S$. In this example, no agent chooses either route $R_3 : S \rightarrow P_1 \rightarrow S$ or $R_4 : S \rightarrow P_2 \rightarrow S$.

5   Conclusions

This work proposes a model to describe the excursionists' movement in historical city centers. The aim of the model is studying how to avoid congestions of pedestrians in the city narrow alleys and of visitors in the city attractions. The results prove that, if agents behave rationally, they can distribute their movements and visits over time and avoid main congestions.

In the light of the above results, the challenge for the city administrators is to promote the excursionist's rational behavior by providing real-time information and congestion forecasts and assisting excursionists in planning their visits. In

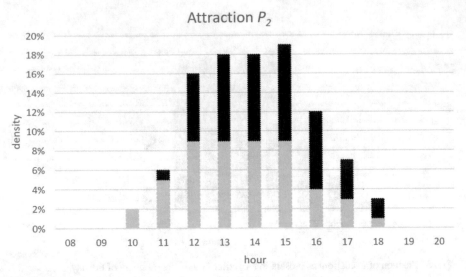

Fig. 6 Fraction of excursionists present in attraction P_2 at different hours of the day

this context, a future development of this work is to study how possible actions undertaken by city administrators can influence excursionists' behavior under different assumptions on their rationality and access to information.

Acknowledgements This research was partially supported by the project Opthysys funded by the University of Trento and by the GNAMPA research project 2017.

References

1. Achdou, Y., Bardi, M., Cirant, M.: Mean field games models of segregation. Math. Models Methods Appl. Sci. **27**, 75 (2017)
2. Bagagiolo, F.: An infinite horizon optimal control problem for some switching systems. Discret. and Contin. Dyn. Syst., Ser. B. **1**, 443–462 (2001)
3. Bagagiolo, F., Bauso, D.: Mean field games and dynamic demand management in power grids. Dyn. Games Appl. **4**, 155–176 (2014)
4. Bagagiolo, F, Benetton, M.: About an optimal visiting problem. Appl. Math. Optim. **65**, 31–51 (2012)
5. Bagagiolo, F., Danieli, K.: Infinite horizon optimal control problems with multiple thermostatic hybrid dynamics. Nonlinear Anal. Hybrid Sys. **6**, 824–838 (2012)
6. Bardi, M., Capuzzo Dolcetta, I.: Optimal Control and Viscosity Solutions of Hamilton-Jacobi-Bellman Equations. Birkhäuser, Boston (1997)
7. Bellman, R.: Dynamic programming treatment of the travelling salesman problem. J. ACM. **9**, 61–63 (1962)
8. Camilli, F., Carlini, E., Marchi, C.: A model problem for mean field games on networks. Discrete Contin. Dyn. Syst. **35**, 4173–4192 (2015)

9. Camilli, F., De Maio R., Tosin, A.: Transport of measures on networks. Netw. Heterog. Media **12**, 191–215 (2017).
10. Cardaliaguet, P.: Notes on mean field games. Unpublished notes. https://www.ceremade. dauphine.fr/cardalia/index.html. Cited 13 July 2017
11. Carmona, R., Delarue, F., Lachapelle, A.: Control of McKean-Vlasov dynamics versus mean field games. Math. Finan. Eco. **7**, 131–166 (2013).
12. Cristiani, E., Priuli, F., Tosin, A.: Modeling rationality to control self-organization of crowds: an environmental approach. SIAM J. Appl. Math. **75**, 605–629 (2015)
13. Hoogendoorn, S., Bovy, P.: Pedestrian route-choice and activity scheduling theory and models. Transportation Research Part B **28**, 169–190 (2004)
14. Huang, L., Wong, S., Zhang, M., Shu, C.W., Lam, W.H.: Revisiting Hughes dynamic continuum model for pedestrian flow and the development of an efficient solution algorithm. Transp. Res. B. **43**, 127–141 (2009)
15. Huang, M., Caines, P., Malhamé, R.: Large population dynamic games; closed-loop McKean-Vlasov systems and the Nash certainty equivalence principle. Commun. Inf. Sys. **6**, 221–252, (2006)
16. Hughes, R.L.: A continuum theory for the flow of pedestrians. Transp. Res. B. **36**, 507–535 (2002)
17. Kachroo, P., Shlayan, N.: Dynamic traffic assignment: A survey of mathematical models and techniques. In: Advances in Dynamic Network Modeling in Complex Transportation Systems, pp. 1–25. Springer (2013)
18. Lasry, J.M., Lions, P.L.: Juex à champ moyen ii. Horizon fini et controle optimal. C. R. Math. **343**, 679–684 (2006)
19. Merchant, D.K., Nemhauser, G.L.: A model and an algorithm for the dynamic traffic assignment problems. Transp. Sci. **12**, 183–199 (1978)
20. Peeta, S., Ziliaskopoulos, A.K.: Foundations of dynamic traffic assignment: The past, the present and the future. Netw. Spat. Econ. **1**, 233–265 (2001).
21. Szeto, W., Wong, S.: Dynamic traffic assignment: model classifications and recent advances in travel choice principles. Centr. Eur. J. Eng. **2**, 1–18 (2012)
22. Vermuyten, H., Beliën, J., De Boeck, L., Reniers, G., Wauters, T.: A review of optimisation models for pedestrian evacuation and design problems. Saf. Sci. **87**, 167–178 (2016)

Limit Game Models for Climate Change Negotiations

Olivier Bahn, Alain Haurie, and Roland Malhamé

Abstract This paper deals with a family of dynamic game models that represent schematically the interaction between groups of countries in achieving the necessary limitation of carbon atmospheric emissions in order to control climate change. We start from a situation where m coalitions of countries exist and behave as m players in a game of sharing a global emission budget through the establishment of an international emissions trading system. We characterize the Nash equilibrium solutions for this game in a deterministic context. Through a simple replication schemes, we increase the number of players, each one becoming infinitesimal, and we characterize the limit games thus obtained. A stochastic version is proposed for this class of models, and the limit games are characterized, using the recently defined concept of mean field games.

1 Introduction

The Paris agreement, negotiated at COP-21 and signed by a majority of nations, is dedicated to limiting to less than 2 °C the surface average temperature (SAT) rise in the twenty-first century. To achieve this goal, the participating countries must reduce their emissions of greenhouse gases (GHG). These emissions are mostly related to the use of fossil energy as an economic production factor. Recent research on

O. Bahn
GERAD and Department of Decision Sciences, HEC Montréal, Montreal, QC, Canada
e-mail: olivier.bahn@gerad.ca

A. Haurie (✉)
ORDECSYS, Chêne-Bougeries, Switzerland;University of Geneva, Geneva, Switzerland;
GERAD, HEC Montréal, Montreal, QC, Canada
e-mail: ahaurie@gmail.com

R. Malhamé
GERAD and Department of Electrical Engineering, POLY Montréal, Montreal, QC, Canada
e-mail: roland.malhame@polymtl.ca

© Springer International Publishing AG 2017
J. Apaloo, B. Viscolani (eds.), *Advances in Dynamic and Mean Field Games*,
Annals of the International Society of Dynamic Games 15,
https://doi.org/10.1007/978-3-319-70619-1_2

mitigation policies have shown that a possible set of backstop technologies could be carbon dioxide removal (CDR), which refers to technologies that reduce the levels of carbon dioxide in the atmosphere. Among such technologies, one finds, in particular, bioenergy with carbon capture and storage (BECCS), direct air capture, ocean fertilization, etc. [10, 17, 18, 20]. On the other hand, recent research on climate modeling tends to show that to limit to $2\,°C$ the SAT rise with sufficiently high probability, one should define a global limiting carbon budget of 1 trillion tons, over the whole period starting from the industrial revolution to the end of the twenty-first century [1, 15]. Up to now the world has emitted over 580 billion tons of carbon (GtC). So there is a remaining budget of a little more than 400 GtC to be shared among all the countries participating in the Paris agreement. Now the fundamental question which arises is "how to share this remaining emission budget among the more than 100 countries that signed the Paris agreement?". This will certainly be a key element of the forthcoming negotiations on climate change. These countries are more or less formally regrouped in coalitions or groups that share similar economic conditions although not necessarily the same size nor the same economic power. Finally, in order to foster economic efficiency, an international emissions trading scheme with full banking and borrowing should be established. When countries are given a share of the remaining emission budget, they will tend to use these emission rights in a strategic way, in order to extract the most welfare benefits, taking into consideration the actions of the other countries that intervene on the international market (see [13] for a discussion of the strategic use of allowances). This implies that one should explore the equilibrium solutions of the dynamic game underlying these forthcoming burden-sharing negotiations to decide what should be a fair allocation of this budget among the different coalitions or groups of countries. This question has been addressed in a stream of papers where a meta-game of burden sharing was built, using statistical emulation of a worldwide, multi-country, general equilibrium model, [3, 11], or an open-loop differential game model based on multi-coalition economic growth models with GHG emissions [5, 6]. In the present paper, we revisit this game theoretic approach with three important modifications introduced to take into account: (i) the fact that each coalition is composed of several countries that will also play noncooperatively in the GHG emission games, (ii) the need to extend the theory to a fully stochastic environment, and (iii) the important role of CDR in achieving the long-term goal of climate sustainability.

We will work with a simple dynamic game paradigm that takes into consideration the fundamental aspects of the climate negotiations environment discussed above. We first introduce a deterministic open-loop differential game model, and, in a second part, we extend the modeling to a stochastic diffusion game framework. In both cases we begin with a formulation of a Nash equilibrium played by a small number of coalitions, and we look at the limit game obtained when each coalition is composed of an ensemble of independent small players who share a similar economy, defined by its productivity and factor rates of substitution. Our approach is strongly influenced by the recent developments of the theory of mean field games (MFG) [7, 8, 14].

2 An *m*-Coalition Deterministic Model

In this section we propose a simple economic growth model with carbon emissions due to the use of fossil energy and possibility to invest in a CDR technology. We assume that the cumulative emission budget is shared among a small number *m* of coalitions of countries, like, e.g., EU-28+Switzerland, USA-Canada-Australia-New Zealand, BRICs, and the rest of the world. To realize efficiency an international emissions trading scheme is supposed to be installed with full banking and borrowing. Each coalition, considered as a "big" player *j*, may then use the supply of emission permits on the market as a strategic variable in order to maximize returns from the emission budget share they control. The other strategic variable will be the investment in CDR technologies in order to reduce the amount of emissions to offset. The model will then be used to assess the welfare loss (WL) for each coalition *j* with reference to a business as usual (BAU) situation where there is no climate policy. We then compare the WLs associated with different possible allocation schemes of the cumulative carbon budget.

2.1 Model Formulation

2.1.1 Economy

Economic output (Y) occurs in the economies according to an extended Cobb-Douglas production function in three inputs, capital (K), labor (L), and fossil energy (the use of which is measured through emission level E):

$$Y(j, t; \cdot) = A(j, t) K(j, t)^{\alpha(j)} (\phi(j, t) E(j, t))^{\theta(j,t)} L(j, t)^{1-\alpha(j)-\theta(j,t)}. \tag{1}$$

Here $A(j, t)$ is the total factor productivity in the economy j at time t, $\alpha(j)$ is the elasticity of output with respect to a malleable stock of capital $K(j, t)$, $\phi(j, t)$ is the energy conversion factor for emissions $E(j, t)$ in the economy j at time t, and $\theta(j, t)$ is the elasticity of output with respect to emissions E. This production function is homogenous of degree 1.

Another stock of capital R is used to remove carbon. In this simple model, we assume that the quantity $\varpi(j, t)$ of carbon removed at time t by coalition j is defined by a function

$$\varpi(j, t) = \Psi(j, t, R(j, t), L(j, t)), \tag{2}$$

which is assumed to be homogenous of degree 1. Economic output is used for consumption (C), investment (I), and the payment of energy costs:

$$Y(j, t) = C(j, t) + I_K(j, t) + I_R(j, t) + \pi(j, t) \phi(j, t) E(j, t), \tag{3}$$

where $\pi(j, t)$ is the energy price at period t in the economy of type j.

2.1.2   Emissions Trading

One assumes that an international market for emissions trading exists. At each
instant of time, given the quantities $\omega(j, t)$ put on the market by each player
$j = 1, \ldots, m$, the emission abatement decisions taken by the players are determined
by the solution of a local optimization problem, where the maximized profit is
given by:

$$Y(j, t) - \pi(j, t)\,\phi(j, t)\,\tilde{E}(j, t) + p(t, \Omega(t))\left[\omega(j, t) - \tilde{E}(j, t, p(t, \Omega(t)))\right],$$

where $p(t, \Omega(t))$ is the market carbon price, which depends on the total permit
supply $\Omega(t) = \sum_{j=1}^{m} \omega(j, t)$, and $\tilde{E}(j, t; p(t, \Omega(t)))$ is the emission response in
coalition j to carbon price $p(t, \Omega(t))$. The price $p(t, \Omega(t))$ is clearing the market,
i.e., the following conditions hold:

$$\Omega(t) \geq \sum_{j=1}^{m} \tilde{E}(j, t, p(t, \Omega(t))), \tag{4}$$

$$0 = p(t, \Omega(t))\left[\Omega(t) - \sum_{j=1}^{m} \tilde{E}(j, t, p(t, \Omega(t)))\right]. \tag{5}$$

At time t the permit market price and emission response of each player j are thus
defined by the equations:

$$0 = \frac{\partial Y(j, t; \cdot)}{\partial \tilde{E}(j, t, p(t, \Omega(t)))} - \pi(j, t)\phi(j, t) - p(t, \Omega(t)) \tag{6}$$

$$0 = \sum_{j=1}^{m} \tilde{E}(j, t, p(t, \Omega(t))) - \Omega(t). \tag{7}$$

Following the same developments as in [13], we can express the marginal influence
of the supply of permits on the emission levels and market price. Taking derivatives
of equation (6) and equation (7) w.r.t. $\Omega(t)$, denoted $'$, we obtain:

$$0 = \tilde{E}(j, t, p(t, \Omega(t)))' - \frac{p(t, \Omega(t))'}{\frac{\partial^2 Y(j, t; \cdot)}{\partial \tilde{E}(j, t, p(\Omega(t)))^2}} \tag{8}$$

$$0 = \sum_{j=1}^{m} \tilde{E}(j, t, p(\Omega(t)))' - 1. \tag{9}$$

Therefore, the derivatives w.r.t. Ω of price and emission levels are given by:

$$p'(t, \cdot) \equiv \frac{d}{d\Omega} p(t, \cdot) = \frac{1}{\sum_{j=1}^{m} \frac{1}{\frac{\partial^2 Y(j,t)}{\partial E(j,t)^2}}} \tag{10}$$

$$\tilde{E}'(j, \cdot) \equiv \frac{d}{d\Omega} \tilde{E}(j, t, \cdot) = \tilde{E}(j, t, p(t, \Omega))'$$

$$= \frac{1}{\sum_{k=1}^{m} \frac{\frac{\partial^2 Y(j,t)}{\partial \tilde{E}(j,t)^2}}{\frac{\partial^2 Y(k,t)}{\partial \tilde{E}(k,t)^2}}} \tag{11}$$

Since $\Omega(t) = \sum_{j=1}^{m} \omega(j, t)$, the partial derivatives w.r.t. $\omega(j, t)$ are the same as the derivatives w.r.t. $\Omega(t)$.

2.1.3 Capital and Emission Budget Dynamics

The dynamics of capital accumulation and GHG budget evolution is defined for $j = 1, \ldots, m$, by the following state equations:

$$\dot{K}(j, t) = I_K(j, t) - \delta_K K(j, t) \tag{12}$$

$$\dot{R}(j, t) = I_R(j, t) - \delta_R R(j, t) \tag{13}$$

$$\dot{B}(j, t) = \Psi(j, t, R(j, t), L(j, t)) - \omega(j, t) \tag{14}$$

where δ_K and δ_R are depreciation rates and initial conditions

$$K(j, 0) = K_j^o \tag{15}$$

$$R(j, 0) = R_j^o \tag{16}$$

$$B(j, 0) = B_j^o \tag{17}$$

are given.

B_j^o is the share of the remaining cumulative emission budget, given to coalition j in the climate negotiations that take place at time $t = 0$. $\Psi(j, t, \cdot, \cdot)$ is a production function, assumed to be concave in both arguments, which determines how much carbon dioxide is removed in region j at time t, given the stock of CDR capital accumulated and the population in that region. In equation (14) we have assumed that the CDR activity increases the emission budget. All state and control variables have to remain nonnegative, $K(j, t) \geq 0$, $R(j, t) \geq 0$, $B(j, t) \geq 0$.

2.1.4 Payoffs

The payoff for player j is the discounted sum of welfare over the planning horizon $[0, T]$, which can be written as follows:

$$\text{WRG}(j) = \int_0^T e^{-\rho t} L(j, t) \log \left[\frac{1}{L(j, t)} \Big(Y(j, t) - \pi(j, t)\phi(j, t)\tilde{E}(j, t, \Omega(t)) \right.$$

$$\left. - I_K(j, t) - I_R(j, t) + p(t, \Omega(t))\big(\omega(j, t) - \tilde{E}(j, t, \Omega(t))\big) \Big) \right] dt$$

$$+ e^{-\rho T} \mathscr{V}(j, T, \mathbf{K}(T), \mathbf{R}(T), \mathbf{B}(T)) \qquad (18)$$

where $\mathscr{V}(j, T, \mathbf{K}(T), \mathbf{R}(T), \mathbf{B}(T))$ is the value at terminal time t for player j associated with the terminal state $(\mathbf{K}(T), \mathbf{R}(T), \mathbf{B}(T))$ of the m players. Here $\tilde{E}(j, t, \Omega(t))$ denotes the emission response of coalition j at time t, given the total supply $\Omega(t)$ of permits on the carbon market.

2.2 Open-Loop Nash Equilibrium Conditions

For the ease of notation we introduce the utility function:

$$U_j(t, \cdot) = L(j, t) \log \left[\frac{1}{L(j, t)} \Big(Y(j, t) - \pi(j, t)\phi(j, t)\tilde{E}(j, t, \Omega(t)) - I_K(j, t) \right.$$

$$\left. - I_R(j, t) + p(t, \Omega(t))\big(\omega(j, t) - \tilde{E}(j, t, \Omega(t))\big) \Big) \right]$$

$$(19)$$

and the new consumption function, taking into consideration the payment for traded permits:

$$C(j, t) = Y(j, t) - \pi(j, t)\phi(j, t)\tilde{E}(j, t, \Omega(t)) - I_K(j, t)$$

$$- I_R(j, t) + p(t, \Omega(t))\big(\omega(j, t) - \tilde{E}(j, t, \Omega(t))\big).$$

$$(20)$$

For each player j define the Hamiltonian:

$$H_j(t, \cdot) = U_j(t, \cdot) + q_K(j, t)\dot{K}(j, t) + q_R(j, t)\dot{R}(j, t) + q_B(j, t)\dot{B}(j, t)$$

$$- \vartheta(j, t)B(j, t), \qquad (21)$$

where q_K, q_R, and q_B are current valued costate (or adjoint) variables and $\vartheta(j, t) \geq 0$ is the multiplier associated with the constraint $B(j, t) \geq 0$. At each time t the equations (6) and (7) define $\tilde{E}(j, t, \Omega(t))$ and $p(t, \Omega(t))$, and the following stationarity conditions must hold for a Nash equilibrium:

$$\omega(j, t) \geq 0$$

$$\omega(j, t) \frac{\partial}{\partial \omega(j, t)} H_j(t, \cdot) = 0$$

$$\frac{\partial}{\partial \omega(j, t)} H_j(t, \cdot) = \frac{\partial}{\partial C(j, t)} U_j(t, \cdot) \left[\frac{\partial Y(j, \cdot)}{\partial \tilde{E}(j, t, \Omega(t))} \tilde{E}(j, t, \Omega(t))' \right.$$

$$- (\pi(j, t)\phi(j, t) + p(t, \Omega(t)))\tilde{E}(j, t, \Omega(t))' + p(t, \Omega(t))$$

$$\left. + p(t, \Omega(t))'(\omega(j, t) - \tilde{E}(j, t, \Omega(t))) \right] - q_B(j, t) \leq 0$$

Using the market equilibrium condition (6), this last equation simplifies as:

$$\omega(j, t) \geq 0$$

$$\omega(j, t) \frac{\partial}{\partial \omega(j, t)} H_j(t, \cdot) = 0$$

$$\frac{\partial}{\partial \omega(j, t)} H_j(t, \cdot) = \frac{\partial}{\partial C(j, t)} U_j(t, \cdot) \left[p(t, \Omega(t)) \right.$$

$$\left. + p(t, \Omega(t))'(\omega(j, t) - E(j, t, \Omega(t))) \right]$$

$$- q_B(j, t) \leq 0 \tag{22}$$

The other equilibrium conditions are:

$$\frac{\partial}{\partial I_K(j, t)} H_j(t, \cdot) = -\frac{\partial}{\partial C(j, t)} U_j(t, \cdot) + q_K(j, t) = 0 \tag{23}$$

$$\frac{\partial}{\partial I_R(j, t)} H_j(t, \cdot) = -\frac{\partial}{\partial C(j, t)} U_j(t, \cdot) + q_R(j, t) = 0 \tag{24}$$

$$\dot{q}_K(j, t) = -\frac{\partial}{\partial K(j, t)} H_j(t, \cdot) + \rho q_K(j, t)$$

$$- \frac{\partial}{\partial K(j, t)} Y(j, t; \cdot) + \rho q_K(j, t) \tag{25}$$

$$\dot{q}_R(j, t) = -\frac{\partial}{\partial R(j, t)} H_j(t, \cdot) + \rho q_R(j, t)$$

$$-q_B(j,t)\frac{\partial}{\partial R(j,t)}\Psi(j,t,\cdot) + \rho q_R(j,t) \qquad (26)$$

$$\dot{q}_B(j,t) = -\frac{\partial}{\partial B(j,t)}H_j(t,\cdot) + \rho q_B(j,t)$$

$$= \vartheta(j,t) + \rho q_B(j,t) \qquad (27)$$

$$\vartheta(j,t)B(j,t) = 0 \qquad (28)$$

$$B(j,t) \geq 0 \qquad (29)$$

$$\vartheta(j,t) \geq 0. \qquad (30)$$

with terminal (transversality) conditions:

$$q_K(j,T) = \frac{\partial}{\partial K(j,T)}\mathcal{V}(j,T;\cdot) \qquad (31)$$

$$q_R(j,T) = \frac{\partial}{\partial R(j,T)}\mathcal{V}(j,T;\cdot) \qquad (32)$$

$$q_B(j,T) = \frac{\partial}{\partial B(j,T)}\mathcal{V}(j,T;\cdot). \qquad (33)$$

Remark 1. In this open-loop Nash equilibrium, each coalition has to take into account the marginal effect of its supply of emission permits on the carbon market, at each time t, as indicated by the derivatives $p(t, \Omega(t))'$ of the carbon price in equation (22). This effect should become smaller if the number of players increases, each player becoming smaller.

2.3   A Limit Game with Many Participants in Each Coalition

We consider the limit game that is obtained when one considers that each coalition is composed of many smaller countries that participate in the equilibrium solution. We propose the simple player replication scheme that has been used in [12] to show that a Wardrop equilibrium can be considered as the limit of a Nash equilibrium when the users of a congested network become infinitesimal. In this player replication scheme, each coalition "size" is divided by n, and these small agents are replicated n times.

The state and control variables for a small player of coalition j are denoted $l = \frac{1}{n}L$, $k = \frac{1}{n}K$, $b = \frac{1}{n}B$, $r = \frac{1}{n}R$, $i_k = \frac{1}{n}I_K$, $i_r = \frac{1}{n}I_R$, $e = \frac{1}{n}E$, $c = \frac{1}{n}C$, $y = \frac{1}{n}Y$ and $\omega_s = \frac{1}{n}\omega$, where:

$$c(j,t) = y(j,t) - \pi(j,t)\phi(j,t)e(j,t) - i_k(j,t) - i_r(j,t)$$

$$+ p(t)\big(\omega_s(j,t) - e(j,t)\big). \qquad (34)$$

The Hamiltonian for a small player is then:

$$h_j(t, \cdot) = l(j, t)U_j(c(j, t)/l(j, t)) + \tilde{q}_k(j, t)\dot{k}(j, t) + \tilde{q}_r(j, t)\dot{r}(j, t) +$$

$$\tilde{q}_b(j, t)\dot{b}(j, t) - \tilde{\vartheta}(j, t)b(j, t), \qquad (35)$$

where $U_j(t, \cdot)$ is still the utility of consumption for one individual, and $\tilde{q}_k = \frac{1}{n}q_K$, $\tilde{q}_r = \frac{1}{n}q_R$, and $\tilde{q}_b = \frac{1}{n}q_B$ are the costate variables for a small player, while $\tilde{\vartheta}$ is the multiplier for the nonnegativity constraint $b \geq 0$. The necessary conditions for an open-loop Nash-equilibrium can be rewritten for a small player in coalition j as follows:

$$\omega_s(j, t)\frac{\partial}{\partial \omega_s(j, t)}h_j(t, \cdot) = 0$$

$$\frac{\partial}{\partial \omega_s(j, t)}h_j(t, \cdot) = -\frac{\partial}{\partial c(j, t)}U_j(t, \cdot)\left[p(t, \Omega(t))\right.$$

$$\left. + p(t, \Omega(t))'\left(\omega_s(j, t) - e(j, t, \Omega(t))\right)\right]$$

$$-\tilde{q}_b(j, t) \leq 0 \qquad (36)$$

$$\frac{\partial}{\partial i_k(j, t)}h_j(t, \cdot) = -\frac{\partial}{\partial c(j, t)}U_j(t, \cdot) + \tilde{q}_k(j, t) = 0 \qquad (37)$$

$$\frac{\partial}{\partial i_r(j, t)}h_j(t, \cdot) = -\frac{\partial}{\partial c(j, t)}U_j(t, \cdot) + \tilde{q}_r(j, t) = 0 \qquad (38)$$

$$\dot{\tilde{q}}_k(j, t) = -\frac{\partial}{\partial k(j, t)}h_j(t, \cdot) + \rho\tilde{q}_k(j, t) \qquad (39)$$

$$\dot{\tilde{q}}_r(j, t) = -\frac{\partial}{\partial r(j, t)}h_j(t, \cdot) + \rho\tilde{q}_r(j, t) \qquad (40)$$

$$\dot{\tilde{q}}_b(j, t) = -\frac{\partial}{\partial b(j, t)}h_j(t, \cdot) + \rho\tilde{q}_b(j, t)$$

$$= \tilde{\vartheta}(j, t) + \rho\tilde{q}_b(j, t) \qquad (41)$$

$$\tilde{\vartheta}(j, t)b(j, t) = 0 \qquad (42)$$

$$b(j, t) \geq 0 \qquad (43)$$

$$\tilde{\vartheta}(j, t) \geq 0. \qquad (44)$$

Using "coalition size" variables in equation (36), one has:

$$\frac{1}{n}\frac{\partial}{\partial \omega(j, t)}H_j(t, \cdot) = \frac{1}{n}\frac{\partial}{\partial C(j, t)}U_j(t, \cdot)\left[p(t, \Omega(t))\right.$$

$$+ p(t, \Omega(t))' \left(\frac{1}{n} \omega(j, t) - \frac{1}{n} E(j, t, \Omega(t)) \right) \Bigg]$$

$$- \frac{1}{n} q_B(j, t)$$

At the limit when $n \to \infty$ the terms $\frac{1}{n} E(j, t, \Omega(t))$ and $\frac{1}{n} \omega(j, t)$ tend to 0. Therefore, equation (36) becomes:

$$0 = \omega(j, t) \left[\frac{\partial}{\partial C(j, t)} U_j(t, \cdot) p(t, \Omega(t)) - q_B(j, t) \right], \tag{45}$$

or equivalently when $\omega(j, t) > 0$:

$$\frac{L(j, t)}{C(j, t)} p(t, \Omega(t)) = q_B(j, t). \tag{46}$$

The other equilibrium conditions (6) and (37)–(40) become:

$$\frac{\partial Y(j, t; \cdot)}{\partial E(j, t)} = \pi(j, t) \phi(j, t) + p(t, \Omega(t)) \tag{47}$$

$$\frac{\partial}{\partial I_K(j, t)} H_j(t, \cdot) = -\frac{\partial}{\partial C(j, t)} U_j(t, \cdot) + q_K(j, t) = 0 \tag{48}$$

$$\frac{\partial}{\partial I_R(j, t)} H_j(t, \cdot) = -\frac{\partial}{\partial C(j, t)} U_j(t, \cdot) + q_R(j, t) = 0 \tag{49}$$

$$\dot{q}_K(j, t) = -\frac{\partial}{\partial K(j, t)} H_j(t, \cdot) + \rho q_K(j, t) \tag{50}$$

$$\dot{q}_R(j, t) = -\frac{\partial}{\partial R(j, t)} H_j(t, \cdot) + \rho q_R(j, t) \tag{51}$$

$$\dot{q}_B(j, t) = -\frac{\partial}{\partial B(j, t)} H_j(t, \cdot) + \rho q_B(j, t)$$

$$= \vartheta(j, t) + \rho q_B(j, t) \tag{52}$$

$$\vartheta(j, t) B(j, t) = 0 \tag{53}$$

$$B(j, t) \geq 0 \tag{54}$$

$$\vartheta(j, t) \geq 0. \tag{55}$$

Remark 2. In this limit-game Nash equilibrium, each coalition does not take into account the marginal effect of its supply of emission permits on the carbon market, at each time t, as indicated by the absence of the terms containing the derivatives $p(t, \Omega(t))'$ of the carbon price in equation (45). This limit equilibrium is similar to the Wardrop equilibrium in a congested network [12].

2.4 The Case Where the Emission Budget Has No Bequest Value

Let us assume that from time T on, the economies should be with net zero emissions. Then, instead of the m state equations (14), there will be m isoperimetric constraints

$$\int_0^T (\Psi(j, t, R(j, t), L(j, t)) - \omega(j, t)) \, dt = B_j \quad j = 1, \ldots, m. \tag{56}$$

We then use a different Hamiltonian

$$\mathcal{H}_j(t, \cdot) = U_j(t, \cdot) + e^{\rho t} v_j \left(\Psi(j, t, R(j, t), L(j, t)) - \omega(j, t) \right)$$
$$+ q_K(j, t)\dot{K}(j, t) + q_R(j, t)\dot{R}(j, t)), \tag{57}$$

where v_j is the multiplier associated with the isoperimetric constraint (isop1). The necessary condition (deriv-H2) becomes now

$$0 = \omega(j, t) \left[\frac{\partial}{\partial C(j, t)} U_j(t, \cdot) p(t, \Omega(t)) - e^{\rho t} v_j \right], \tag{58}$$

or equivalently when $\omega(j, t) > 0$:

$$\frac{L(j, t)}{C(j, t)} p(t, \Omega(t)) = e^{\rho t} v_j. \tag{59}$$

When one optimizes the weighted sum of payoffs with weights $\alpha_j, j = 1, \ldots m$, with the combined constraint

$$\sum_{j=1}^m \int_0^T (\Psi(j, t, R(j, t), L(j, t)) - \omega(j, t)) \, dt = B \tag{60}$$

one obtains the first-order condition

$$\alpha_j \frac{\partial}{\partial C(j, t)} U_j(t, \cdot) \left[p(t, \Omega(t)) \right.$$
$$\left. + \sum_{\ell=1}^m \alpha_\ell \frac{\partial}{\partial C(\ell, t)} U_\ell(t, \cdot) p(t, \Omega(t))' (\omega(j, t) - E(j, t, \Omega(t))) \right] = e^{\rho t} v, \tag{61}$$

where v is the multiplier associated with the global isoperimetric constraint (60). From (61) one easily obtains that the following conditions hold true

$$\alpha_j \frac{L(j, t)}{C(j, t)} p(t, \Omega(t)) = e^{\rho t} v \quad j = 1, \ldots, m. \tag{62}$$

This is the same condition as in (59) if we define $v_j = \frac{v}{\alpha_j}$.

Indeed, from (61) one obtains

$$\alpha_j \frac{\partial}{\partial C(j,t)} U_j(t,\cdot) p(t, \Omega(t)) =$$

$$-\sum_{\ell=1}^{m} \alpha_\ell \frac{\partial}{\partial C(\ell,t)} U_\ell(t,\cdot) p(t, \Omega(t))' \big(\omega(j,t) - E(j,t,\Omega(t))\big) + e^{\rho t} v, \qquad (63)$$

which implies that all the terms $\alpha_j \frac{\partial}{\partial C(j,t)} U_j(t,\cdot)$, $j = 1,\ldots,m$, are equal. This implies that

$$\sum_{\ell=1}^{m} \alpha_\ell \frac{\partial}{\partial C(\ell,t)} U_\ell(t,\cdot) p(t, \Omega(t))' \big(\omega(j,t) - E(j,t,\Omega(t))\big) = 0,$$

hence the result.

Therefore, when the budget has no bequest value, the limit game equilibrium is equivalent to an optimization of the weighted sum of payoffs for the m coalitions.

2.5 The Use of the Limit Game to Assess Fair Climate Agreement

For each coalition j we may compute a BAU payoff, using m decoupled optimal control problems, without restrictions on emissions. Call $\Phi(j,0)$ the discounted sum of welfare over the planning horizon. Now allocate a share $\bar{B}_j^0$ of the safe cumulative emission budget to coalition $j, j = 1,\ldots,m$.

Solving the limit game, as defined above, one obtains an expected welfare $W(j,0) \leq \Phi(j,0)$ and an expected price schedule $p(t, \Omega(t))$, expected emission schedule $E(j,t)$, and expected investment schedules in both types of capital and in supply of emission permits on the carbon market. We compute for each coalition the relative welfare loss (W_L):

$$W_L(j,0;\bar{B}(\cdot)) = \frac{\Phi(j,0) - W(j,0)}{\Phi(j,0)}. \qquad (64)$$

Now, changing the initial allocation $\bar{B}_j^0$, one solves the Rawlsian [19] distributive justice "meta-game":

$$\min_{\bar{B}(\cdot)} \max_j W_L(j,0;\bar{B}(\cdot)). \qquad (65)$$

Remark 3. Through this approach we will define a budget allocation for the different coalitions that will have a similar implication in terms of relative welfare loss, when all the players in all coalitions use their allocations in a noncooperative way, as price takers on the international carbon market.

3 A Stochastic Nash Equilibrium Model with m Coalitions of Players

As usual, when considering a dynamic game, one has to look at the possibility to define a subgame perfect equilibrium solution, for example, by introducing feedback or closed-loop strategies for the players. The use of such strategies is mandatory if the dynamic system is stochastic. In this section we reformulate the model in a full probabilistic framework, and we characterize a limit game, using a concept introduced in [14] and further developed under the name of mean field games [8, 16].

The dynamics of capital accumulation and GHG budget evolution is defined for $j = 1, \ldots, m$, as controlled stochastic diffusions:

$$dK(j, t) = \big(I_K(j, t) - \delta_K K(j, t)\big)dt + \sigma_K d\varepsilon_K(t), \tag{66}$$

$$dR(j, t) = \big(I_R(j, t) - \delta_R R(j, t)\big)dt + \sigma_R d\varepsilon_R(t), \tag{67}$$

$$dB(j, t) = \big(\Psi(j, t, R(j, t), L(j, t)) - \omega(j, t)\big)dt + \sigma d\varepsilon_B(t), \tag{68}$$

where $\varepsilon_K(t)$, $\varepsilon_R(t)$ and $\varepsilon_B(t)$ are independent Wiener processes (white noise).

The control variables must remain nonnegative, $I_K(j, t) \geq 0$, $I_R(j, t) \geq 0$, $\omega(j, t) \geq 0$. For the capital stocks $K(j, t), R(j, t)$, we may assume a reflecting boundary at 0. The budget $B(j, t)$ is not constrained to remain positive. A negative value indicates an over emission that will be penalized in the terminal value function.

The payoff for player j can thus be written as the expected value:

$$\mathrm{WRG}(j) = \mathrm{E}_\gamma \Bigg[\int_0^T e^{-\rho t} L(j, t) \log \Bigg[\frac{1}{L(j, t)} \Big\{ Y(j, t) - \pi(j, t)\phi(j, t)\tilde{E}(j, t, \Omega(t))$$

$$- I_K(j, t) - I_R(j, t)$$

$$+ p(t, \Omega(t))\big(\omega(j, t) - \tilde{E}(j, t, \Omega(t))\big) \Big\} \Bigg] dt$$

$$+ e^{-\rho T} \mathcal{W}(j, T; \mathbf{K}(T), \mathbf{R}(T), \mathbf{B}(T)) \Bigg] \tag{69}$$

where $\mathcal{W}(j, T; \mathbf{K}(T), \mathbf{R}(T), \mathbf{B}(T))$ is the value at terminal time t for player j associated with the terminal state $(\mathbf{K}(T), \mathbf{R}(T), \mathbf{B}(T))$ of the m players. Here the expected value is taken with respect to the probability measure induced by the strategies γ chosen by all players. We shall assume that these strategies are state feedbacks, that is, the investment decisions and the supply of permits of a coalition are functions of the state variables of all coalitions and of time. Indeed the players are interdependent through the emissions trading scheme, since their emissions $E(j, t, \Omega(t))$ and the carbon price $p(t, \Omega(t))$ are both depending on the total supply of permits:

$$\Omega(t) = \sum_{j=1}^{m} \omega(j, t). \tag{70}$$

This implies that the value function for each player will be a function of the state variables of all players.

Remark 4. We have assumed that the utility functions, as well as the production functions, are homogenous of degree 1. Therefore, the value functions should also be of this type.

3.1 Nash Equilibrium for the m Coalitions

3.1.1 HJB Equations

As shown in [21] and [4], a Nash equilibrium for this stochastic game will be characterized by a set of m coupled HJB equations that we summarize below. For all $j = 1, \ldots, m$ the current-valued payoff function $V(j, t; \mathbf{K}, \mathbf{R}, \mathbf{B})$ from time t, given state $(\mathbf{K}, \mathbf{R}, \mathbf{B})$ for all coalitions, satisfies the functional equations:

$$\rho V(j, t; \cdot) - \frac{\partial V(j, t; \cdot)}{\partial t} = \max_{\{I_K(j,t), I_R(j,t), \omega(j,t)\}} \left\{ L(j, t) \log \left[\frac{1}{L(j, t)} \left\{ Y(j, t) \right. \right. \right.$$

$$- \pi(j, t)\phi(j, t)\tilde{E}(j, t, \Omega(t)) - I_K(j, t) - I_R(j, t)$$

$$\left. + p(t, \Omega(t))(\omega(j, t) - \tilde{E}(j, t, \Omega(t))) \right\} \right]$$

$$+ \sum_{\ell=1}^{m} \left(\frac{\partial V(j, t; \cdot)}{\partial B(\ell, t)} (\Psi(\ell, t, R(\ell, t), L(j, t)) - \omega(\ell, t)) \right.$$

$$+ \frac{\partial V(j, t; \cdot)}{\partial K(\ell, t)} (I_K(\ell, t) - \delta_i K_i(\ell, t)) + \frac{\partial V(j, t; \cdot)}{\partial R(\ell, t)} (I_R(\ell, t) - \delta_R R(\ell, t))$$

$$\left. + \frac{\sigma_K^2}{2} \frac{\partial^2 V(j, t; \cdot)}{\partial K(\ell, t)^2} + \frac{\sigma_R^2}{2} \frac{\partial^2 V(j, t; \cdot)}{\partial R(j, t)^2} + \frac{\sigma_{B_\ell}^2}{2} \frac{\partial^2 V(j, t; \cdot)}{\partial B(\ell, t)^2} \right) \right\} \tag{71}$$

with initial conditions:

$$K(j, 0) = \bar{K}_j^0$$

$$B(j, 0) = \bar{B}_j^0,$$

$$j = 1, \ldots, m,$$

$$\sum_{j=1}^{m} \bar{B}_j^0 = \bar{B};$$

and terminal condition:

$$V(j, T; \mathbf{K}(T), \mathbf{R}(T), \mathbf{B}(T)) = \mathscr{W}(j, T; \mathbf{K}(T), \mathbf{R}(T), \mathbf{B}(T)), \quad j = 1, \ldots, m. \quad (72)$$

3.1.2 Equilibrium Strategies

The equilibrium feedback strategy is determined by the conditions of maximization of the RHS of the HJB equations. The optimality conditions are

Deriving w.r.t. $I_K(j, t)$:

$$0 = -\frac{L(j, t)}{C(j, t)} + \frac{\partial V(j, t; \cdot)}{\partial K(j, t)} \quad (73)$$

Deriving w.r.t. $I_R(j, t)$:

$$0 = -\frac{L(j, t)}{C(j, t)} + \frac{\partial V(j, t; \cdot)}{\partial R(j, t)} \quad (74)$$

Deriving w.r.t. $\omega(j, t)$:

$$0 = \frac{L(j, t)}{C(j, t)} \left[p(t, \Omega(t)) + p(t, \Omega(t))' (\omega(j, t) - E(j, t, \Omega(t))) \right]$$
$$- \frac{\partial V(j, t; \cdot)}{\partial B(j, t)}. \quad (75)$$

Notice that the players are interdependent through the determination of the price of carbon and the emission levels as shown by the derivatives (10) and (11), which enter into equation (75) determining the permit supply. They are also interdependent through the bequest function, which depends on all state variables for the m players.

Remark 5. In this stochastic Nash equilibrium, the equilibrium strategies are feedback rules over the whole set of state variables and time, because the value functions are themselves dependent on all state variables and time.

3.2 MFG Limit

The HJB equations for a Nash equilibrium will be extremely difficult to solve, due to the curse of dimensionality in dynamic programming. As the number of players grows, the difficulty increases also. However, if the players become "small," or infinitesimal, the Nash equilibrium can be approximated by an equilibrium in an MFG as shown below.

3.2.1 Replication Schemes and Measure Space of Agents

In the deterministic case, we used a simple replication scheme where each coalition j would be composed of N_j players of size $1/N_j$. Then we looked at the limit when $N_j \to \infty$. Notice that we could assimilate the weight $1/N_j$ to a probability, and the whole state of the coalition j would be the average of the states of each small player, when normalized to a coalition full value.

This probabilistic interpretation is essential in the stochastic case. Indeed, a country of size $1/N_j$ at time 0 will not remain at the same ratio when time passes, and the state dynamics is affected by random perturbations. Therefore, we shall work with a description of the countries that compose a coalition as a measure space of agents. In each coalition j, one considers a measure space of countries defined as follows.

Let $x_j(t) \equiv (L_j, K_j, R_j, B_j)(t)$ describe the state of a country at time t, once normalized at the coalition level. An elementary player associated with this state value will have a weight (or elementary probability) $dm_j(x_j(t); t)$, where $m_j(x_j(t); t)$ is a probability measure on the space of possible state values X_j when normalized at a coalition j full value. We may assume that a probability density function (pdf) $\Theta_j(L_j, K_j, R_j, B_j; t)$ defines the joint distribution of (L_j, K_j, R_j, B_j), at time t. The initial distribution $\Theta_j(L_j, K_j, R_j, B_j, 0)$ is given. This defines in particular a distribution of the initial endowment B_j of emission permits for each category in coalition j.

3.2.2 The Equilibrium Conditions for an Elementary Player

We suppose that, in each coalition, all the infinitesimal players are competing. Indeed because the number of actors becomes very large (infinite), each player will base its decision at time t on the observation of its own state and the information provided by the m pdf's $\Theta_j(K_j, R_j, B_j; t)$.

An elementary player in coalition j is characterized at time t by elementary state

$$(dL_j(t), dK_j(t), dR_j(t), dB_j(t)) = dm_j(x_j(t); t)(L_j(t), K_j(t), R_j(t), B_j(t)),$$

where $x_j(t) \equiv (L_j(t), K_j(t), R_j, B_j(t))$ is the state normalized at coalition level, and $dm_j(x(t); t)$ is the elementary probability of $x(t)$ at time t. Let us write the Nash equilibrium condition for this elementary player.

1. Taking into consideration the homogeneity of degree 1 of the value functions, the necessary condition corresponding to (73) is now

$$0 = -\frac{dm_j(x_j(t); t)L(j, t)}{dm_j(x_j(t); t)C(j, t)} + \frac{\partial V(j, t; dm_j(x_j(t); t)x_j(t))}{\partial dm_j(x_j(t))K(j, t)}$$

$$0 = -\frac{L(j, t)}{C(j, t)} + \frac{\partial V(j, t; \cdot)}{\partial K(j, t)},$$

which is precisely the same as (73). The same reasoning will apply to conditions (74), although for the condition (75), we obtain

$$
0 = \frac{L(j,t)}{C(j,t)}\left[p(t,\Omega(t)) + p(t,\Omega(t))'dm_j(x_j(t);t)x_j(t))(\omega(j,t) - E(j,t,\Omega(t)))\right]
$$
$$
- \frac{\partial V(j,t;\cdot)}{\partial B(j,t)}
$$
$$
= \frac{L(j,t)}{C(j,t)}p(t,\Omega(t)) - \frac{\partial V(j,t;\cdot)}{\partial B(j,t)} + O(dm_j(x_j(t))). \qquad (76)
$$

where $O = (dm_j(x_j(t));t))$ tends to 0 when $dm_j(x_j(t);t);t)$ tends to 0,

2. The total supply of permits on the market $\Omega(t)$ is thus defined as the average of supply for coalition j,

$$
\Omega(t) = \sum_{j=1}^{m} \int_{X_j} \tilde{\omega}_j(x_j(t);t)\, dm_j(x_j(t);t);t) + O(dm_j(x_j(t);t)), \qquad (77)
$$

where $\tilde{\omega}_j(x(t);t)$ is the supply of permit by all the elementary players as defined by their respective feedback decision rules and $O = (dm_j(x_j(t);t))$ tends to 0 when $dm_j(x_j(t));t)$ tends to 0.

3.3 The Limit MFG

Hence, for each country (player) of coalition j, the value function will satisfy a "local" HJB equation:

$$
\rho V(j,t;\cdot) - \frac{\partial V(j,t;\cdot)}{\partial t} = \max_{\{I_K(j,t),I_R(j,t),\omega(j,t)\}} \left\{ L(j,t)\log\left[\frac{1}{L(j,t)}\left\{Y(j,t)\right.\right.\right.
$$
$$
- \pi(j,t)\phi(j,t)\tilde{E}(j,t,\Omega(t)) - I_K(j,t) - I_R(j,t)
$$
$$
\left.\left.+ p(t,\Omega(t))(\omega(j,t) - \tilde{E}(j,t,\Omega(t)))\right\}\right]
$$
$$
+ \frac{\partial V(j,t;\cdot)}{\partial B(j,t)}(\Psi(j,t,R(j,t),L(j,t)) - \omega(j,t)) + \frac{\partial V(j,t;\cdot)}{\partial K(j,t)}(I_K(\ell,t) - \delta_i K_i(\ell,t))
$$
$$
+ \frac{\partial V(j,t;\cdot)}{\partial R(j,t)}(I_R(j,t) - \delta_R R(j,t))
$$
$$
+ \frac{\sigma_K^2}{2}\frac{\partial^2 V(j,t;\cdot)}{\partial K(j,t)^2} + \frac{\sigma_R^2}{2}\frac{\partial^2 V(j,t;\cdot)}{\partial R(j,t)^2} + \frac{\sigma_{B_j}^2}{2}\frac{\partial^2 V(j,t;\cdot)}{\partial B(j,t)^2}\right\}. \qquad (78)
$$

At time t, given a global supply of permits $\Omega(t)$, the carbon market price and emission response $\mathscr{E}(j, t, , \Omega(t); K_j, R_j, B_j)$ of the players of coalition j are defined by the equations:

$$0 = \frac{\partial Y(j, t; \cdot)}{\partial \tilde{E}(j, t, \cdot)} - \pi(j, t)\phi(j, t) + p(t, \Omega(t)) \tag{79}$$

$$\Omega(t) = \sum_{j=1}^{m} \int \Theta_j(K_j, R_j, B_j, t)\tilde{E}(j, t, \Omega(t); K_j, R_j, B_j)\, d\Theta. \tag{80}$$

Therefore, both the carbon price and the emission level of the players are now functionals of the pdfs $\Theta_j(K_j, R_j, B_j, t)$ for $j = 1, \ldots, m$. Now, a given player (country), being infinitesimal, has no influence on the carbon market equilibrium by changing its infinitesimal level of supply $\omega(j, t)$.

Hence, the equilibrium strategy defining the permit supply of a player of type j in equation (75) for a Nash equilibrium is now defined by the simpler condition:

$$\frac{L(j, t)}{C(j, t)}p(t, \Omega(t)) = \frac{\partial V(j, t; \cdot)}{\partial B(j, t)}, \tag{81}$$

which corresponds to equation (46) in the deterministic case.

As $p(t, \Omega(t))$ and $E(j, t; \Omega(t))$ are defined by the market-clearing conditions (79)-(80), we can summarize the optimal strategies as functions:

$$\varphi_{I_K}(K(j, t), R(j, t), B(j, t); \Theta(\cdot, t))$$

$$\varphi_{I_R}(K(j, t), R(j, t), B(j, t); \Theta(\cdot, t))$$

and

$$\psi_{\omega_j}(K(j, t), R(j, t), B(j, t)); \Theta(\cdot, t)).$$

Finally, for each coalition j, the distribution $\Theta_j(x, t)$ evolves according to a Kolmogorov F-P (KFP) equation, where x stands for the vector of all state variables (K, R, B) and Q denotes the covariance matrix:

$$\frac{\sigma^2}{2}\partial_{xx}^2\left[\|x\|_Q^2 \Theta_j(x, t)\right] = \partial_t \Theta(x, t)$$

$$+ \partial_{K_i(j,t)}\left[\Theta_j(x, t)\left(\varphi_{I_K}(K(j, t), R(j, t), B(j, t); \Theta(\cdot, t)) - \delta_i K_i(j, t)\right)\right]$$

$$+ \partial_{R(j,t)}\left[\Theta_j(x, t)\left(\varphi_{I_R}(K(j, t), R(j, t), B(j, t); \Theta(\cdot, t)) - \delta_i K_i(j, t)\right)\right]$$

$$- \partial_{B(j,t)}\left[\Theta_j(x, t)\psi_{\omega_j}(K(j, t), R(j, t), B(j, t)); \Theta(\cdot, t))\right]. \tag{82}$$

The terminal conditions must be reformulated as follows:

$$V(j, T; K(j, T), R(j, T), B(j, T); \Theta(\cdot, T)) =$$

$$\mathscr{W}(j, T; K(j, T), R(j, T), B(j, T); \Theta(\cdot, T)). \qquad (83)$$

This means that the bequest function depends for each coalition on its own state variables at time T and the distribution of all players at final time T.

In summary, this MFG equilibrium is defined by a fixed-point condition involving the pdfs $\Theta_j(x, t)$. Given a nominal set of m pdfs $\Theta_j(x, t)$, for each t, one can characterize, through the HJB equations (78) and the carbon market conditions (8)-(9), a set of equilibrium strategies, which, injected in the KFP equations, define a new set of pdfs $\bar{\Theta}_j(x, t)$. The equilibrium is reached when $\Theta_j(x, t) \equiv \bar{\Theta}_j(x, t)$ for $j = 1, \ldots, m$.

Remark 6. In this mean field game limit for the stochastic Nash equilibrium, we have used the concept of a measure space of agents in each coalition j. This is a useful abstraction introduced initially by Robert Auman [2]. The challenge for the modeler will be to determine, from the economic data describing the coalitions, the pdf's at the initial time, which are part of the initial conditions for this system of dynamic equations. As the HJB equations are decoupled, the complexity for computing the equilibrium is much reduced. Once the value functions are obtained, the meta-game for a fair allocation of the emission budget to each coalition can be formulated as in Section 2.5.

4 Conclusion

In this paper we have proposed a dynamic game paradigm to assess the fairness of the allocation of the safe cumulative emission budget to the different coalitions of countries that are engaged in climate negotiations. We have treated both the deterministic and fully stochastic cases. By relying on the limit game formulations obtained using the MFG paradigms, we propose an approach which circumvents the obstacle of dimensionality for computing equilibria.

Without the possibility to use CDR, the exploitation of the share of the emission budget by a country or a group of countries would be very similar to the exploitation of a nonrenewable resource, like oil, for example. We would therefore rediscover a form of the Hotelling rule in the deterministic limit games discussed in the paper. In a stochastic context, the developments would be very close to those of Giraud et al. [9] dealing with oil field exploitation.

The numerical solution for the stochastic case remains a challenge although one that can be managed, unlike the computation of many player stochastic Nash equilibria, which are notoriously still out of reach of existing numerical methods. The numerics will be the object of a further development of this research on the game theoretic dimension of climate change negotiations. The developments in this paper have shown that the MFG concepts could have interesting applications in the assessment of negotiations for the design of fair climate change policies.

Acknowledgements This research has been supported by the Natural Sciences and Engineering Research Council of Canada (O. Bahn and R. Malhamé).

References

1. M.R. Allen, D.J. Frame, C. Huntingford, C.D. Jones, J.A. Lowe, M. Meinshausen, and N. Meinshausen. Warming caused by cumulative carbon emissions towards the trillionth tonne. *Nature*, 458:1163–1166, 2009.
2. R. J. Aumann. Markets with a continuum of traders. *Econometrica*, 32:39–50, 1964.
3. F. Babonneau, A. Haurie, and M. Vielle. A robust meta-game for climate negotiations. *Computational Management Science*, 13:1–31, January 2013.
4. T. Başar and G.J. Olsder. *Dynamic Noncooperative Game Theory*. Academic Press, New York, 1995.
5. O. Bahn and A. Haurie. A class of games with coupled constraints to model international ghg emission agreements. *International Game Theory Review*, Vol. 10:337–362, 2008.
6. O. Bahn and A. Haurie. A cost-effectiveness differential game model for climate agreements. *Dynamic Games and Applications*, 6(1):1–19, 2016.
7. P.-N. Giraud, O. Guéant, J.-M. Lasry, and P.-L. Lions. A mean field game model of oil production in presence of alternative energy producers. mimeo 2010?
8. O. Guéant. A reference case for mean field games models. *J. Math.Pures Appl.*, 92:276–294, 2009.
9. O. Guéant, J. M. Lasry, and P. L. Lions. Mean field games and oil production. Technical report, HAL, 2010.
10. S. Hallegatte, J. Rogelj, M. Allen, L. Clarke, O. Edenhofer, C.B. Field, P. Friedlingstein, L. van Kesteren, R. Knutti, K.J. Mach, M. Mastrandrea, A. Michel, J. Minx, M. Oppenheimer, G.-K. Plattner, K. Riahi, M. Schaeffer, T.F. Stocker, and D.P. van Vuuren. Mapping the climate change challenge. *Nature Clim. Change*, 6(7):663–668, 07 2016.
11. A. Haurie, F. Babonneau, N. Edwrads, P. Holden, A. Kanudia, M. Labriet, M. Leimbach, B. Pizzileo, and M. Vielle. *Macroeconomics of Global Warming*, chapter Fairness in Climate Negotiations : a Meta-Game Analysis Based on Community Integrated Assessment. Oxford Handbook, 2014.
12. A. Haurie and P. Marcotte. On the relationship between Nash-Cournot and Wardrop equilibria. *Networks*, 15:295–308, 1985.
13. C. Helm. International emissions trading with endogenous allowance choices. *Journal of Public Economics*, 87:2737–2747, 2003.
14. M. Huang, R. Malhamé, and P.E. Caines. Large population stochastic dynamic games: closed-loop Mckean-Vlasov systems and the nash certainty equivalence principle. *Commun. Inf. Syst.*, 6(3):221–252, 2006.
15. R. Knutti, J. Rogelj, J. Sedlacek, and E.M.Fischer. A scientific critique of the two-degree climate change target. *Nature Geoscience*, 9(1):13–18, January 2016.
16. J.-M. Lasry and P.-L. Lions. Mean field games. *Japanese Journal of Mathematics*, 2(1):229–260, 2007.
17. S. Mathesius, M. Hofmann, K. Caldeira, and Schellnhuber H.-J. Long-term response of oceans to co2 removal from the atmosphere. *Nature Climate Change*, 5(12):1107–1113, 2015.
18. J. Meadowcroft. Exploring negative territory carbon dioxide removal and climate policy initiatives. *Climatic Change*, 118(1):137–149, 2013.

19. J. Rawls. *A theory of justice*. Harvard University Press, 1971.
20. M. Tavoni and R. Socolow. Modeling meets science and technology: an introduction to a special issue on negative emissions. *Climatic Change*, 118(1):1–14, 2013.
21. K. Uchida. On the existence of Nash equilibrium point in n-person nonzero-sum stochastic differential games. *SIAM Journal on Control and Optimization*, 16(1):142–149.

A Segregation Problem in Multi-Population Mean Field Games

Pierre Cardaliaguet, Alessio Porretta, and Daniela Tonon

Abstract We study a two-population mean field game in which the coupling between the two populations becomes increasingly singular. In the case of a quadratic Hamiltonian, we show that the limit system corresponds a partition of the space into two components in which the players have to solve an optimal control problem with state constraints and mean field interactions.

Mean field games (MFG) formalize noncooperative dynamic games with an infinite number of players. The interest for such problems started with the pioneering works of Lasry and Lions [21–23] and of Huang, Caines, and Malhamé [16, 17]. Since then the theory has grown very fast: see, for instance, the survey papers [4, 14] or the monographs [2, 15].

We consider here MFG problems with several populations. Let us recall that the analysis of multi-population mean field games goes back to the early work of Huang, Malhamé, and Caines [16]. More recently, Lachapelle and Wolfram [19] discuss a multi-population time-dependent MFG problem. Feleqi [11] investigates the derivation of the ergodic multi-population model from an ergodic N-player differential game as the number N of players tends to infinity (hence generalizing Lasry-Lions result [23]). Cirant [8] analyzes multi-population models in bounded domains with Neumann-type boundary conditions and provides examples of MFG where multiplicity of solutions arises (see also Kolokoltsov, Li, and Yang [18] for very general diffusions). An interesting issue, first discussed in an example in [8] and then investigated in more details by Cirant and Verzini [9], is the vanishing viscosity framework, which, since the two populations are antagonist, leads to "segregation phenomena," i.e., to situations where the support of the two population densities is disjoint. Cacace and Camilli [3] and Achdou, Bardi, and Cirant [1] present numerical evidence of such a phenomenon.

P. Cardaliaguet (✉) • D. Tonon
Université Paris-Dauphine, PSL Research University, CNRS, Ceremade, 75016 Paris, France
e-mail: cardaliaguet@gmail.com

A. Porretta
Dipartimento di Matematica, Università di Roma "Tor Vergata", Roma, Italy

© Springer International Publishing AG 2017
J. Apaloo, B. Viscolani (eds.), *Advances in Dynamic and Mean Field Games*,
Annals of the International Society of Dynamic Games 15,
https://doi.org/10.1007/978-3-319-70619-1_3

In this note we investigate a model similar to the one of [8, 9], but, instead of assuming that the viscosity of the population vanishes, we suppose that the coupling between the two populations becomes increasingly singular. More precisely, our MFG system takes the form

$$
\begin{cases}
\lambda_i - \Delta u_i(x) + H(x, Du_i(x)) = f_i(x, m_i(x)) + \dfrac{m_j(x)}{\epsilon} & \text{in } \mathbb{T}^d, \ i,j = 1,2, i \neq j \\
-\Delta m_i(x) - \operatorname{div}(m_i D_p H(x, Du_i(x))) = 0 & \text{in } \mathbb{T}^d, \ i = 1,2 \\
m_i(x) \geq 0, \ \displaystyle\int_{\mathbb{T}^d} m_i(x)dx = 1
\end{cases}
$$

(1)

where $\epsilon > 0$ is a small parameter (for simplicity, we work with periodic boundary conditions, i.e., in the torus $\mathbb{T}^d = \mathbb{R}^d/\mathbb{Z}^d$). The singular term m_j/ϵ describes a strong cost for a player of population i to be in places where the population j has a positive density. The interesting point is that the diffusion of both populations is fixed and positive: so both populations have a natural tendency to be spread in space. The players have therefore to take a special care to ensure that their individual state does not intersect the support of the other population density. At least formally, one expects that the limit problem should have the structure of two state constraint stochastic control problems (one for each population): namely, the state space should be split into two domains, and each population should have an ergodic control problem with state constraint inside of each domain. Moreover, one expects the partition of each domain to be optimal for some cost criterium.

Note that in general, the solution to (1) is not unique and one cannot expect this behavior to hold for any solution. For instance, if $H(x, 0) = 0$ and the f_i, $i = 1, 2$, do not depend on x, then an obvious solution to (1) is given by $u_i = 0$, $m_i = 1$, and $\lambda_i = f_i(1) + 1/\epsilon$, which has no segregated limit as $\epsilon \to 0$. It turns out that (1) has a variational structure and that the segregation result is expected to hold for minimizers of the variational problem. In this setting, we show that the pair of minimizers $(m_1^\epsilon, m_2^\epsilon)$ converges to a minimizer of a limit functional which involves the exclusion constraint $m_1 m_2 = 0$ (see Proposition 3.2 for details). Unfortunately, a detailed qualitative description of minimizers of this limit functional, and in particular the derivation of the related mean field game system, seems out of reach in full generality for the moment.

As in [9], but in a different framework, we manage to give a complete description of the limit problem in the case of quadratic Hamiltonian ("the model case") and when $f_1 = f_2$: indeed, in this case, a classical change of variable (Hopf-Cole transform) yields to an optimal partition problem in the flavor of Conti, Terracini, and Verzini [10] (see also [24]). Using techniques from this paper, we show that the (minimal) solution to (1) converges to a couple $(\bar{m}_i, \hat{u}_i)$ which is a classical solution of the state constraint ergodic control problem

$$
\begin{cases}
\lambda_i - \Delta u_i + \dfrac{1}{2}|Du_i|^2 = f(x, m_i) & \text{in } \mathbb{T}^d \cap \{m_i > 0\}, \ i = 1, 2 \\
-\Delta m_i - \operatorname{div}(m_i Du_i) = 0 & \text{in } \mathbb{T}^d, \ i = 1, 2 \\
m_i \geq 0, \ \displaystyle\int_{\mathbb{T}^d} m_i = 1, \ m_1 m_2 = 0.
\end{cases}
$$

In addition, if, for a (sufficiently smooth) subset Ω of $\mathbb{T}^d$, we set

$$\mu(\Omega) := \min\left\{\int_\Omega [|Dv|^2 + \frac{1}{2}F(x, v^2)], \quad v \in H_0^1(\Omega) : \quad \int_\Omega v^2 = 1\right\},$$

(where F is an antiderivative of f in the last variable), then the partition $\{m_1 > 0\}, \{m_2 > 0\}$ of $\mathbb{T}^d$ is optimal for the problem:

$$\inf\left\{\mu(\Omega_1) + \mu(\Omega_2), \quad \Omega_1, \Omega_2 \text{ open sets: } \overline{\Omega}_1 \cup \overline{\Omega}_2 = \mathbb{T}^d, \Omega_1 \cap \Omega_2 = \emptyset\right\}.$$

1 Assumptions

Throughout the paper, we assume that, for $i = 1, 2, f_i : \mathbb{T}^d \times [0, +\infty) \to \mathbb{R}$ is continuous, increasing with respect to the second variable and satisfies the following inequality for some $q > 1$

$$\frac{1}{C_0}|m|^{q-1} - C_0 \le f_i(x, m) \le C_0|m|^{q-1} + C_0 \quad \forall m > 0. \tag{2}$$

Let us define

$$F_i(x, m) := \begin{cases} \int_0^m f_i(x, \rho)d\rho & \text{if } m \ge 0 \\ +\infty & \text{otherwise} \end{cases}.$$

Then F_i is continuous on $\mathbb{T}^d \times [0, +\infty)$, strictly convex and differentiable in the second variable, and satisfies

$$\frac{1}{qC_0}|m|^q - C_0 \le F_i(x, m) \le \frac{C_0}{q}|m|^q + C_0. \tag{3}$$

The Hamiltonian $H : \mathbb{T}^d \times \mathbb{R}^d \to \mathbb{R}$ is supposed to be strictly convex, continuous in both variables, and differentiable in the second variable, with $D_p H$ continuous in both variables. Moreover, the Hamiltonian is supposed to verify, for some $r > 1$,

$$\frac{1}{rC_0}|p|^r - C_0 \le H(x, p) \le \frac{1}{r}C_0|p|^r + C_0. \tag{4}$$

We note that, due to the above assumption, H^*, the Fenchel conjugate with respect to p of H, satisfies the inequalities

$$\frac{1}{r'C_0}|p|^{r'} - C_0 \le H^*(x, p) \le \frac{1}{r'}C_0|p|^{r'} + C_0, \tag{5}$$

with r' solving $\frac{1}{r} + \frac{1}{r'} = 1$.

Let us define the energy of the problem as

$$\mathcal{E}^\epsilon(m_1, w_1, m_2, w_2) := \int_{\mathbb{T}^d} m_1 H^*(x, -\frac{w_1}{m_1}) + m_2 H^*(x, -\frac{w_2}{m_2}) + F_1(x, m_1) + F_2(x, m_2) + \frac{m_1 m_2}{\epsilon}$$

where (m_1, w_1) and (m_2, w_2) satisfy the constraints

$$- \Delta m_i + \text{div}(w_i) = 0 \qquad \text{in } \mathbb{T}^d, \ i = 1, 2. \tag{6}$$

2   Existence of a Solution with Minimal Energy

Let $\alpha := r'q/(r' + q - 1) > 1$.

Definition 2.1 *Let, for $i = 1, 2$, $(\lambda_i, u_i, m_i) \in \mathbb{R} \times W^{1,r}(\mathbb{T}^d) \times W^{1,\alpha}(\mathbb{T}^d) \cap L^q(\mathbb{T}^d)$, we say that $(\lambda_1, u_1, m_1, \lambda_2, u_2, m_2)$ is a weak solution for the ergodic MFG system (1) if:*

i) $m_i \geq 0$, $\int_{\mathbb{T}^d} m_i = 1$, $\int_{\mathbb{T}^d} u_i = 0$ *and* $m_i D_p H(x, Du) \in L^\alpha(\mathbb{T}^d; \mathbb{R}^d)$;
ii)

$$\lambda_i - \Delta u_i + H(x, Du_i) \leq f_i(x, m_i) + \frac{m_j}{\epsilon} \qquad \text{in } \mathbb{T}^d \qquad i, j = 1, 2, i \neq j$$

holds in the sense of distributions;
iii)

$$-\Delta m_i - \text{div}(m_i D_p H(x, Du_i)) = 0 \quad \text{in } \mathbb{T}^d, \ i = 1, 2$$

holds in the sense of distributions;
iv)

$$\lambda_i = \int_{\mathbb{T}^d} m_i \left(f(x, m_i) + \frac{m_j}{\epsilon} + H^*(x, -\frac{w_i}{m_i}) \right) \qquad i, j = 1, 2, i \neq j.$$

Remark 2.2 *Note that (iv) roughly says that there is an equality in inequality (ii) on $\{m_i > 0\}$. Indeed, this is exactly what one obtains when testing (ii) against m_i and one uses (iii). Although we will not address this question in the paper, we can expect to obtain stronger solutions for $1 < r \leq 2$: in this case one should expect to obtain that the m_i are positive and that equality holds in (ii).*

Theorem 2.3 *For any $\epsilon > 0$, there exists at least one weak solution $(\lambda_1^\epsilon, u_1^\epsilon, m_1^\epsilon, \lambda_2^\epsilon, u_2^\epsilon, m_2^\epsilon)$ to the MFG system (1), such that $(m_1^\epsilon, -m_1^\epsilon D_p H(\cdot, Du_1^\epsilon), m_2^\epsilon, -m_2^\epsilon D_p H(\cdot, Du_2^\epsilon))$ minimizes the energy $\mathcal{E}^\epsilon$ under the constraints (6).*

Proof Let (m_1, w_1, m_2, w_2) be smooth with finite energy $\mathcal{E}^\epsilon$, $m_i > 0$ and satisfying the equation (6). From the definition of $\mathcal{E}^\epsilon$ and the growth conditions (3) and (5), it follows that

$$\mathscr{E} \geq \int_{\mathbb{T}^d} \frac{m_1}{C_0} \left| \frac{w_1}{m_1} \right|^{r'} + \int_{\mathbb{T}^d} \frac{m_2}{C_0} \left| \frac{w_2}{m_2} \right|^{r'} + \int_{\mathbb{T}^d} \frac{1}{C_0} |m_1|^q + \int_{\mathbb{T}^d} \frac{1}{C_0} |m_2|^q + \frac{m_1 m_2}{\epsilon} - 4C_0 \,.$$

Therefore, for $i = 1, 2$,

$$\|m_i\|_{L^q(\mathbb{T}^d)}^q \leq C \left(1 + \mathscr{E}^\epsilon(m_1, w_1, m_2, w_2) \right)$$

and

$$\int_{\mathbb{T}^d} \left(\frac{|w_i|}{m_i^{1/r}} \right)^{r'} = \int_{\mathbb{T}^d} \frac{|w_i|^{r'}}{m_i^{r'-1}} \leq C \left(1 + \mathscr{E}^\epsilon(m_1, w_1, m_2, w_2) \right) \,.$$

Moreover, using Holder inequality, we obtain that, for $i = 1, 2$,

$$\int_{\mathbb{T}^d} |w_i|^{\frac{r'q}{r'+q-1}} \leq \left(\int_{\mathbb{T}^d} m_i^q \right)^{\frac{r'-1}{r'+q-1}} \left(\int_{\mathbb{T}^d} \frac{|w_i|^{r'}}{m_i^{r'-1}} \right)^{\frac{q}{r'+q-1}} \leq C \left(1 + \mathscr{E}^\epsilon(m_1, w_1, m_2, w_2) \right)$$

where $\alpha := \frac{r'q}{r'+q-1} > 1$ since $r' > 1$ and $q > 1$.

Suppose $\alpha \geq 2$. Then, since, for $i = 1, 2$, $(m_i, w_i,)$ satisfies (6), and $w_i \in L^\alpha(\mathbb{T}^d)$, a classical theorem on elliptic partial differential equations (see, e.g., Theorem 7.1 of [12]), ensures that $m_i \in W^{1,\alpha}(\mathbb{T}^d)$ and

$$\|Dm_i\|_{L^\alpha(\mathbb{T}^d)} \leq C \|w_i\|_{L^\alpha(\mathbb{T}^d)} \,.$$

The same result can be obtained through a duality argument for $1 < \alpha < 2$.

Accordingly, if $(m_1^n, w_1^n, m_2^n, w_2^n)$ is a minimizing sequence for $\mathscr{E}^\epsilon$, then a subsequence (denoted in the same way) converges weakly in $W^{1,\alpha} \times L^\alpha \times W^{1,\alpha} \times L^\alpha$ to a quadruple $(\bar{m}_1, \bar{w}_1, \bar{m}_2, \bar{w}_2)$ which solves (6). As the functional

$$(m_1, w_1, m_2, w_2) \rightarrow \int_{\mathbb{T}^d} m_1 H^*(x, -\frac{w_1}{m_1}) + m_2 H^*(x, -\frac{w_2}{m_2})$$

is convex, we obtain

$$\int_{\mathbb{T}^d} \bar{m}_1 H^*(x, -\frac{\bar{w}_1}{\bar{m}_1}) + \bar{m}_2 H^*(x, -\frac{\bar{w}_2}{\bar{m}_2}) \leq \liminf \int_{\mathbb{T}^d} m_1^n H^*(x, -\frac{w_1^n}{m_1^n}) + m_2^n H^*(x, -\frac{w_2^n}{m_2^n}) \,.$$

Moreover, (m_1^n, m_2^n) converges strongly in L^α and a.e., so that by Fatou,

$$\int_{\mathbb{T}^d} F_1(x, \bar{m}_1) + F_2(x, \bar{m}_2) + \frac{\bar{m}_1 \bar{m}_2}{\epsilon} \leq \liminf \int_{\mathbb{T}^d} F_1(x, m_1^n) + F_2(x, m_2^n) + \frac{m_1^n m_2^n}{\epsilon} \,.$$

Therefore the quadruplet $(\bar{m}_1, \bar{w}_1, \bar{m}_2, \bar{w}_2)$ minimizes $\mathscr{E}^\epsilon$.

It remains to associate with $(\bar{m}_1, \bar{w}_1, \bar{m}_2, \bar{w}_2)$ a weak solution to the MFG system (1). We note that if we freeze $(\bar{m}_2, \bar{w}_2)$, then $(\bar{m}_1, \bar{w}_1)$ is the unique minimizer of the strictly convex functional

$$(m_1, w_1) \rightarrow \mathscr{E}^\epsilon(m_1, w_1, \bar{m}_2, \bar{w}_2). \tag{7}$$

Let us assume for a while that $\bar{m}_2$ is L^∞: we remove this assumption at the end of the proof. Following [5–7], the minimization problem in (7) can be seen as the dual problem—in the sense of Fenchel-Rockafellar—of the convex problem:

$$\inf_{(\lambda, \alpha, u)} \int_{\mathbb{T}^d} F_1^*(x, \alpha(x))dx - \lambda$$

where the infimum is taken over triplets $(\lambda, \alpha, u) \in \mathbb{R} \times L^{q'}(\mathbb{T}^d) \times W^{1,r}(\mathbb{T}^d)$ where u satisfies the constraint (in the sense of distributions)

$$\lambda - \Delta u(x) + H(x, Du(x)) \leq \alpha(x) + \frac{\bar{m}_2(x)}{\epsilon} \quad \text{in } \mathbb{T}^d \quad \text{and} \quad \int_{\mathbb{T}^d} u = 0.$$

Here we have set

$$F_1^*(x, \alpha) := \sup_{m \geq 0} m\alpha - F_1(x, m),$$

so that $F_1^*(x, \alpha) = 0$ for $\alpha \leq 0$. One easily checks that this problem has at least one solution $(\bar{\lambda}_1, \bar{u}_1, \bar{\alpha}_1)$ with $\bar{\alpha}_1 \geq 0$ (see, for instance, [6] and [5]).

Our next claim is that the constraint inequality for $(\bar{\lambda}_1, \bar{u}_1, \bar{\alpha}_1)$ can be tested against $(\bar{m}_1, \bar{w}_1)$ in the following sense:

$$\bar{\lambda}_1 + \int_{\mathbb{T}^d} \bar{w}_1 \cdot D\bar{u}_1 + \bar{m}_1 \left(H(x, D\bar{u}_1) + H^*(x, -\frac{\bar{w}_1}{\bar{m}_1}) \right) \leq \int_{\mathbb{T}^d} \bar{m}_1 \left(\bar{\alpha}_1 + \frac{\bar{m}_2}{\epsilon} + H^*(x, -\frac{\bar{w}_1}{\bar{m}_1}) \right). \tag{8}$$

Note that the integrand on the left-hand side of the inequality is nonnegative, while the right-hand side is finite. We prove (8) by regularizing the equation for $(\bar{m}_1, \bar{w}_1)$ by convolution: let $(\bar{m}_1^\theta, \bar{w}_1^\theta) = \rho_\theta * (\bar{m}_1, \bar{w}_1)$, where ρ_θ is a standard convolution kernel. As

$$-\Delta \bar{m}_1^\theta + \text{div}(\bar{w}_1^\theta) = 0,$$

multiplying by $\bar{m}_1^\theta$ the inequality satisfied by $(\bar{\lambda}_1, \bar{u}_1, \bar{\alpha}_1)$ and integrating lead to

$$\bar{\lambda}_1 + \int_{\mathbb{T}^d} \bar{w}_1^\theta \cdot D\bar{u}_1 + \bar{m}_1^\theta \left(H(x, D\bar{u}_1) + H^*(x, -\frac{\bar{w}_1^\theta}{\bar{m}_1^\theta}) \right) \leq \int_{\mathbb{T}^d} \bar{m}_1^\theta \left(\bar{\alpha}_1 + \frac{\bar{m}_2}{\epsilon} + H^*(x, -\frac{\bar{w}_1^\theta}{\bar{m}_1^\theta}) \right).$$

By Fatou Lemma we can pass to the limit in the nonnegative left-hand side. For the right-hand side, we use the fact that $\bar{m}_2$ is in L^∞ while $\bar{\alpha}_1$ is in $L^{q'}$ to get

$$\lim_{\theta \to 0} \int_{\mathbb{T}^d} \bar{m}_1^\theta \left(\bar{\alpha}_1 + \frac{\bar{m}_2}{\epsilon} \right) = \int_{\mathbb{T}^d} \bar{m}_1 \left(\bar{\alpha}_1 + \frac{\bar{m}_2}{\epsilon} \right).$$

(Note that this would be an issue for a general $\bar{m}_2$). The convergence

$$\lim_{\theta \to 0} \int_{\mathbb{T}^d} \bar{m}_1^\theta H^* (x, -\frac{\bar{w}_1^\theta}{\bar{m}_1^\theta}) = \int_{\mathbb{T}^d} \bar{m}_1 H^* (x, -\frac{\bar{w}_1}{\bar{m}_1})$$

can be checked as in [6]. This gives (8).

Next we use the duality between the two problems. One has

$$
\begin{aligned}
0 &= \int_{\mathbb{T}^d} \left(\bar{m}_1 H^* (x, -\frac{\bar{w}_1}{\bar{m}_1}) + F_1(x, \bar{m}_1) + \frac{\bar{m}_1 \bar{m}_2}{\epsilon} + F_1^*(x, \bar{\alpha}_1) \right) dx - \bar{\lambda}_1 \\
&\geq \int_{\mathbb{T}^d} \bar{m}_1 \left(H^* (x, -\frac{\bar{w}_1}{\bar{m}_1}) + \bar{\alpha}_1 + \frac{\bar{m}_2}{\epsilon} \right) dx - \bar{\lambda}_1 \\
&\geq \int_{\mathbb{T}^d} \bar{w}_1 \cdot D\bar{u}_1 + \bar{m}_1 \left(H(x, D\bar{u}_1) + H^* (x, -\frac{\bar{w}_1}{\bar{m}_1}) \right) \geq 0.
\end{aligned}
$$

where we used (8) in the second inequality. So there is an equality in each inequality, which means that (8) is an equality, that

$$\bar{\alpha}_1 \in \partial_m F_1(x, \bar{m}_1(x)) \qquad \text{a.e.,} \qquad (9)$$

and that

$$\bar{w}_1(x) = -\bar{m}_1(x) D_p H(x, D\bar{u}_1(x)) \qquad \bar{m}_1 - \text{a.e..} \qquad (10)$$

Note that (9) means that $\bar{\alpha}_1(x) = f_1(x, \bar{m}_1(x))$ in $\{\bar{m}_1 > 0\}$ and $\bar{\alpha}_1(x) = 0$ in $\{\bar{m}_1 = 0\}$, so that

$$\bar{\alpha}_1(x) = f_1(x, \bar{m}_1(x)) \qquad \text{a.e..}$$

Moreover, (10) can be rewritten as

$$\bar{w}_1(x) = -\bar{m}_1(x) D_p H(x, D\bar{u}_1(x)) \qquad \text{a.e..}$$

Thus the equality in (8) becomes

$$\bar{\lambda}_1 = \int_{\mathbb{T}^d} \bar{m}_1 \left(f(x, \bar{m}_1) + \frac{\bar{m}_2}{\epsilon} + H^* (x, -\frac{\bar{w}_1}{\bar{m}_1}) \right).$$

It remains to remove the assumption $\bar{m}_2 \in L^\infty$. For this we first note that, in the minimization problem (7), $\bar{w}_2$ plays no role, nor does the relation between $\bar{m}_2$ and $\bar{w}_2$, nor the fact that $\int_{\mathbb{T}^d} \bar{m}_2 = 1$. So we simply approximate $\bar{m}_2$ by $\bar{m}_2^\delta :=$ $\min(\bar{m}_2, 1/\delta)$ which is in L^∞. Let $(\bar{m}_1^\delta, \bar{w}_1^\delta)$ be the minimizer of the map

$$\tilde{\mathscr{E}}_\delta^\epsilon(m_1, w_1) := \int_{\mathbb{T}^d} m_1 H^*(x, -\frac{w_1}{m_1}) + \bar{m}_2 H^*(x, -\frac{\bar{w}_2}{\bar{m}_2}) + F_1(x, m_1) + F_2(x, \bar{m}_2) + \frac{m_1 \bar{m}_2^\delta}{\epsilon}$$

Note that

$$\lim_{\delta \to 0} \tilde{\mathscr{E}}_\delta^\epsilon(\bar{m}_1^\delta, \bar{w}_1^\delta) \leq \lim_{\delta \to 0} \tilde{\mathscr{E}}_\delta^\epsilon(\bar{m}_1, \bar{w}_1) = \mathscr{E}^\epsilon(\bar{m}_1, \bar{w}_1, \bar{m}_2, \bar{w}_2)$$

because, by construction,

$$\lim_{\delta \to 0} \int_{\mathbb{T}^d} \bar{m}_1 \bar{m}_2^\delta = \int_{\mathbb{T}^d} \bar{m}_1 \bar{m}_2.$$

In particular, the family $(\bar{m}_1^\delta, \bar{w}_1^\delta)$ is bounded in $(L^q \cap W^{1,\alpha}) \times L^\alpha$ and therefore converges, up to a subsequence $\delta \to 0$, to some $(\hat{m}_1, \hat{w}_1)$ strongly in L^α and a.e. for the first variable and weakly in L^α for the second one. By Fatou and the convexity of H^*, we get

$$\int_{\mathbb{T}^d} \hat{m}_1 H^*(x, -\frac{\hat{w}_1}{\hat{m}_1}) + \bar{m}_2 H^*(x, -\frac{\bar{w}_2}{\bar{m}_2}) + F_1(x, \hat{m}_1) + F_2(x, \bar{m}_2) + \frac{\hat{m}_1 \bar{m}_2}{\epsilon}$$
$$\leq \lim_{\delta \to 0} \tilde{\mathscr{E}}_\delta^\epsilon(\bar{m}_1^\delta, \bar{w}_1^\delta) \leq \mathscr{E}^\epsilon(\bar{m}_1, \bar{w}_1, \bar{m}_2, \bar{w}_2).$$

As $(\hat{m}_1, \hat{w}_1)$ satisfies the constraint, we infer that $(\hat{m}_1, \hat{w}_1)$ is equal to the unique minimum of $\mathscr{E}^\epsilon(\cdot, \cdot, \bar{m}_2, \bar{w}_2)$, i.e., to $(\bar{m}_1, \bar{w}_1)$. By strict convexity of problem (7), one also infers that $(\bar{m}_1^\delta, \bar{w}_1^\delta)$ converges to $(\bar{m}_1, \bar{w}_1)$ in $(L^q \cap W^{1,\alpha}) \times L^\alpha$ while

$$\lim_{\delta \to 0} \int_{\mathbb{T}^d} \bar{m}_1^\delta H^*(x, -\frac{\bar{w}_1^\delta}{\bar{m}_1^\delta}) = \int_{\mathbb{T}^d} \bar{m}_1 H^*(x, -\frac{\bar{w}_1}{\bar{m}_1})$$

and

$$\lim_{\delta \to 0} \int_{\mathbb{T}^d} \bar{m}_1^\delta \bar{m}_2^\delta = \int_{\mathbb{T}^d} \bar{m}_1 \bar{m}_2.$$

Let $(\bar{\lambda}_1^\delta, \bar{u}_1^\delta, \bar{\alpha}_1^\delta)$ be the solution of the primal problem associated with $\tilde{\mathscr{E}}_\delta^\epsilon$. We know that

$$\bar{\lambda}_1^\delta - \Delta \bar{u}_1^\delta(x) + H(x, D\bar{u}_1^\delta(x)) \leq f(x, \bar{m}_1^\delta(x)) + \frac{\bar{m}_2^\delta(x)}{\epsilon} \quad \text{in } \mathbb{T}^d, \qquad \int_{\mathbb{T}^d} \bar{u}_1^\delta = 0,$$
$$\tag{11}$$

$$\bar{w}_1^\delta(x) = -\bar{m}_1^\delta(x) D_p H(x, D\bar{u}_1^\delta(x)) \quad \text{a.e.,} \tag{12}$$

and

$$\bar{\lambda}_1^\delta = \int_{\mathbb{T}^d} \bar{m}_1^\delta \left(f(x, \bar{m}_1^\delta) + \frac{\bar{m}_2^\delta}{\epsilon} + H^*(x, -\frac{\bar{w}_1^\delta}{\bar{m}_1^\delta}) \right).$$

As $\delta \to 0$, we have

$$\bar{\lambda}_1 := \lim_{\delta \to 0} \bar{\lambda}_1^\delta = \int_{\mathbb{T}^d} \bar{m}_1 \left(f(x, \bar{m}_1) + \frac{\bar{m}_2}{\epsilon} + H^*(x, -\frac{\bar{w}_1}{\bar{m}_1}) \right).$$

By (11) and the coercivity of H, $\bar{u}_1^\delta$ is bounded in $W^{1,r}$, so that (still along a subsequence) it converges weakly to some $\bar{u}_1$ which satisfies

$$\bar{\lambda}_1 - \Delta \bar{u}_1(x) + H(x, D\bar{u}_1(x)) \leq f(x, \bar{m}_1(x)) + \frac{\bar{m}_2(x)}{\epsilon} \qquad \text{in } \mathbb{T}^d, \qquad \int_{\mathbb{T}^d} \bar{u}_1 = 0,$$

since H is convex in p. Finally, note that (12) is equivalent to

$$\bar{m}_1^\delta \left(H(x, D\bar{u}_1^\delta) + H^*(x, -\frac{\bar{w}_1^\delta}{\bar{m}_1^\delta}) \right) + D\bar{u}_1^\delta \cdot \bar{w}_1^\delta = 0 \qquad \text{a.e..}$$

We integrate this equality over $\{M^{-1} \leq \bar{m}_1^\delta \leq M\}$ and pass to the limit to get

$$\int_{\{M^{-1} \leq \bar{m}_1 \leq M\}} \bar{m}_1 \left(H(x, D\bar{u}_1) + H^*(x, -\frac{\bar{w}_1}{\bar{m}_1}) \right) + D\bar{u}_1 \cdot \bar{w}_1 \leq 0$$

because of the weak convergence of $D\bar{u}_1^\delta$ in L^r, the convexity of H, and the strong convergence of $\bar{w}_1^\delta \mathbf{1}_{\{M^{-1} \leq \bar{m}_1^\delta \leq M\}}$ in $L^{r'}$. Then we let $M \to +\infty$ to obtain

$$\int_{\mathbb{T}^d} \bar{m}_1 \left(H(x, D\bar{u}_1) + H^*(x, -\frac{\bar{w}_1}{\bar{m}_1}) \right) + D\bar{u}_1 \cdot \bar{w}_1 \leq 0,$$

so that

$$\bar{w}_1(x) = -\bar{m}_1(x) D_p H(x, D\bar{u}_1(x)) \qquad \text{a.e..}$$

Let us collect for later use the estimates established in the proof:

Lemma 2.4 *For any quadruple (m_1, w_1, m_2, w_2) satisfying (6), we have, for $\alpha := r'q/(r'+q-1) > 1$ and $i = 1, 2$, $m_i \in W^{1,\alpha}(\mathbb{T}^d) \cap L^q(\mathbb{T}^d)$, $w_i \in L^\alpha(\mathbb{T}^d, \mathbb{R}^d)$, and*

$$\|m_i\|_{L^q(\mathbb{T}^d)}^q \leq C \left(1 + \mathscr{E}^\epsilon(m_1, w_1, m_2, w_2) \right),$$

$$\|m_i\|_{W^{1,\alpha}(\mathbb{T}^d)}^\alpha \leq C \left(1 + \mathscr{E}^\epsilon(m_1, w_1, m_2, w_2) \right),$$

$$\|w_i\|^\alpha_{L^\alpha(\mathbb{T}^d,\mathbb{R}^d)} \le C\left(1 + \mathscr{E}^\epsilon(m_1, w_1, m_2, w_2)\right),$$

$$\left\|\frac{w_i}{m_i^{1/r}}\right\|^{r'}_{L^{r'}(\mathbb{T}^d,\mathbb{R}^d)} \le C\left(1 + \mathscr{E}^\epsilon(m_1, w_1, m_2, w_2)\right).$$

Lemma 2.5 *Let $i = 1, 2$. For $1 < r \le 2$,*

$$\left\|m_i^{1/r}\right\|^{r'}_{H^1(\mathbb{T}^d)} \le C\left(1 + \mathscr{E}^\epsilon(m_1, w_1, m_2, w_2)\right).$$

Proof Let $i = 1, 2$.

Case $r = 2$: Multiplying formally (6) by $\log(m_i)$ and integrating, we have

$$\int_{\mathbb{T}^d} \frac{|Dm_i|^2}{m_i} - \int_{\mathbb{T}^d} \frac{\langle Dm_i, w_i\rangle}{m_i} = 0,$$

thus, using Cauchy-Schwarz inequality,

$$\int_{\mathbb{T}^d} \frac{|Dm_i|^2}{m_i} \le \left(\int_{\mathbb{T}^d} \frac{|Dm_i|^2}{m_i}\right)^{\frac{1}{2}} \left(\int_{\mathbb{T}^d} \frac{|w_i|^2}{m_i}\right)^{\frac{1}{2}},$$

so that

$$\left\|Dm_i^{1/2}\right\|_{L^2(\mathbb{T}^d)} \le C\left\|\frac{w_i}{m_i^{1/2}}\right\|_{L^2(\mathbb{T}^d)}.$$

Therefore,

$$\left\|m_i^{1/2}\right\|^2_{H^1(\mathbb{T}^d)} \le C\left(1 + \mathscr{E}^\epsilon(m_1, w_1, m_2, w_2)\right).$$

Case $1 < r < 2$: Multiplying (6) by $m_i^{-\frac{2}{r}+1}$, and integrating, we have

$$\int_{\mathbb{T}^d} m_i^{-\frac{2}{r}}|Dm_i|^2 - \int_{\mathbb{T}^d} \langle m_i^{-\frac{2}{r}}Dm_i, w_i\rangle = 0.$$

Thus, using Cauchy-Schwarz inequality,

$$\int_{\mathbb{T}^d} m_i^{-\frac{2}{r}}|Dm_i|^2 \le \left(\int_{\mathbb{T}^d} m_i^{-\frac{2}{r}}|Dm_i|^2\right)^{\frac{1}{2}} \left(\int_{\mathbb{T}^d} \left(\frac{|w_i|}{m_i^{1/r}}\right)^2\right)^{\frac{1}{2}}.$$

Using the space embedding $L^r \subset L^2$, which holds for $1 < r < 2$,

$$\left(\int_{\mathbb{T}^d} \left(\frac{|w_i|}{m_i^{1/r}} \right)^2 \right)^{\frac{1}{2}} \leq C \left(\int_{\mathbb{T}^d} \left(\frac{|w_i|}{m_i^{1/r}} \right)^r \right)^{\frac{1}{r}},$$

then,

$$\left\| Dm_i^{1/r} \right\|_{L^2(\mathbb{T}^d)} \leq C \left\| \frac{w_i}{m_i^{1/r}} \right\|_{L^r(\mathbb{T}^d)}.$$

Therefore

$$\left\| m_i^{1/r} \right\|_{H^1(\mathbb{T}^d)}^{r} \leq C \left(1 + \mathscr{E}^\epsilon(m_1, w_1, m_2, w_2) \right).$$

3 Singular Limit in the General Case

3.1 The Segregation Problem

We now describe the limit of the variational problem as $\epsilon \to 0$. Let the energy of the limit problem be defined as

$$\mathscr{E}(m_1, w_1, m_2, w_2) := \int_{\mathbb{T}^d} m_1 H^*(x, -\frac{w_1}{m_1}) + m_2 H^*(x, -\frac{w_2}{m_2}) + F_1(x, m_1) + F_2(x, m_2),$$

where (m_1, w_1) and (m_2, w_2) satisfy the constraints (6) and the exclusion constraint

$$m_1 m_2 = 0 \qquad \text{a.e. in } \mathbb{T}^d \tag{13}$$

We discuss the existence of segregate configurations.

Lemma 3.1 *There exists a smooth quadruplet (m_1, w_1, m_2, w_2) which satisfies the continuity equation (6) in the sense of distributions and the exclusion condition (13) and such that $\mathscr{E}(m_1, w_1, m_2, w_2)$ is finite.*

Note that in particular,

$$\inf \mathscr{E}^\epsilon \leq \mathscr{E}(m_1, w_1, m_2, w_2),$$

where the right-hand side is finite and does not depend on ϵ.

Proof It is enough to check that, for any $\epsilon > 0$ small, there exists a pair (m, w) supported in $B_\epsilon(0)$ such that $-\Delta m + \mathrm{div} w = 0$ and $\int_{\mathbb{T}^d} m \left| -\frac{w}{m} \right|^{r'} + |m|^q$ is finite. In view of the growth assumption on H^* and F, this will ensure that the quadruplet $(m_1, w_1, m_2, w_2)(\cdot) := ((m, w)(\cdot), (m, w)(\cdot + x))$ has a finite energy $\mathcal{E}$ provided ϵ is small enough and $|x| \geq 2\epsilon$. Let us set

$$
m(x) = \begin{cases} C(\epsilon^2 - |x|^2))^{r'} & \text{on } B(0, \epsilon) \\ m(x) := 0 & \text{otherwise} \end{cases}, \quad w(x) := Dm(x) = \begin{cases} -2Cr'(\epsilon^2 - |x|^2))^{r'-1}x & \text{on } B(0, \epsilon) \\ w(x) = 0 & \text{otherwise} \end{cases}
$$

where C is such that m has mass 1. Then the quantity $\int_{B(0,\frac{1}{2})} \frac{|w|^{r'}}{m^{r'-1}}$ is finite and therefore $\mathcal{E}(m, w)$ is finite. Moreover, as $r' > 1$, we also have that $-\Delta m + \mathrm{div}(w) = 0$ in $\mathbb{T}^d$ in the sense of distribution.

3.2 The Singular Limit

Let $(\lambda_1^\epsilon, u_1^\epsilon, m_1^\epsilon, \lambda_2^\epsilon, u_2^\epsilon, m_2^\epsilon)$ be a solution of (1) as in Theorem 2.3: setting $w_i^\epsilon := -m_i^\epsilon D_p H(\cdot, Du_i^\epsilon)$ (for $i = 1, 2$), the quadruplet $(m_1^\epsilon, w_1^\epsilon, m_2^\epsilon, w_2^\epsilon)$ minimizes the energy $\mathcal{E}^\epsilon$ under the constraint (6).

Proposition 3.2 *As $\epsilon \to 0$, there exists a subsequence, again denoted $(m_1^\epsilon, w_1^\epsilon, m_2^\epsilon, w_2^\epsilon)$, which converges weakly in $W^{1,\alpha} \times L^\alpha \times W^{1,\alpha} \times L^\alpha$ to a quadruple $(\bar{m}_1, \bar{w}_1, \bar{m}_2, \bar{w}_2)$, where $(\bar{m}_1, \bar{w}_1, \bar{m}_2, \bar{w}_2)$ minimizes $\mathcal{E}$ under the constraints (6) and (13). Moreover,*

$$
\lim \int_{\mathbb{T}^d} \frac{m_1^\epsilon m_2^\epsilon}{\epsilon} = 0.
$$

Proof In view of Lemma 3.1, there exists C such that

$$
\mathcal{E}^\epsilon(m_1^\epsilon, w_1^\epsilon, m_2^\epsilon, w_2^\epsilon) \leq C.
$$

Thus, by Lemma 2.4, (m_1^ϵ) and (m_2^ϵ) are bounded in $W^{1,\alpha}$ and (w_1^ϵ) and (w_2^ϵ) are bounded in L^α, where $\alpha := r'q/(r' + q - 1) > 1$. Up to a subsequence, (m_1^ϵ) and (m_2^ϵ) converge strongly in L^α and a.e., while (w_1^ϵ) and (w_2^ϵ) converge weakly in L^α to $\bar{m}^1$, $\bar{m}^2$, $\bar{w}^1$, and $\bar{w}^2$, respectively. By lower semicontinuity of $\mathcal{E}$,

$$
\mathcal{E}(\bar{m}^1, \bar{w}^1, \bar{m}^2, \bar{w}^2) \leq \liminf \mathcal{E}(m_1^\epsilon, w_1^\epsilon, m_2^\epsilon, w_2^\epsilon) \leq \liminf \mathcal{E}^\epsilon(m_1^\epsilon, w_1^\epsilon, m_2^\epsilon, w_2^\epsilon).
$$

Moreover, $(\bar{m}_1, \bar{w}_1, \bar{m}_2, \bar{w}_2)$ satisfies (6). Since

$$
\limsup \int_{\mathbb{T}^d} \frac{m_1^\epsilon m_2^\epsilon}{\epsilon} \leq \limsup \mathcal{E}^\epsilon(m_1^\epsilon, w_1^\epsilon, m_2^\epsilon, w_2^\epsilon) \leq C,
$$

we have at the limit $\bar{m}_1 \bar{m}_2 = 0$ a.e., so that $(\bar{m}_1, \bar{w}_1, \bar{m}_2, \bar{w}_2)$ satisfies also (13).

On the other hand, for any (m^1, w^1, m^2, w^2) which satisfies (6) and (13), we have

$$\mathscr{E}(m_1, w_1, m_2, w_2) = \mathscr{E}^\epsilon(m_1, w_1, m_2, w_2) \geq \mathscr{E}^\epsilon(m_1^\epsilon, w_1^\epsilon, m_2^\epsilon, w_2^\epsilon),$$

Therefore

$$\mathscr{E}(\bar{m}^1, \bar{w}^1, \bar{m}^2, \bar{w}^2) = \inf \mathscr{E}(m_1, w_1, m_2, w_2) = \lim \mathscr{E}(m_1^\epsilon, w_1^\epsilon, m_2^\epsilon, w_2^\epsilon) = \lim \mathscr{E}^\epsilon(m_1^\epsilon, w_1^\epsilon, m_2^\epsilon, w_2^\epsilon).$$

and

$$\lim \int_{\mathbb{T}^d} \frac{m_1^\epsilon m_2^\epsilon}{\epsilon} = \lim \mathscr{E}^\epsilon(m_1^\epsilon, w_1^\epsilon, m_2^\epsilon, w_2^\epsilon) - \mathscr{E}(m_1^\epsilon, w_1^\epsilon, m_2^\epsilon, w_2^\epsilon) = 0.$$

4 The Model Case

We consider here the model case where H is purely quadratic, normalized as

$$H(x, Du) = \frac{1}{2}|Du|^2,$$

and the functions f_i coincide with the same increasing and continuous function $f : \mathbb{T}^d \times \mathbb{R} \to \mathbb{R}$.

In this case, the system (1) reads as

$$\begin{cases} \lambda_i - \Delta u_i + \frac{1}{2}|Du_i|^2 = f(x, m_i) + \frac{m_j}{\epsilon} & \text{in } \mathbb{T}^d, \ i, j = 1, 2, i \neq j \\ -\Delta m_i - \text{div}(m_i Du_i) = 0 & \text{in } \mathbb{T}^d, \ i = 1, 2 \\ m_i \geq 0, \ \int_{\mathbb{T}^d} m_i = 1 \end{cases} \tag{14}$$

which can be transformed through the Hopf-Cole change of unknown $u \mapsto e^{-\frac{u}{2}}$. Note that the density equation gives $m_i = e^{-u_i}/\|e^{-u_i}\|_{L^1(\mathbb{T}^d)}$. Therefore, it is convenient to set

$$v_i := \frac{e^{-\frac{u_i}{2}}}{\|e^{-\frac{u_i}{2}}\|_{L^2(\mathbb{T}^d)}}, \quad m_i = v_i^2 \quad i = 1, 2.$$

Hence (14) is transformed into

$$\begin{cases} -2\Delta v_i + f(x, v_i^2)v_i = \lambda_i v_i - \frac{v_i v_j^2}{\epsilon} & \text{in } \mathbb{T}^d, \ i, j = 1, 2, i \neq j \\ v_i \geq 0, \ \|v_i\|_{L^2(\mathbb{T}^d)} = 1 \quad i = 1, 2 \end{cases} \tag{15}$$

It is convenient to consider the optimization problems for $\epsilon = 0$, both for the MFG system and for its semilinear equivalent version. For this latter, it takes the form

$$\inf \left\{ \int_{\mathbb{T}^d} [|Dv_1|^2 + |Dv_2|^2] + \int_{\mathbb{T}^d} \frac{1}{2}[F(x, v_1^2) + F(x, v_2^2)], \right.$$
$$\left. v_i \in H^1(\mathbb{T}^d), \ v_i \geq 0 : \quad \int_{\mathbb{T}^d} v_i^2 = 1, \ v_1 v_2 = 0 \quad \text{in } \mathbb{T}^d, \ i = 1, 2 \right\} \tag{16}$$

Notice that the problem can also be rephrased as a partition problem: in each Ω_i (given by the support of v_i), there is a constrained minimization:

$$\mu(\Omega_i) := \min \left\{ \int_{\Omega_i} [|Dv|^2 + \frac{1}{2}F(x, v^2)], \quad v \in H_0^1(\Omega_i) : \quad \int_{\Omega_i} v^2 = 1 \right\}$$

and the associated optimal partition problem is

$$\inf \left\{ \mu(\Omega_1) + \mu(\Omega_2), \quad \Omega_1, \Omega_2 \text{ open sets: } \overline{\Omega}_1 \cup \overline{\Omega}_2 = \mathbb{T}^d, \ \Omega_1 \cap \Omega_2 = \emptyset \right\} \tag{17}$$

It is even more relevant to consider the optimal partition problem for the segregation model. Here, the limit problem can be written as

$$\inf \left\{ \frac{1}{2} \int_{\mathbb{T}^d} [|\alpha_1|^2 m_1 + |\alpha_2|^2 m_2] + \int_{\mathbb{T}^d} [F(x, m_1) + F(x, m_2)], \right.$$
$$m_i \in L^1(\mathbb{T}^d)_+, \ \alpha_i \in L^2(m_i \, dx) : \quad -\Delta m_i - \text{div}(\alpha_i m_i) = 0 \quad \text{in } \mathbb{T}^d \tag{18}$$
$$\left. \int_{\mathbb{T}^d} m_i = 1, \ m_1 m_2 = 0 \quad \text{in } \mathbb{T}^d, \ i = 1, 2 \right\}$$

and the optimal partition problem is defined in terms of the eigenvalue

$$\lambda(\Omega_i) := \min \left\{ \int_{\Omega_i} [\frac{1}{2}|\alpha|^2 m + F(x, m)], \quad m \in L^1(\mathbb{T}^d) : \quad \int_{\mathbb{T}^d} m = 1, \right.$$
$$\left. -\Delta m - \text{div}(\alpha m) = 0 \quad \text{in } \mathbb{T}^d, m = 0 \text{ in } \Omega_i^c \right\}$$

It is not by chance that we call λ an eigenvalue, since it is actually the first eigenvalue of the operator $w \mapsto -\Delta w + F(x, m)w$ in $H_0^1(\Omega_i)$.

Moreover, this partition problem admits an interpretation (and a reformulation) in terms of mean field games with state constraints. To be more precise, let α be a continuous function on Ω_i and X_t be the solution (on a standard probability space $(\mathscr{X}, \mathscr{F}, \mathscr{F}_t, \mathbb{P})$) of the SDE

$$\begin{cases} dX_t = \alpha(X_t)dt + \sqrt{2}dB_t \\ X_0 = x \in \Omega_i \end{cases}$$

starting in Ω_i and adapted to a Brownian motion B_t. Given a fixed function m, one can consider the ergodic state constraint problem (as in [20]):

$$\inf_{\alpha \in \mathscr{A}} \lim_{T \to \infty} \frac{1}{T} \mathbb{E} \int_0^T [\frac{1}{2}|\alpha(X_t)|^2 + f(m(X_t))] \, dt$$
$$\mathscr{A} = \{\alpha \in C^0(\Omega_i) : X_t \in \Omega_i \text{ a.s. } \forall t > 0\} \, .$$

The associated mean field game problem consists in finding a couple $(\bar{\alpha}, \bar{m})$ such that, for fixed $\bar{m}$, $\bar{\alpha}$ is an optimal feedback for the above state constraint problem and, in turn, $\bar{m}$ is the invariant measure associated to the dynamics controlled by $\bar{\alpha}$. Whenever such a fixed point exists, we have

$$\lambda(\Omega_i) = \inf_{\alpha \in \mathscr{A}} \lim_{T \to \infty} \mathbb{E} \frac{1}{T} \int_0^T \frac{1}{2}[\frac{1}{2}|\alpha(X_t)|^2 + f(\bar{m}(X_t))] \, dt \, .$$

Finally, the optimal partition problem for state constraint mean field games reads as

$$\inf \{\lambda(\Omega_1) + \lambda(\Omega_2), \quad \Omega_1, \Omega_2 \text{ open sets: } \overline{\Omega}_1 \cup \overline{\Omega}_2 = \mathbb{T}^d, \ \Omega_1 \cap \Omega_2 = \emptyset\} \, .$$
$$(19)$$

We now analyze the semilinear problem, and we start with a property of minimizers of (16). We follow here [10] (see in particular Lemma 3.2 of [10]).

Lemma 4.1 *Let* (v_1, v_2) *be an optimal couple for (16). Then*

$$-2\Delta(v_1 - v_2) + f(x, (v_1 - v_2)^2)(v_1 - v_2) = \lambda_1 v_1 - \lambda_2 v_2 \qquad \text{in } \mathbb{T}^d \qquad (20)$$

holds in the sense of distributions.

Proof For any smooth $\varphi \geq 0$, and small $t > 0$, one considers the couple (w_1, w_2) given by

$$w_1 = (\sigma v_1 - \delta v_2 + t\varphi)^+, \quad w_2 = (\sigma v_1 - \delta v_2 + t\varphi)^-,$$

where, for fixed $t > 0$, $\sigma < 1$ and $\delta > 1$ are chosen so that $\|w_i\|_{L^2(\mathbb{T}^d)} = 1$ for $i = 1, 2$. Of course, by construction we have $w_1 w_2 = 0$. Notice that

$$\|w_1\|^2_{L^2(\mathbb{T}^d)} = \sigma^2 + 2\sigma t \int_{\mathbb{T}^d} v_1 \varphi + O(t^2)$$

so $\sigma = 1 - t \int_{\mathbb{T}^d} v_1 \varphi + O(t^2)$; similarly one finds $\delta = 1 + t \int_{\mathbb{T}^d} v_2 \varphi + O(t^2)$.
Next, by computing

$$J(w_1, w_2) = \int_{\mathbb{T}^d} [|Dw_1|^2 + |Dw_2|^2] + \int_{\mathbb{T}^d} \frac{1}{2}[F(x, w_1^2) + F(x, w_2^2)],$$

and using that

$$\lambda_i = \int_{\mathbb{T}^d} 2|Dv_i|^2 + f(x, v_i^2)v_i^2$$

one verifies that

$$J(w_1, w_2) \le J(v_1, v_2) + t \left\{ \int_{\mathbb{T}^d} 2Dv_1 D\varphi - 2Dv_2 D\varphi + \int_{\mathbb{T}^d} f(x, (v_1)^2) v_1 \varphi - f(x, (v_2)^2) v_2 \varphi \right.$$
$$\left. - (\lambda_1 \int_{\mathbb{T}^d} v_1 \varphi - \lambda_2 \int_{\mathbb{T}^d} v_2 \varphi) \right\} + O(t^2)$$

Therefore, if (v_1, v_2) is a minimizing couple, this implies that

$$\int_{\mathbb{T}^d} 2Dv_1 D\varphi - 2Dv_2 D\varphi + \int_{\mathbb{T}^d} f(x, (v_1)^2) v_1 \varphi - f(x, (v_2)^2) v_2 \varphi \ge \lambda_1 \int_{\mathbb{T}^d} v_1 \varphi - \lambda_2 \int_{\mathbb{T}^d} v_2 \varphi$$

hence

$$-2\Delta(v_1 - v_2) + f(x, (v_1)^2) v_1 - f(x, (v_2)^2) v_2 \ge \lambda_1 v_1 - \lambda_2 v_2$$

holds in the sense of distributions.

Reversing the roles of v_1, v_2, one obtains the reverse inequality and the conclusion.

We also have

Lemma 4.2 *Let (v_1, v_2) be an optimal couple for (16). Then, for $i = 1, 2$, we have $v_i \in W^{1,\infty}$ and*

$$-2\Delta v_i + f(x, v_i^2) v_i \le \lambda_i v_i \qquad \text{in } \mathbb{T}^d$$

holds in the sense of distributions, with equality in $\{v_i > 0\}$.

Proof Indeed, as (v_1, v_2) satisfies (20), $v_1 = (v_1 - v_2)_+$ is a subsolution to (20). Hence the following inequality holds in the sense of distributions:

$$-2\Delta v_1 + f(x, v_1^2) v_1 \le \lambda_1 v_1 - \lambda_2 v_2 \le \lambda_1 v_1.$$

Because of the bound from below of f, this implies that

$$-2\Delta v_1 - (\lambda_1 + C_0) v_1 \le 0.$$

As $\|v_1\|_2 = 1$, we deduce from classical elliptic regularity (see Theorem 8.17 in [13], for instance) that $\|v_1\|_\infty$ is bounded. In particular, $(v_1 - v_2)$ solves a linear uniformly elliptic equation with bounded coefficients and is therefore $C^{1,\alpha}$ for some $\alpha \in (0, 1)$. In particular, $v_1 = (v_1 - v_2)_+$ is Lipschitz continuous. The symmetric argument applies to v_2.

Finally, the semilinear problem admits a solution.

Theorem 4.3 *Problem (16) admits a solution (v_1, v_2). Moreover this problem coincides with problem (17) where the minimum is attained for $\Omega_1 = supp(v_1), \Omega_2 = supp(v_2)$.*

Proof It is clear that the infimum in (17) is above the infimum in (16). On the other hand, one can prove as in Section 3 that Problem (16) admits a solution (v_1, v_2). Then Lemma 4.2 states that (v_1, v_2) is continuous, so that the sets $\Omega_i = \{v_i > 0\}$ are open. Obviously, $\Omega_1 \cap \Omega_2 = \emptyset$. Recalling the linear elliptic equation satisfied by $v_1 - v_2$ (Lemma 4.1), it is well known that, as $v_1 - v_2$ is not identically equal to zero, it cannot vanish on an open set by a unique continuation argument. Thus $\overline{\Omega}_1 \cup \overline{\Omega}_2 = \mathbb{T}^d$. So (Ω_1, Ω_2) is a competitor in problem (17), and therefore a minimum.

Now we deduce a solution for the segregation problem. Indeed, by setting $m_i = v_i^2$ and $\alpha_i = \frac{Dm_i}{m_i}$, we construct a solution to problem (18); and by setting $u_i = -\log m_i$ in Ω_i, we can characterize the optimal MFG system with state constraint in each Ω_i.

4.1 The Singular Limit

Let $(v_1^\epsilon, v_2^\epsilon)$ be a solution of (15), and then $(v_1^\epsilon, v_2^\epsilon)$ is a minimizer of the following problem:

$$\inf \left\{ \int_{\mathbb{T}^d} 2[|Dv_1|^2 + |Dv_2|^2] + \int_{\mathbb{T}^d} [F(x, v_1^2) + F(x, v_2^2)] + \frac{(v_1 v_2)^2}{\epsilon} , \right.$$
$$\left. v_i \in H^1(\mathbb{T}^d), \ v_i \ge 0 : \ \int_{\mathbb{T}^d} v_i^2 = 1 \quad \text{in } \mathbb{T}^d, \ i = 1, 2 \right\}. \tag{21}$$

Proposition 4.4 *As $\epsilon \to 0$, there exists a subsequence, again denoted $(v_1^\epsilon, v_2^\epsilon)$, which converges strongly in $H^1 \times H^1$ to a couple $(\bar{v}_1, \bar{v}_2)$, where $(\bar{v}_1, \bar{v}_2)$ minimizes problem (16). Moreover,*

$$\lim_{\epsilon \to 0} \int_{\mathbb{T}^d} \frac{(v_1^\epsilon v_2^\epsilon)^2}{\epsilon} = 0.$$

Proof Define

$$\mathscr{E}^\epsilon(v_1, v_2) = \int_{\mathbb{T}^d} 2[|Dv_1|^2 + |Dv_2|^2] + \int_{\mathbb{T}^d} [F(x, v_1^2) + F(x, v_2^2)] + \frac{(v_1 v_2)^2}{\epsilon}$$

and

$$\mathscr{E}(v_1, v_2) = \int_{\mathbb{T}^d} 2[|Dv_1|^2 + |Dv_2|^2] + \int_{\mathbb{T}^d} [F(x, v_1^2) + F(x, v_2^2)].$$

Note that, for any $(v_1, v_2) \in H^1(\mathbb{T}^d) \times H^1(\mathbb{T}^d)$, $v_i \ge 0$, $\int_{\mathbb{T}^d} v_i^2 = 1$ in $\mathbb{T}^d$, $i = 1, 2$, such that $v_1 v_2 = 0$, we have

$$\inf \mathscr{E}^\epsilon \le \mathscr{E}^\epsilon(v_1, v_2) = \mathscr{E}(v_1, v_2).$$

Hence, it exists $C > 0$ such that $\inf \mathscr{E}^\epsilon \le C$.

Let $(v_1^\epsilon, v_2^\epsilon)$ be a smooth minimizer of (21) satisfying equation (15). Due to the fact that $(v_1^\epsilon, v_2^\epsilon)$ minimizes (21), we have that $(Dv_1^\epsilon, Dv_2^\epsilon)$ is bounded in $L^2(\mathbb{T}^d) \times L^2(\mathbb{T}^d)$ and therefore $(v_1^\epsilon, v_2^\epsilon)$ is bounded in $H^1(\mathbb{T}^d) \times H^1(\mathbb{T}^d)$.

Thus, up to a subsequence, $(v_1^\epsilon, v_2^\epsilon)$ converges weakly in $H^1(\mathbb{T}^d) \times H^1(\mathbb{T}^d)$ and strongly in $L^2(\mathbb{T}^d) \times L^2(\mathbb{T}^d)$ to a couple $(\bar{v}_1, \bar{v}_2)$. Using Fatou and lower semicontinuity

$$\int_{\mathbb{T}^d} 2[|D\bar{v}_1|^2 + |D\bar{v}_2|^2] + \int_{\mathbb{T}^d} [F(x, \bar{v}_1^2) + F(x, \bar{v}_2^2)]$$
$$\leq \liminf_{\epsilon \to 0} \int_{\mathbb{T}^d} 2[|Dv_1^\epsilon|^2 + |Dv_2^\epsilon|^2] + \int_{\mathbb{T}^d} [F(x, (v_1^\epsilon)^2) + F(x, (v_2^\epsilon)^2)] + \frac{(v_1^\epsilon v_2^\epsilon)^2}{\epsilon}.$$

Moreover, at the limit, we have that, $\bar{v}_1 \bar{v}_2 = 0$ a.e., indeed

$$\limsup_{\epsilon \to 0} \int_{\mathbb{T}^d} \frac{(v_1^\epsilon v_2^\epsilon)^2}{\epsilon} \leq \limsup_{\epsilon \to 0} \int 2[|Dv_1^\epsilon|^2 + |Dv_2^\epsilon|^2] + \int_{\mathbb{T}^d} [F(x, (v_1^\epsilon)^2) + F(x, (v_2^\epsilon)^2)] + \frac{(v_1^\epsilon v_2^\epsilon)^2}{\epsilon} \leq C.$$

Furthermore, since $\bar{v}_1 \bar{v}_2 = 0$ a.e., we have

$$\mathscr{E}(\bar{v}_1, \bar{v}_2) = \mathscr{E}^\epsilon(\bar{v}_1, \bar{v}_2) \geq \mathscr{E}^\epsilon(v_1^\epsilon, v_2^\epsilon).$$

Hence,

$$\mathscr{E}(\bar{v}_1, \bar{v}_2) = \inf \mathscr{E}(v_1, v_2) = \lim_{\epsilon \to 0} \mathscr{E}^\epsilon(v_1^\epsilon, v_2^\epsilon)$$

and $(\bar{v}_1, \bar{v}_2)$ minimizes (16).

Moreover, $\int_{\mathbb{T}^d} |Dv_i^\epsilon|^2$ converges to $\int_{\mathbb{T}^d} |D\bar{v}_i|^2$ for $i = 1, 2$. Hence, $(v_1^\epsilon, v_2^\epsilon)$ converges strongly in $H^1(\mathbb{T}^d) \times H^1(\mathbb{T}^d)$.

Let us come back to the original problem (14).

Corollary 4.5 *For $i = 1, 2$, as $\epsilon \to 0$, there exists a subsequence, again denoted m_i^ϵ, which converges strongly in $L^{\frac{2^*}{2}}(\mathbb{T}^d)$ to $\bar{m}_i$. Moreover, there exists a sequence $c_i^\epsilon = 2\ln(\|e^{-\frac{u_i^\epsilon}{2}}\|_{L^2(\mathbb{T}^d)})$ such that $u_i^\epsilon + c_i^\epsilon$ converges a.e., up to a subsequence, to $\hat{u}_i$. The couple $(\bar{m}_i, \hat{u}_i)$ is a classical solution of*

$$\begin{cases} \lambda_i - \Delta u_i + \frac{1}{2}|Du_i|^2 = f(x, m_i) & \text{in } \mathbb{T}^d \cap \{m_i > 0\}, \ i = 1, \ldots, 2 \\ -\Delta m_i - \text{div}(m_i Du_i) = 0 & \text{in } \mathbb{T}^d, \ i = 1, \ldots, 2 \\ m_i \geq 0, \ \int_{\mathbb{T}^d} m_i = 1, \ m_1 m_2 = 0 \end{cases} \tag{22}$$

Proof Fix $i = 1, 2$. Since, up to a subsequence, v_i^ϵ strongly converges in $H^1(\mathbb{T}^d)$ to $\bar{v}_i$ and $m_i^\epsilon := (v_i^\epsilon)^2$, we have that m_i^ϵ strongly converges in $L^1(\mathbb{T}^d)$ to $\bar{m}_i := (\bar{v}_i)^2$, or using Sobolev inequality, there exists $\beta = 2^* > 2$ such that $\|v_i\|_{L^\beta} \leq C \|v_i\|_{H^1(\mathbb{T}^d)}$; hence m_i^ϵ strongly converges in $L^{\frac{\beta}{2}}(\mathbb{T}^d)$.

Let $c_i^\epsilon = 2\ln(\|e^{-\frac{u_i^\epsilon}{2}}\|_{L^2(\mathbb{T}^d)})$. Then,

$$v_i^\epsilon = e^{-\frac{u_i^\epsilon + c_i^\epsilon}{2}} \quad \text{and} \quad u_i^\epsilon = -2\ln(v_i^\epsilon) - c_i^\epsilon.$$

Hence, $u_i^\epsilon + c_i^\epsilon$ converges a.e. to $-2\ln(\bar{v}_i) =: \hat{u}_i$, in the set $\{\bar{m}_i > 0\}$. Moreover, $Du_i^\epsilon = -2\frac{Dv_i^\epsilon}{v_i^\epsilon}$ converges a.e. to $-2\frac{D\bar{v}_i}{\bar{v}_i} = D\hat{u}_i$, in the set $\{\bar{m}_i > 0\}$. Therefore, we have that $\bar{m}_i, \hat{u}_i$ verifies

$$-\Delta\bar{m}_i - \text{div}(\bar{m}_i D\hat{u}_i) = 0.$$

Thanks to Lemma 4.1, $\bar{v}_i$ is a classical solution of

$$-2\Delta v_i + f(x, v_i^2)v_i = \lambda_i v_i,$$

in the set $\{\bar{m}_i > 0\}$. Therefore, using the fact that $\bar{v}_i = e^{-\frac{\hat{u}_i}{2}}$, we have that $\hat{u}_i$ is a classical solution of

$$\lambda_i - \Delta u_i + \frac{1}{2}|Du_i|^2 = f(x, m_i),$$

for $m_i = \bar{v}_i^2$.

Hence, $(\bar{m}_i, \hat{u}_i)$ is a classical solution of (22), in the set $\{\bar{m}_i > 0\}$.

4.2 The n-Population Case

When considering the case of n populations, the system (1) reads as

$$\begin{cases} \lambda_i - \Delta u_i + \frac{1}{2}|Du_i|^2 = f(x, m_i) + \dfrac{\Pi_{j\neq i} m_j}{\epsilon} & \text{in } \mathbb{T}^d, \ i = 1,\ldots,n \\ -\Delta m_i - \text{div}(m_i Du_i) = 0 & \text{in } \mathbb{T}^d, \ i = 1,\ldots,n \\ m_i \geq 0, \ \displaystyle\int_{\mathbb{T}^d} m_i = 1 \end{cases} \tag{23}$$

As before Hopf-Cole transform can be used to obtain the system

$$\begin{cases} -2\Delta v_i + f(x, v_i^2)v_i = \lambda_i v_i - \dfrac{v_i \Pi_{j\neq i} v_j^2}{\epsilon} & \text{in } \mathbb{T}^d, \ i,j = 1,\ldots,n, i \neq j \\ v_i \geq 0, \ \|v_i\|_{L^2(\mathbb{T}^d)} = 1 & i = 1,\ldots,n \end{cases} \tag{24}$$

for the function

$$v_i := \frac{e^{-\frac{u_i}{2}}}{\|e^{-\frac{u_i}{2}}\|_{L^2(\mathbb{T}^d)}}, \quad m_i = v_i^2 \quad i = 1,\ldots,n.$$

In this case the optimization problem for $\epsilon = 0$ in the semilinear version takes the form

$$\inf \left\{ \sum_{i=1}^{n} \int_{\mathbb{T}^d} |Dv_i|^2 + \frac{1}{2} \sum_{i=1}^{n} \int_{\mathbb{T}^d} F(x, v_i^2), \right.$$
$$\left. v_i \in H^1(\mathbb{T}^d) \ v_i \geq 0 : \quad \int_{\mathbb{T}^d} v_i^2 = 1, \ v_i v_j = 0 \quad \text{in } \mathbb{T}^d, \ i,j = 1,\ldots,n, \ i \neq j \right\}$$

(25)

As in the two-population case, the problem can also be rephrased as a partition problem: in each Ω_i (given by the support of v_i), there is a constrained minimization

$$\mu(\Omega_i) := \min \left\{ \int_{\Omega_i} [|Dv|^2 + \frac{1}{2} F(x, v^2)], \quad v \in H_0^1(\Omega_i) : \quad \int_{\Omega_i} v^2 = 1 \right\}$$

and the associated optimal partition problem

$$\inf \left\{ \sum_{i=1}^{n} \mu(\Omega_i), \quad \Omega_i \text{ open sets: } \overline{\Omega}_i \cup \overline{\Omega}_j = \mathbb{T}^d, \ \Omega_i \cap \Omega_j = \emptyset, \ i = 1,\ldots,n, \ i \neq j \right\}$$

(26)

Moreover, the limit segregation problem can be written as

$$\inf \left\{ \frac{1}{2} \sum_{i=1}^{n} \int_{\mathbb{T}^d} |\alpha_i|^2 m_i + \sum_{i=1}^{n} \int_{\mathbb{T}^d} F(x, m_i), \right.$$
$$m_i \in L^1(\mathbb{T}^d)_+, \ \alpha_i \in L^2(m_i \, dx) : \quad -\Delta m_i - \text{div}(\alpha_i m_i) = 0 \quad \text{in } \mathbb{T}^d$$
$$\left. \int_{\mathbb{T}^d} m_i = 1, \ m_i \, m_j = 0 \quad \text{in } \mathbb{T}^d, \ i,j = 1,\ldots,n, \ i \neq j \right\}$$

(27)

and the optimal partition problem is defined in terms of the eigenvalue

$$\lambda(\Omega_i) := \min \left\{ \int_{\Omega_i} [\frac{1}{2} |\alpha|^2 m + F(x, m)], \quad m \in L^1(\mathbb{T}^d) : \quad \int_{\mathbb{T}^d} m = 1, \right.$$
$$\left. -\Delta m - \text{div}(\alpha m) = 0 \quad \text{in } \mathbb{T}^d, m = 0 \text{ in } \Omega_i^c \right\}.$$

$$\inf \left\{ \sum_{i=1}^{n} \lambda(\Omega_i), \quad \Omega_i \text{ open sets: } \overline{\Omega}_i \cup \overline{\Omega}_j = \mathbb{T}^d, \ \Omega_i \cap \Omega_j = \emptyset, \ i = 1,\ldots,n, \ i \neq j \right\}.$$

(28)

Generalizing the proofs presented in [10], Section 3, we can analyze the semilinear problem obtaining some properties of the minimizers, which extend Lemma 4.1 and Lemma 4.2 to the n-population case.

Lemma 4.6 *Let $(v_1, \ldots, v_n)$ be an optimal n-uple for (25). For all $i = \{1, \ldots, n\}$ fixed, let*

$$w_i = v_i - \sum_{j \neq i} v_j.$$

Then, for each i,

$$-2\Delta w_i + f(x, w_i^2)w_i \geq \lambda_1 v_1 - \sum_{j \neq i} \lambda_j v_j \quad \text{in } \mathbb{T}^d$$

holds in the sense of distributions.

Lemma 4.7 *Let* $(v_1, \ldots, v_n)$ *be an optimal n-uple for (25). For all* $i = \{1, \ldots, n\}$ *fixed,*

$$-2\Delta v_i + f(x, v_i^2)v_i \leq \lambda_i v_i \quad \text{in } \mathbb{T}^d$$

holds in the sense of distributions.

These two Lemmas and the monotonicity formulae from [10], Section 4, can be used to prove Lipschitz regularity of the solution $(v_1, \ldots, v_n)$ following [10], Section 5. This implies in particular that the sets $\Omega_i = \text{supp}(v_i)$ are open. Thus we have the following theorem.

Theorem 4.8 *Problem (25) admits a solution* $(v_1, \ldots, v_n)$. *Moreover this problem coincides with problem (26) where the minimum is attained for* $\Omega_i = \text{supp}(v_i)$, *for all* $i = \{1, \ldots, n\}$.

Acknowledgement The first and last authors are partially supported by the ANR (Agence Nationale de la Recherche) project ANR-16-CE40-0015-01.

References

1. Achdou, Y., Bardi, M. and Cirant, M. *Mean Field Games models of segregation.* Mathematical Models and Methods in Applied Sciences, 27 (01), (2017), 1–39.
2. Bensoussan, A., Frehse, J. and Yam, S.C.P. MEAN FIELD GAMES AND MEAN FIELD TYPE CONTROL THEORY. New York: Springer, (2016).
3. Cacace, S. and Camilli, F. *A Generalized Newton Method for Homogenization of Hamilton–Jacobi Equations.* SIAM Journal on Scientific Computing, 38 (6), (2016), A3589–A3617.
4. Cardaliaguet, P. Notes on mean field games (from P.-L. Lions lectures at College de France), (2013).
5. Cardaliaguet, P., Graber, J., Porretta, A. and Tonon, D. *Second order mean field games with degenerate diffusion and local coupling.* Nonlinear Differ. Equ. Appl., 22 (5), (2015), 1287–1317.
6. Cardaliaguet, P. and Graber, J. *Mean field games systems of first order.* ESAIM: : Control, Optimisation and Calculus of Variations, 21 (3), (2015), 690–722.
7. Cardaliaguet, P. *Weak solutions for first order mean field games with local coupling.* In "Analysis and Geometry in Control Theory and its Applications." ed. Bettiol, P., Cannarsa, P., Colombo, G., Motta, M., & Rampazzo, F. Springer INdAM Series 11, (2015), 111–158.
8. Cirant, M. *A generalization of the Hopf-Cole transformation for stationary Mean Field Games systems.* Comptes Rendus Mathematique, 353 (9), (2015), 807–811.
9. Cirant, M. and Verzini, G. *Bifurcation and segregation in quadratic two-populations Mean Field Games systems.* ESAIM: COCV 23 (2017), 1145–1177

10. Conti, M., Terracini, S. and Verzini, G. *An optimal partition problem related to nonlinear eigenvalues.* J. Funct. Anal., 198 (1), (2003), 160–196.
11. Feleqi, E. *The derivation of ergodic mean field game equations for several populations of players.* Dynamic Games and Applications, 3 (4), (2013), 523–536.
12. Giaquinta, M. and Martinazzi, L. AN INTRODUCTION TO THE REGULARITY THEORY FOR ELLIPTIC SYSTEMS, HARMONIC MAPS AND MINIMAL GRAPHS. Lecture Notes. Scuola Normale Superiore di Pisa (New Series), 11, Edizioni della Normale, Pisa, (2012).
13. Gilbarg, D. and Trudinger, N.S. ELLIPTIC PARTIAL DIFFERENTIAL EQUATIONS OF SECOND ORDER. Springer, (2015).
14. Gomes, D.A. and Saúde, J. *Mean field games models - a brief survey.* Dynamic Games and Applications, 4 (2), (2014), 110–154.
15. Gomes, D.A., Pimentel, E. and Voskanyan, V. REGULARITY THEORY FOR MEAN-FIELD GAME SYSTEMS. Springer Briefs in Mathematics. Springer, (2016).
16. Huang, M., Malhamé, R.P. and Caines, P.E. *Large population stochastic dynamic games: closed-loop McKean-Vlasov systems and the Nash certainty equivalence principle.* Communication in information and systems, 6 (3), (2006), 221–252.
17. Huang, M., Caines, P.E. and Malhamé, R.P. *Large-Population Cost-Coupled LQG Problems With Nonuniform Agents: Individual-Mass Behavior and Decentralized ϵ-Nash Equilibria.* IEEE Transactions on Automatic Control, 52 (9), (2007), 1560–1571.
18. Kolokoltsov, V. N., Li, J. and Yang, W. *Mean field games and nonlinear Markov processes.* Preprint arXiv:1112.3744 (2011).
19. Lachapelle, A. and Wolfram, M.-T. *On a mean field game approach modeling congestion and aversion in pedestrian crowds,* Transportation research part B: methodological, 45 (10), (2011), 1572–1589.
20. Lasry, J.-M. and Lions, P.-L. *Nonlinear elliptic equations with singular boundary conditions and stochastic control with state constraints.* Mathematische Annalen, 283 (4), (1989), 583–630.
21. Lasry, J.-M. and Lions, P.-L. *Jeux à champ moyen. I. Le cas stationnaire.* C. R. Math. Acad. Sci. Paris, 343 (9), (2006), 619–625.
22. Lasry, J.-M. and Lions, P.-L. *Jeux à champ moyen. II. Horizon fini et contrôle optimal.* C. R. Math. Acad. Sci. Paris, 343 (10), (2006), 679–684.
23. Lasry, J.-M. and Lions, P.-L. *Mean field games.* Jpn. J. Math., 2 (1), (2007), 229–260.
24. Soave, N., Tavares, H., Terracini, S. and Zilio, A. *Hölder bounds and regularity of emerging free boundaries for strongly competing Schrödinger equations with nontrivial grouping.* Nonlinear Analysis: Theory, Methods & Applications, 138, (2016), 388–427.

Evolutionary Game of Coalition Building Under External Pressure

Alekos Cecchin and Vassili N. Kolokoltsov

Abstract We study the fragmentation-coagulation, or merging and splitting, model as introduced in Kolokoltsov (Math Oper Res, 2016, in press, doi:10.1287/moor.2016.0838), where N small players can form coalitions to resist to the pressure exerted by the principal. It is a Markov chain in continuous time, and the players have a common reward to optimize. We study the behavior as N grows and show that the problem converges to a (one player) deterministic optimization problem in continuous time, in the infinite dimensional state space ℓ^1. We apply the method developed in Gast et al. (IEEE Trans Autom Control 57:2266–2280, 2012), adapting it to our different framework. We use tools involving dynamics in ℓ^1, generators of Markov processes, martingale problems, and coupling of Markov chains.

Keywords Fragmentation-coagulation • Merging and splitting • Evolutionary coalition formation • Markov decision process • Major agent • Mean field limit

1 Introduction

In this paper, we study dynamic optimization problems on Markov decision processes composed of a large number of interacting agents; in particular, we investigate the so-called fragmentation-coagulation, or merging and splitting, model. Our aim is to analyze its limit as the number of players tends to infinity.

Following Kolokoltsov [16], we describe a model which is a Markov chain in continuous time. A natural reaction of the society of small players to the pressure exerted by the principal can be executed by forming stable groups that can confront this pressure in an effective manner (but possibly imposing certain obligatory

A. Cecchin (✉)
Department of Mathematics, University of Padua, Via Trieste 63, Padova, Italy
e-mail: acecchin@math.unipd.it

V.N. Kolokoltsov
Department of Statistics, University of Warwick, Coventry CV4 7AL, UK
e-mail: v.kolokoltsov@warwick.ac.uk

© Springer International Publishing AG 2017
J. Apaloo, B. Viscolani (eds.), *Advances in Dynamic and Mean Field Games*,
Annals of the International Society of Dynamic Games 15,
https://doi.org/10.1007/978-3-319-70619-1_4

regulations for the members of the group). Analysis of such possibility leads one naturally to models of mean field enhanced coagulation processes under external pressure. The major player can change her strategy only in discrete deterministic time.

Coagulation fragmentation processes are well studied in statistical physics; see e.g., [19]. In particular, general mass exchange processes, which in our social environment become general coalition-forming processes preserving the total number of participants, were analyzed in [13] and [14] with their law of large number limits for discrete and general state spaces. In the same way, problems in economics, like merging banks or firms on the market, were studied in [21] and [22], while an application to scientific citation networks or the network of internet links is discussed in [17]. Some simple situation of nonlinear Markov games on a finite state space was analyzed in [15], proving the convergence of Nash equilibria for finite games to equilibria of a limiting deterministic differential game.

Very recently, several authors have studied games of coalition formation. A notion of core equilibrium is proposed in [9] and found via a fixed point method. An application to contracts and networks is analyzed in [10]. A study of the incentives offered by the government to municipalities to merge into larger groups is provided in [24]. Players' preferences over winning coalitions are derived by applying strongly monotonic power indices on the game in [12], where the author also investigates whether there are core stable coalitions. An application of systems of coalition formation to the climate change problem is discussed in [7], where also numerical simulations are performed.

Here we are interested in the response of such systems to external parameters that may be set by the principal who has her own agenda. Thus, we add to the analysis a major player fitting the model to a more general framework. There are two main difficulties in studying this model. Firstly, the total number of coalitions is not constant in time, as they can merge or split. Secondly, the dynamics both of the system of small players and of the limiting system are supposed to lie on the infinite dimensional space ℓ^1, which can be viewed also as a space of measures, instead of a fixed $\mathbb{R}^d$. In fact the dimension of the state space for the system of coalitions grows as the number N of small players tends to infinity. If the system is in the state $x \in \ell^1$, then $x_k = h n_k$ where n_k is the number of coalitions of size k and h is a suitable parameter depending on N, for instance, the inverse of the initial number of coalitions.

Our main result is to show that this problem converges, as N grows, to a one-player deterministic optimization problem in continuous time, in the infinite-dimensional state space ℓ^1, the so-called mean field limit. We prove convergence of the value functions and provide also an approximated optimal policy for the system of small players. Such optimal policy is usually found by using the dynamic programming algorithm for the finite horizon case, but this approach suffers from the curse of dimensionality, which makes the algorithm impractical when the state space is too large. Solving the HJB equation for the limiting system numerically is sometimes rather easy. It provides a deterministic optimal policy whose reward is remarkably close to the optimal reward.

We apply the method developed by Gast, Gaujal, and Le Boudec in [8], where the authors obtained the same kind of results, but in a different setting. They consider discrete-time Markov chains as prelimit systems whose state space is finite and fixed. Their proofs are in line with classic mean field arguments and use stochastic approximation techniques. Moreover their approach is algorithmic; they construct two intermediate systems: one with a finite number of agents controlled by a limit policy and one with a limit system controlled by a stochastic policy induced by the finite system.

Several papers in the literature are concerned with the problem of mixing the limiting behavior of a large number of objects with optimization. In [5], the value function of the Markov decision process (MDP) is approximated by a linearly parameterized class of functions, and a fluid approximation of the MDP is used. It is shown that a solution of the HJB equation is a value function for a modification of the original MDP problem. In [23], the curse of dimensionality of dynamic programming is circumvented by approximating the value function by linear regression. In [8] they use instead a mean field limit approximation and prove the asymptotic optimality in N of the limit policy. Actually, most of the papers dealing with mean field limits of optimization problems over large systems are set in a game theory framework, leading to the concept of mean field games, introduced by Lasry and Lions [18] and P.E. Caines, M. Huang, and R.P. Malhamé [2].

Notice finally that in this paper, we analyze only a preliminary step for a full game setting with major and minor players, namely, the response of the minor players to the action of the major one. The full analysis (not developed here) would include the reaction of the major on the behavior of the minor players and the search for the corresponding equilibrium. However, this development does not seem to present serious difficulty, since our analysis reduces it effectively to a two-player game: the major and the pool of small players.

Contribution and Structure of the Paper

In [16] Kolokoltsov shows the convergence of the optimization problems related to the system of small players to an optimization problem in discrete time for the limiting system (this will be similar to Theorem 3). His proof is based on an argument that is focused on the generators of the Markov chains and shows their convergence. In this paper, we want to show the convergence to an optimization problem in continuous time, so we apply the ideas from [8] where they used a completely different argument for the proof, focusing on trajectories and constructing two auxiliary systems.

In Section 2, we describe properly the fragmentation coagulation model starting from [16], the limiting system, and the related optimization problems. We define the state space where all the dynamics considered lie, which is a compact set $S \subset \ell^1$, and in the end we state the assumptions we need to obtain the convergence. In Section 3, we present our main results and define the two auxiliary systems.

Then we show how to construct an approximated optimal policy starting from an optimal action function for the mean field limit. Moreover we consider a class of applications in which a particular choice of the rate functions allows to reduce the limiting problem to an optimization problem in one dimension, providing an explicit solution in a simplified case and a more effective numerical scheme in general. Finally, in Section 4, we complete the proofs, showing that the general requirements for convergence expressed in [8] can fit to our model, with some modification. We use theorems about semigroups and generators of Markov processes and related martingale problems. Moreover we apply the notion of coupling of Markov chains and also a particular Markovian coupling.

2   Model and Assumptions

2.1   The Space $B_+(L, R]$

We denote, as usual, the space of measures

$$\ell^1(\mathbb{N}) := \left\{ x = (x_1, x_2, \ldots) : x_k \in \mathbb{R}, \quad \sum_k |x_k| < \infty \right\}. \tag{1}$$

Denote by ℓ^1_+ the space of positive measures on $\mathbb{N}$: $\ell^1_+(\mathbb{N}) := \{ x \in \ell^1 : x_k \geq 0 \}$. The usual norm in ℓ^1 is $||x||_{\ell^1} := \sum_k |x_k|$.

Let $L : \mathbb{N} \longrightarrow \mathbb{R}$ be the identity function, which means $L(k) = k$. We define a new norm $||x||_{\ell^1(L)} := \sum_k L(k)|x_k|$, so that we can consider the subset of ℓ^1

$$\ell^1(L) := \{ x \in \ell^1 : ||x||_{\ell^1(L)} < \infty \}$$

which is a Banach space equipped with this norm.

Let us denote by $B(L, R)$ the ball of radius R in $\ell^1(L)$, centered in 0, and $\ell^1_+(L) := \ell^1(L) \cap \ell^1_+$, $B_+(L, R) := B(L, R) \cap \ell^1_+$.

Lemma 1 *The set $B_+(L, R)$ is relatively compact in the norm topology of ℓ^1.*

Proof By Prohorov's compactness criterion, a family of measures is relatively compact in the weak topology if and only if it is tight. We have $\sum_k kx_k < R$ for any $x \in B_+(L, R)$. Thus, for any $n \in \mathbb{N}$ and any $x \in B_+(L, R)$

$$n \sum_{k \geq n} x_k \leq \sum_{k \geq n} kx_k \leq \sum_k kx_k < R$$

which gives $\sum_{k \geq n} x_k < \frac{R}{n}$ for any $n \in \mathbb{N}$ and any $x \in B_+(L, R)$. So for any $\epsilon > 0$, there exists $n \in N$ such that

$$x(\mathbb{N} \setminus [0, n]) = \sum_{k \geq n} x_k < \epsilon$$

for all $x \in B_+(L, R)$, which means that the tightness condition is satisfied.

By *Schur*'s theorem, any weakly convergent sequence in ℓ^1 is actually convergent in the norm of ℓ^1. Therefore the set $B_+(L, R)$ is relatively compact in the topology of ℓ^1.

We denote by $B_+(L, R]$ the closure of $B_+(L, R)$ in the norm topology of ℓ^1, which is compact in ℓ^1, although not in the topology of $\ell^1(L)$. The set

$$S := B_+(L, R]$$

will be the state space for the dynamics considered.

We will assume that the functions defined on S have some regularity. Let Z be a closed convex subset of a normed space Y and $f : Z \longrightarrow Y$ a function. Recall that the *directional derivative $Df(x) : Y \longrightarrow Y$* of f in the point $x \in Z$ is a linear form that calculated in a vector $\xi \in Y$ is defined as $Df(x).\xi := \lim_{t \to 0} \frac{f(x+t\xi)-f(x)}{t}$. The second-order derivative $D^2 f(x)$ is a bilinear form defined as $D^2 f(x).[\xi, \eta] := D(Df(x).\xi).\eta$. Thus the norms of the derivatives in Y are defined as norms of linear maps:

$$||Df(x)||_Y := \sup_{||\xi||=1} ||Df(x).\xi||_Y, \tag{2}$$

$$||D^2 f(x)||_Y := \sup_{||\xi||=||\eta||=1} ||D^2 f(x).[\xi, \eta]||_Y. \tag{3}$$

We say that $f \in \mathscr{C}^1(Z)$ if the function $(x, \xi) \mapsto Df(x).\xi$ is continuous from $Z \times Y$ to Y. Similarly, $f \in \mathscr{C}^2(Z)$ if the function $(x, \xi, \eta) \mapsto D^2 f(x).[\xi, \eta]$ is continuous from $Z \times Y^2$ to Y. These are subsets of $\mathscr{C}(Z)$ and Banach spaces under the norms

$$||f||_{\mathscr{C}^1(Z)} := \sup_{x \in Z} \{||Df(x)||_Y + ||f(x)||_Y\} \tag{4}$$

$$||f||_{\mathscr{C}^2(Z)} := \sup_{x \in Z} \{||D^2 f(x)||_Y + ||f(x)||_Y\}. \tag{5}$$

We will use these definitions for the sets $Z = S$, which is convex and compact, and Y to be either ℓ^1 or $\ell^1(L)$.

2.2 System of Small Players

We describe a so-called *fragmentation-coagulation*, or *merging and splitting*, model in which there are N indistinguishable small players that form coalitions to resist to the pressure exerted by a major player, following [16].

The state space is

$$\mathbb{N}^{fin} := \{n = (n_1, n_2, \ldots) : \text{there is only a finite number of nonzero entries}\} \tag{6}$$

where $n_k \in \mathbb{N}$ denotes the total number of coalitions of size k, so the total number of small players is $N = \sum_k kn_k$ and the total number of coalitions is $\sum_k n_k$. The dynamics will be better described in the rescaled space

$$h\mathbb{N}^{fin} = \{x = hn = (x_1, x_2, \ldots)\} \tag{7}$$

where h can be taken, for instance, as the inverse of the initial number of coalitions. We want to study the limit as $h \to 0$. All this $h\mathbb{N}^{fin}$ spaces, as h changes, can be viewed as subspaces of the space ℓ^1. The total number of players is conserved: this motivates the choice of $L(k) = k$ in the previous section, since $||x||_{\ell^1(L)} = hN$ for any state x; we will return to this in Section 2.6.

The dynamics evolves in continuous time as a Markov chain. It is described as follows:

- to any randomly chosen pair of coalitions of size i and j is attached a random exponential clock of parameter $hC_{ij}(x, b)$ so that they merge if it rings;
- to any randomly chosen coalition of size i is attached an exponential clock of parameter $F_{ij}(x, b)$ such that, if it rings, the coalition splits into two coalitions of size j and $i - j$.

Here the functions C and F may depend on the whole composition x, and b is a control parameter which lies in a compact metric space (E, d).

The minimum of all these exponential random variables is an exponential random variable with the parameter

$$s(x, b) := \sum_{i,j} n_i n_j h C_{ij}(x, b) + \sum_i \sum_{j<i} n_i F_{ij}(x, b). \tag{8}$$

When this minimum clock rings, the system goes from the state n to either $n - e_i - e_j + e_{i+j}$ (two coalitions merge) or $n - e_i + e_j + e_{i-j}$ (a coalition splits). The sequence $(e_i)_{i=1}^\infty$ denotes the standard basis in $\mathbb{R}^\infty$. The first case happens if the minimum holds for the clock of parameter $hC_{ij}(x, b)$, thus with probability given by $\frac{hC_{ij}(x,b)n_i n_j}{s(x,b)}$, while the second case happens with probability $\frac{F_{ij}(x,b)n_i}{s(x,b)}$.

Hence the infinitesimal generator of this Markov chain on the space $\mathbb{N}^{fin}$ is

$$\Lambda_{b,n} G(n) = s(x, b) \sum_{i,j} \frac{hC_{ij}(x, b)n_i n_j}{s(x, b)} \left[G(n - e_i - e_j + e_{i+j}) - G(n) \right] \tag{9}$$

$$+ s(x, b) \sum_i \sum_{j<i} \frac{F_{ij}(x, b)n_i}{s(x, b)} \left[G(n - e_i + e_j + e_{i-j}) - G(n) \right].$$

Equation (9) can be equivalently presented as the infinitesimal generator

$$\Lambda_{b,h} G(x) = \frac{1}{h} \sum_{i,j} C_{ij}(x, b)x_i x_j \left[G(x - he_i - he_j + he_{i+j}) - G(x) \right] \tag{10}$$

$$+ \frac{1}{h} \sum_i \sum_{j<i} F_{ij}(x, b)x_i \left[G(x - he_i + he_j + he_{i-j}) - G(x) \right]$$

of the Markov chain describing the system of small players on the space $h\mathbb{N}^{fin} \subset \ell^1(\mathbb{N})$, for every $G \in \mathscr{C}(S)$.

Notation 1 $X^h(t, x, b)$ *is the state (in S) at time t of the Markov Chain given by this generator (10) which is in x at $t = 0$, under the control parameter b given by the major player.*

The process $X^h(t, x, b)$ describes the evolution of the coalitions of small players, which will be also called the system with N agents.

2.3 Limiting System

The limiting deterministic evolution, the *mean field limit*, is described by the so-called *Smoluchowski equation*. For every x in the compact subset $S \subset \ell^1$, the ODE for the component i is

$$\dot{x}_i = f_i(x, b) := \sum_{j<i} C_{j,i-j}(x, b) x_j x_{i-j} - 2 \sum_j C_{ij}(x, b) x_i x_j \qquad (11)$$

$$+ 2 \sum_{j>i} F_{ji}(x, b) x_j - \sum_{j<i} F_{ij}(x, b) x_i.$$

Notation 2 $X(t, x, b)$ *is the* flow *at time t of the ODE*

$$\dot{x} = f(x, b) \qquad (12)$$

starting in x at $t = 0$ under the control parameter $b \in E$, where f is given by (11). In integral form

$$X(t, x, b) = x + \int_0^t f(X(s, x, b), b) ds. \qquad (13)$$

We view the dynamics given by a deterministic ODE as a Markov process. The semigroup is

$$U_t G(x) = G(X(t, x)) \qquad (14)$$

for every $G \in \mathscr{C}(S)$, and its generator is given by

$$\Lambda G(x) := \sum_i f_i(x) \frac{\partial G}{\partial x_i}(x), \qquad (15)$$

for any $G \in \mathscr{C}^1(S)$. The first-order partial differential operator defined in (15) has characteristics which solve equation (12).

So, for the limiting ODE given by (11), the corresponding infinitesimal generator given by (15) is

$$
\Lambda_b G(x) = \sum_{i,j} C_{ij}(x,b) x_i x_j \left[\frac{\partial G}{\partial x_{i+j}} - \frac{\partial G}{\partial x_i} - \frac{\partial G}{\partial x_j} \right] \tag{16}
$$
$$
+ \sum_i \sum_{j<i} F_{ij}(x,b) x_i \left[\frac{\partial G}{\partial x_{i-j}} + \frac{\partial G}{\partial x_j} - \frac{\partial G}{\partial x_i} \right]
$$

for any $G \in \mathscr{C}^1(S)$ and $b \in E$.

We can thus deduce that pointwise convergence of the generators holds. Namely, for the generators of the Markov chains defined by (10) and the generator of the deterministic limit defined by (16), we obtain

$$
\lim_{h \to 0} \Lambda_{b,h} G(x) = \Lambda_b G(x) \tag{17}
$$

for every $G \in \mathscr{C}^1(S)$, $x \in S$ and every $b \in E$.

We show moreover the convergence in law, for any fixed parameter $b \in E$, of the processes X^h to X in the Skorokhod space $D([0,T], S)$ of cadlag functions, which is the right space where to study these processes. The convergence is then also in probability, as the limit is deterministic, and hence a constant in the Skorokhod space.

Proposition 1 *Let all the functions C_{ij} and F_{ij} be in $\mathscr{C}^1(S)$. Suppose that the initial points $x(h)$ converge in ℓ^1 to x_0, as $h \to 0$. Then, the processes $X^h(\cdot, x(h), b)$ converge in law on the Skorokhod space $D([0,T], S)$ to $X(\cdot, x_0, b)$, as $h \to 0$, for any $b \in E$, whereas the processes are defined in Notations 1 and 2.*

Proof Let $b \in E$ be fixed. The set $\mathscr{C}^1(S)$ is a dense linear subspace of $\mathscr{C}(S)$, and under the assumption of smooth C_{ij} and F_{ij}, it is invariant under the limiting semigroup (U_t) defined in (14), since the function f defined in (11) turns out to be in $\mathscr{C}^1(S)$. So by ([11], proposition 17.9) the set $\mathscr{C}^1(S)$ is a core for the generator Λ_b defined in (16).

Then expanding G in Taylor series, we have that

$$
\lim_{h \to 0} \Lambda_{b,h} G = \Lambda_b G \tag{18}
$$

uniformly for every $G \in \mathscr{C}^1(S)$. Thus, the claim follows from ([11], theorem 17.25) which characterizes the convergence of processes in $D([0,T], S)$.

2.4 Controlled Systems

Here we deal with the system of small players under some control, i.e., a strategy given by the major player. We assume that this major player focuses in finite horizon time $n\tau$ and can update her strategy only in discrete times

$$k\tau \qquad k = 0, 1, \ldots, n-1.$$

The constants $\tau > 0$ and $n \in \mathbb{N}$ are fixed, and in each time step, the controller can change the control parameter b regarding what has happened in the time interval.

The starting point of the Markov chain is $x_0 = x(h) \in h\mathbb{Z}^{fin}$, with the control parameter b_0. After the first time step, the Markov chain is in the state $x_1 = X_h(\tau, x_0, b_0)$. Now the major player can change the parameter, so it becomes $b_1 = b_1(x_1)$ that may depend on the current state of the system. She repeats the same procedure at each time step, and therefore, in the end, we get what is called a *policy*.

Notation 3 A policy *is a sequence of decision rules*

$$\pi = (\pi_0, \pi_1, \ldots, \pi_{n-1}) \tag{19}$$

that specify the action of the mayor player at each time step, with

$$\pi_k = \pi_k(x_k) \tag{20}$$

and

$$x_k = X^h(\tau, x_{k-1}, \pi_{k-1}). \tag{21}$$

Let $X_\pi^h(t, x_0)$ *denote the state of the system at time t when the controller applies policy* π, *starting from the initial point* x_0. *To shorten the notation, we shall sometimes write* $X_\pi^h(t)$ *instead of* $X_\pi^h(t, x_0)$. *It is called the controlled system of small players.*

Equation (21) can be also written as $x_k = X_\pi^h(k\tau)$. At each time step $k\tau$, the controller has an *instantaneous reward* $B(x_k, b_k)$, and in the end, she has a *final reward* $V_0(x_n)$. Our goal is to find a strategy that maximizes

$$V_{\pi,n}^h(x(h)) := E[\tau B(x_0, \pi_0) + \ldots + \tau B(x_{n-1}, \pi_{n-1}) + V_0(x_n)] \tag{22}$$

$$= E\left[\sum_{k=0}^{n-1} \tau B(X_\pi^h(k\tau), \pi(X_\pi^h(k\tau)) + V_0(X_\pi^h(n\tau)) \,\middle|\, X_\pi^h(0) = x_0\right]$$

where B and V_0 are given continuous functions. It is called the *value* for the system with N players. The maximum over all possible policies is then the *optimal value* for the system with N agents

$$V_n^h(x(h)) := \sup_\pi V_{\pi,n}^h(x(h)). \tag{23}$$

We may want to find this optimum value via the usual *dynamic programming* method. First of all we define the *Shapley operator*

$$S[h]V(x) := \sup_{b \in E}[\tau B(x, b) + E(V(X_h(\tau, x, b)))] \tag{24}$$

and then by backward recurrence

$$V_k^h = S[h]V_{k-1}, \tag{25}$$

hence we get

$$V_n^h = S[h]^n V_0. \tag{26}$$

However, this procedure might be unfeasible to calculate practically when the number of players increases. So we will consider the optimum of the limit and then study how close these optima are.

2.4.1 Controlled Limiting System

We want to study the mean field limit system, given by equation (11), in a classical control theory setup. Recall that (E, d) is a compact metric space, the one where the parameter b lies.

Notation 4 *We define an* action function *to be a piecewise Lipschitz function from finite horizon time to E*

$$\alpha : [0, T] \to E.$$

We note that an action function is different from a policy, because the latter depends on the state of the system at each step, while the former does not. Thus in this context, we rewrite equation (12) where f is defined in (11) as

$$\dot{x} = f(x, \alpha), \tag{27}$$

meaning $\dot{x}(t) = f(x(t), \alpha(t))$ for every $t \geq 0$, considering hence $b = \alpha(t)$, i.e., the control parameter is a function of the time.

Notation 5 $X(t, x, \alpha)$ *is the* flux *of the ODE (27), i.e., the solution at time t that is in x at $t = 0$ under the control parameter $b = \alpha = \alpha(s)$. In integral form*

$$X(t, x_0, \alpha) = x_0 + \int_0^t f(X(s, x_0, \alpha), \alpha(s))ds. \tag{28}$$

We are in a finite horizon time T, and now we want to maximize

$$v_\alpha(x) := \int_0^T B(X(s,x,\alpha),\alpha(s))ds + V_0(X(T,x,\alpha)) \tag{29}$$

where B and V_0 are the same as in (22). This is the *value* of the limiting system. The *optimal value* is then

$$v(x) = \sup_\alpha v_\alpha(x). \tag{30}$$

Our aim is to study how and under what assumptions we have the convergence of the optimum of the system of small players (23) to this optimum (30). In fact we want both h and τ tend to 0. To achieve this goal, we need further auxiliary systems, to get also the convergence for every policy and every action function.

2.5 Stability of S

We show that the state space $S := B_+(L, R]$ is stable for all the dynamics considered. We need some regularity for the functions involved in the model. We require that all the functions $C_{ij}(x, b)$ and $F_{ij}(x, b)$ are positive and in $\mathscr{C}^2(S)$ in the variable x, for any b, i.e., twice continuously differentiable on the compact subspace $S \subset \ell^1$, in the topology of ℓ^1. Since S is convex, we can take the directional derivatives in every direction, so we have

$$C := \sup_{i,j} C_{ij}(x,b) < \infty, \quad F = \sup_i \sum_{j<i} F_{ij}(x,b) < \infty \tag{31}$$

$$C(1) := \sup_{i,j,k} \left| \frac{\partial C_{ij}}{\partial x_k}(x,b) \right| < \infty, \quad F(1) := \sup_{i,k} \sum_{j<i} \left| \frac{\partial F_{ij}}{\partial x_k}(x,b) \right| < \infty \tag{32}$$

$$C(2) := \sup_{i,j,k,l} \left| \frac{\partial^2 C_{ij}}{\partial x_k \partial x_l}(x,b) \right| < \infty, \quad F(2) := \sup_{i,k,l} \sum_{j<i} \left| \frac{\partial^2 F_{ij}}{\partial x_k \partial x_l}(x,b) \right| < \infty. \tag{33}$$

These constants are all finite as S is compact.

Let us recall that L is the identity, i.e., $L(k) = k$. Therefore, using the above equalities in (11), we get that also $f : S \to \ell^1$ is twice continuously differentiable (in $\mathscr{C}^2(S)$) as a map both in ℓ^1 and in $\ell^1(L)$ with the following bounds

$$\|f(x)\|_{\ell^1} \le 3C\|x\|_{\ell^1}^2 + 3F\|x\|_{\ell^1} \tag{34}$$

$$\|f(x)\|_{\ell^1(L)} \le 3(C\|x\|_{\ell^1} + 3F)\|x\|_{\ell^1(L)} \tag{35}$$

$$\|Df(x)\|_{\ell^1} \le 6C\|x\|_{\ell^1} + 3F + 3[C(1)\|x\|_{\ell^1} + F(1)]\|x\|_{\ell^1} \tag{36}$$

$$\|Df(x)\|_{\ell^1 (L)} \leq 8C\|x\|_{\ell^1 (L)} + 3F + 3[2C(1)\|x\|_{\ell^1} + F(1)]\|x\|_{\ell^1 (L)} \qquad (37)$$

$$\|D^2 f(x)\|_{\ell^1} \leq 6[C + F(1) + [C(1) + F(2)]\|x\|_{\ell^1} + C(2)\|x\|_{\ell^1}^2] \qquad (38)$$

$$\|D^2 f(x)\|_{\ell^1 (L)} \leq 9[C + F(1) + [C(1) + F(2)]\|x\|_{\ell^1 (L)} + C(2)\|x\|_{\ell^1 (L)}^2]. \qquad (39)$$

We recall now some fact about ODEs in Banach space of measures. In the Markovian dynamics of the system of small players, every state represents the number of coalitions of different sizes, which is of course positive. Hence we are interested in an evolution $f : S \longrightarrow \ell^1$ for the dynamic (12) $\dot{x} = f(x)$ that preserves positivity, i.e., such that for any initial point $x \in \ell^1_+$, the solution $X(t, x)$ belongs to ℓ^1_+ for any $t \geq 0$. We say that f must be *conditionally positive*, in the following sense:

Definition 1 A function $f : \ell^1 \longrightarrow \ell^1$ is said to be *conditionally positive* if for any $x \in \ell^1_+$ with $x_k = 0$ one has $f_k(x) \geq 0$.

Further, we need the following definitions.

Definition 2 A function $f : \ell^1_+ \longrightarrow \ell^1$ is called *L-subcritical* if

$$\sum_k L(k) f_k(x) \leq 0. \qquad (40)$$

As a motivation, we observe that $\frac{d}{dt}\|x\|_{\ell^1 (L)} \leq 0$ if f is *L*-subcritical and $\dot{x} = f(x)$.

Definition 3 A function $f : \ell^1_+ \longrightarrow \ell^1$ is said to satisfy the *Lyapunov condition* if

$$\sum_k L(k) f_k(x) \leq a \sum_k L(k) x_k + b \qquad (41)$$

for some constant a and b, for all $x \in \ell^1_+$.

The main result concerning the dynamics in ℓ^1 is the following lemma.

Lemma 2 *Assume that the function f is conditionally positive, satisfies the Lyapunov condition, and is Lipschitz continuous in the norm of $\ell^1(L)$ on any bounded set of $\ell^1(L)_+$. Then for any $x \in \ell^1(L)_+$, the Cauchy problem (12) has a unique global (defined for all times) solution $X(t, x)$ in $\ell^1_+(L)$. Moreover,*

$$X(t, x) \in B_+(L, e^{at}(\|x\|_{\ell^1 (L)} + bt)). \qquad (42)$$

In particular if f is L-subcritical, then any $B_+(L, R)$ is invariant.

Proof By local Lipschitz continuity and conditional positivity, evolution (12) is locally well-posed and preserves positivity. Moreover, by the Lyapunov condition,

$$(L, X(t, x)) \leq (L, x) + \int_0^t [a(L, X(s, x)) + b] ds$$

where $(L, x) = \sum_k L(k) x_k$ is the duality between functions and measures. So by Gronwall's lemma and the preservation of positivity,

$$0 \le (L, X(t, x)) \le e^{at}[(L, x) + bt],$$

implying that the solution can be extended to all times with required bounds.

We have found that, if the assumptions of the lemma are satisfied, the set $B_+(L, x_0]$ is invariant under an L-subcritical evolution f, and the ODE has a unique global solution, starting from x_0.

Let us check that the assumptions of Lemma 2 are satisfied for f defined in (11). The function f is conditionally positive because its domain is S, which is a subset of ℓ^1_+, and the functions $C_{ij}(x, b)$ and $F_{ij}(x, b)$ are positive. Further, if we consider the function $G \in \mathscr{C}^1(S)$ defined by $G(x) = ||x||_{\ell^1(L)} = \sum_k k x_k$ and apply the generator (16) to this function, then we have $\sum_k k f_k(x) = 0$, since the derivatives of G are $\frac{\partial G}{\partial x_k}(x) = k$. This implies that f is L-subcritical.

Considering equation (37) and thanks to the boundedness of S in $\ell^1(L)$, we obtain that f is Lipschitz continuous as a map in $\ell^1(L)$. So all the assumptions of Lemma 2 are satisfied, showing the well posedness of the problem and the invariance of S. We would like this set to be invariant also for the system of small players.

2.6 State Space for the Small Players

For any h, $X^h(t, x)$ is a continuous-time Markov chains on $h\mathbb{N}^{fin} \cap \ell^1_+$. Let us say that $X^h(t, x)$ are L-non-increasing, if any jump of $X^h(t, x)$ cannot increase L. In this case, a trajectory $X^h(t, x)$ stays forever in $B_+(L, R)$ whenever the initial point $x \in B_+(L, R)$. Moreover, X^h is L-subcritical in the sense that its generator $\Lambda_{h,b}$ satisfies the inequality $\Lambda_{h,b}(L) \le 0$.

The Markov chains $X^h(t, x(h))$ are L-non-increasing and have bounded generators. In fact the state space of the coalitions of small players is actually finite for any fixed h. Indeed we recall that if $X^h(t, x(h))$ is in the state x, then x_k is h times the number of coalitions of size k. The total number of small players is fixed $N = N(h)$ for any h, so $x_k = 0$ for any $k \ge N$ and $x_k \le N/h$ for any $k \le N$, meaning that the state space is finite.

The total number of small players N is of course constant. So the norm in $\ell^1(L)$ of the states x of the Markov chain $X^h(t)$ is conserved:

$$||x||_{\ell^1(L)} = \sum_{k=1}^{N} k x_k = h \sum_{k=1}^{N} k n_k = hN(h). \tag{43}$$

Hence, if the initial point $x(h)$ is in $S = B_+(L, R]$, then any state is in S, i.e., the set $B_+(L, R]$ is invariant for the dynamics of X^h.

Notation 6 $S(h)$ *is the finite state space of the Markov chain X^h, the system of $N = N(h)$ small players. It is a subset of the compact $B_+(L, R]$ in ℓ^1 and a subset of $\mathbb{R}^N$ and of the set $h\mathbb{N}^{fin}$. Denote by $M(h)$ the number of elements of $S(h)$*

$$S(h) := h\mathbb{N}^{fin} \cap B_+(L, R].$$

So we can define $S := B_+(L, R]$ for a suitable R such that this set contains all the initial data $x(h)$ and x_0. Such an R exists because we will consider $\lim_{h \to 0} x(h) = x_0$ and then the sequence is bounded. S is the compact set in ℓ^1 invariant for all the dynamics considered. Thanks to (43), we have

$$N(h) \leq \frac{R}{h} \tag{44}$$

for any h. Further $\lim_{h \to 0} M(h) = \lim_{h \to 0} N(h) = +\infty$ and the finite spaces $S(h)$ are decreasing, i.e., if $h > l$, then $S(h) \subset S(l)$, and $\bigcup_{h>0} S(h) \subseteq S$. Moreover,

$$S(h) \subset S \subset \ell^1(L) \subset \ell^1 \subset \ell^2. \tag{45}$$

2.7 Assumptions in the Model

Let us summarize the assumptions we make on our model. Recall that in the system of N small players, the controller acts at time steps $k\tau$ for $k = 0, 1, \ldots, n - 1$.

- **(H1)** $S = B_+(L, R]$ is the state space of all the dynamics considered, and the initial states lie in S;
- **(H2)** The functions $C_{ij}(x, b)$ and $F_{ij}(x, b)$ are positive and in $\mathscr{C}^2(S)$ in the variable x, for any $b \in E$, and Lipschitz continuous in the variable b, for any $x \in S$;
- **(H3)** The rewards $B(x, b)$ and $V_0(x)$ are Lipschitz continuous in x in the ℓ^2-norm, uniformly in b, and B is bounded;
- **(H4)** The time step $\tau = \tau(h)$ depends on h, as well as $N = N(h)$ does, and

$$\lim_{h \to 0} \tau(h) = \lim_{h \to 0} \tau(h)\sqrt{N(h)} = 0;$$

- **(H5)** The horizon T is fixed and the number of steps is

$$n(h) := \left\lfloor \frac{T}{\tau(h)} \right\rfloor \tag{46}$$

for any h, which tends to infinity, as h tends to 0;
- **(H6)** The rescaling parameter h of the model is such that

$$\lim_{h \to 0} h(N(h))^2 = \lim_{h \to 0} \tau(h)(N(h))^2 = 0. \tag{47}$$

Equivalently, we can think of studying the limit as N tends to infinity. So the parameter $h = h(N)$ has to satisfy the latter conditions, and it represents the rescaling parameter for the system of N players.

3 Mean Field Convergence

In this section, we present our main results. We follow the ideas in [8]; hence, we firstly introduce the two auxiliary systems.

3.1 First Auxiliary System

This is a system with N agents controlled by an action function borrowed from the mean field limit. More precisely, let α be an action function that specifies the action to be taken at time t. Although α has been defined for the limiting system, it can also be used for the system with N players. In this case, the action function α can be seen as a policy that does not depend on the state of the system.

At step k, the controller applies action

$$\alpha_k := \alpha(k\tau),$$

so (20) gives a policy $(\alpha_0, \ldots, \alpha_{n-1})$ as in (19), but independent of the state of the system.

By abuse of notation, we denote by $X_\alpha^h(t)$ as in (3) the state of the system at time t when applying the policy derived from the action function α as explained above. In what follows, policies will always be denoted by π and action functions by α. Here (21) becomes

$$x_k = X^h(\tau, x_{k-1}, \alpha_{k-1}) = X_\alpha^h(k\tau),$$

starting from initial point x_0 with control parameter $\alpha_0 = \alpha(0)$.

The value for this system, similar to (22), is defined by

$$V_{\alpha,n}^h(x_0) := E[\tau B(x_0, \alpha_0) + \ldots + \tau B(x_{n-1}, \alpha_{n-1}) + V_0(x_n)] \tag{48}$$

$$= E\left[\sum_{k=0}^{n-1} \tau B(X_\alpha^h(k\tau), \alpha(k\tau)) + V_0(X_\alpha^h(n\tau)) \,\middle|\, X_\alpha^h(0) = x_0\right].$$

3.2 Second Auxiliary System

The method of proof uses a second auxiliary system in which trajectories are considered. This is a limiting system controlled by an action function derived from the policy of the original system with N agents.

Consider the system with N players under policy π. The stochastic process $X_\pi^h = X_{\pi,n(h)}^h$ is defined on some probability space Ω. To every $\omega \in \Omega$, there corresponds a trajectory $X_\pi^h(\omega)$, and for every $\omega \in \Omega$, we define the piecewise constant action function $A_\pi^h(\omega)$, as explained in the following

Notation 7

$$A_\pi^h(\omega) : [0, T] \to E$$

is an action function such that

- *this random function is constant on each interval* $[k\tau, (k + 1)\tau[$ *for any* $k = 0, 1, \ldots, n - 1$;
- $A_\pi^h(\omega)(k\tau) := \pi_k = \pi_k(X_\pi^h(k\tau))$ *is the action taken by the major player of the system with N agents at time slot $k\tau$, under policy π.*

Recall from Notation 5 that for any $x_0 \in S \subset \ell^1(\mathbb{N})$ and any action function α, $X(x_0, \alpha)$ is the solution of the ODE (27). For every ω, $X(t, x_0, A_\pi^h(\omega))$ is the solution of the limiting system with action function $A_\pi^h(\omega)$, as in (28), i.e.,

$$X(t, x_0, A_\pi^h(\omega)) = x_0 + \int_0^t f(X(s, x_0, A_\pi^h(\omega)), A_\pi^h(\omega)(s))ds. \tag{49}$$

The value function for this system is as in (29).

When ω is fixed, $X(t, x_0, A_\pi^h(\omega))$ is a continuous time deterministic process corresponding to one trajectory $X_\pi^h(\omega)$. When considering all possible realizations of X_π^h, $X(t, x_0, A_\pi^h)$ is a random, continuous time function *coupled* to X_π^h, i.e., a stochastic process. Its randomness comes only from the action term A_π^h, in the ODE (27). In the following, we omit the dependence on ω in our writing. A_π^h and X_π^h will always designate the processes corresponding to the same ω.

3.3 Main Results

The main result establishes the convergence of the optimization problem for the system with N players to the optimization problem for the mean field limit, through their value functions.

Theorem 1 *Under assumptions (H1)–(H6), if $\lim_{h\to 0} x(h) = x_0$ almost surely, respectively in probability, then*

$$\lim_{h\to 0} V^h(x(h)) = v(x_0) \tag{50}$$

almost surely, respectively in probability, where V^h and v are the optimal values defined in (23) and (30).

The second result states that an optimal action function for the mean field limit provides an asymptotically optimal strategy for the system with N agents. Let us denote by $(\hat{X}^h_\alpha(t))_{t\geq 0}$ the continuous time process which is the affine interpolation of $X^h_\alpha(t)$ (the first auxiliary system) in the points $k\tau$ and similarly by $\hat{X}^h_\pi(t)$ the affine interpolation of $X^h_\pi(t)$ under policy π.

Theorem 2 *Under assumptions (H1)–(H6), let α be a piecewise Lipschitz continuous action function on $[0, T]$, of Lipschitz constant K_α, and with at most p discontinuity points. Then there exist functions J, I'_0, and B' satisfying*

$$\lim_{h\to 0} I'_0(h, \alpha) = \lim_{h\to 0} J(h, T) = 0, \quad \lim_{\substack{h\to 0 \\ \delta\to 0}} B'(h, \delta) = 0$$

such that for all $\epsilon > 0$

$$P\left\{ \sup_{0\leq t\leq T} \left\| \hat{X}^h_\alpha(t) - X(t, x_0, \alpha) \right\| > \left[\left\| X^h(0) - x_0 \right\| + I'_0(h, \alpha)T + \epsilon \right] e^{L_1 T} \right\} \leq \frac{J(h, T)}{\epsilon^2} \tag{51}$$

and

$$\left| V^h_\alpha(X^h(0)) - v_\alpha(x_0) \right| \leq B'\left(h, \left\| X^h(0) - x_0 \right\| \right). \tag{52}$$

Inequality (52) implies also that if

$$\lim_{h\to 0} X^h(0) = x_0 \tag{53}$$

almost surely, respectively in probability, then

$$\lim_{h\to 0} V^h_\alpha(X^h(0)) = v_\alpha(x_0) \tag{54}$$

almost surely, respectively in probability.

The following Corollary combines Theorems 1 and 2. It states that an optimal action function for the limiting system is asymptotically optimal for the system of small players.

Corollary 1 *If α_* is an optimal action function for the limiting system and if $\lim_{h\to 0} X^h(0) = x_0$ almost surely, respectively in probability, then*

$$\lim_{h\to 0} \left| V^h_{\alpha_*}(X^h(0)) - V^h(X^h(0)) \right| = 0 \tag{55}$$

almost surely, respectively in probability.

Proof The assumption says that there exists an action function α_* that maximizes v, i.e., $v(x_0) = v_{\alpha_*}(x_0) = \max_\alpha v_\alpha(x_0)$. Hence, we have

$$\left| V^h_{\alpha_*}(X^h(0)) - V^h(X^h(0)) \right| \leq \left| V^h_{\alpha_*}(X^h(0)) - v_{\alpha^*}(x_0) \right| + \left| V^h(X^h(0)) - v(x_0) \right|.$$

So, if (53) holds, the first modulus goes to 0 by (54) and the second by (50). Therefore, (55) is given combining Theorems 1 and 2.

3.3.1 Auxiliary Results

In order to prove the main theorems, we need two auxiliary results.

Theorem 3 *Under assumptions (H1)–(H6), there exist functions I_0 and J satisfying* $\lim_{h\to 0} I_0(h, \alpha) = \lim_{h\to 0} J(h, T) = 0$ *such that for any $\epsilon > 0$, $h > 0$ and any policy π*

$$P\left\{ \sup_{0 \leq t \leq T} \left\| \hat{X}^h_\pi(t) - X(t, x_0, A^h_\pi) \right\| > \left[\left\| X^h(0) - x_0 \right\| + I_0(h)T + \epsilon \right] e^{L_1 T} \right\} \leq \frac{J(h, T)}{\epsilon^2}.$$
(56)

If (53) holds, this theorem shows the convergence in probability of the controlled system with N agents, with explicit bounds.

The second statement deals with the convergence of the value for the controlled system of small players to the value of the second auxiliary system. Let π be a policy and A^h_π be the sequence of actions corresponding to a trajectory of X^h_π, as in Notation 7. Equation (29) defines the value for the deterministic limit, whereas α is an action function. When applying the random action function A^h_π, this defines a random variable $v_{A^h_\pi}(x_0)$. A consequence of Theorem 3 is the convergence of $V^h_\pi(X^h(0))$ to the expectation of this random variable.

Theorem 4 *Let A^h_π be the random action function associated with X^h_π as in Notation 7. Under assumptions (H1)–(H6), there exists a function B satisfying*

$$\lim_{\substack{h\to 0 \\ \delta \to 0}} B(h, \delta) = 0$$
(57)

such that

$$\left| V^h_\pi(X^h(0)) - E[v_{A^h_\pi}(x_0)] \right| \leq B\left(h, \left\| X^h(0) - x_0 \right\| \right).$$
(58)

This implies that if (53) holds almost surely, respectively in probability, then

$$\lim_{h\to 0} \left| V^h_\pi(X^h(0)) - E[v_{A^h_\pi}(x_0)] \right| = 0$$
(59)

almost surely, respectively in probability.

The proofs of the main results use the two auxiliary systems. The first auxiliary system provides a strategy for the system with N agents derived from an action function of the mean field limit. It cannot do better than the optimal value of the system of small players, and it is close to the optimal value of the mean field limit. Therefore, the optimal value for the system with N players is lower bounded by the optimal value for the mean field limit.

The second auxiliary system is used in the opposite direction: it shows that for large N, the two optimal values are the same.

3.4 Requirements for Convergence

Let us denote by $||x||$ the ℓ^2-norm of x and define the *drift* of the model as

$$F^h(x, b) := E[X^h(\tau, x, b) - x]. \tag{60}$$

Due to Theorems 2–6 in [8], in order to prove Theorems 1–4, it is sufficient to show that

- **(A1)** There exist some nonrandom functions $I_1(h)$ and $I_2(h)$ such that

$$\lim_{h \to 0} I_1(h) = \lim_{h \to 0} I_2(h) = 0$$

and that for all x and all policies π, the number of coalitions $\Delta_\pi^h(k)$ that perform a transition between time step $k\tau$ and $(k + 1)\tau$ satisfies

$$E(\Delta_\pi^h(k)|X_\pi^h(k\tau) = x) \leq \frac{1}{h}I_1(h) \tag{61}$$

$$E(\Delta_\pi^h(k)^2|X_\pi^h(k\tau) = x) \leq \frac{1}{h^2}\tau I_2(h); \tag{62}$$

- **(A2)** There exists a function $I_0(h)$ such that $\lim_{h \to 0} I_0(h) = 0$ and

$$\left\| \frac{F^h(x, b)}{\tau(h)} - f(x, b) \right\| \leq I_0(h) \tag{63}$$

for every $x \in S$ and $b \in E$ where the function f is the one in (11), and moreover f is defined on $S \times E$, and there exists a constant L_2 such that

$$|f(x, b)| \leq L_2; \tag{64}$$

- **(A3)** There exist constants L_1, K and K_B such that for all $x, y \in S$ and $a, b \in E$

$$||F^h(x, b) - F^h(y, b)|| \leq L_1||x - y||I(h) \tag{65}$$

$$\|f(x, b) - f(y, a)\| \leq K(\|x - y\| + d(a, b)) \tag{66}$$

$$\|B(x, b) - B(y, b)\| \leq K_B \|x - y\| \tag{67}$$

$$\|V_0(x) - V_0(y)\| \leq K_B \|x - y\| \tag{68}$$

and the reward is bounded:

$$\sup_{\substack{x \in S \\ b \in E}} \max \{|B(x, b)|, |V_0(x, b)|\} =: \|B\|_\infty. \tag{69}$$

We will show in the next section that if our assumptions (H1)–(H3) are satisfied, then (A1)–(A3) hold.

Let us now fix the functions appearing in (A1), (A2), and (A3):

$$K := 6CR + 3F + 3R[C(1)R + F(1)], \tag{70}$$

$$L_2 := 3CR^2 + 3FR, \tag{71}$$

$$I_0(h) := \sqrt{N(h)}\frac{\tau}{2}(R_1 + hR_2), \tag{72}$$

$$I_1(h) := \tau(CR^2 + F), \tag{73}$$

$$I_2(h) := (CR^2 + F)[\tau(CR^2 + F) + h], \tag{74}$$

where

$$R_1 := 3(CR^2 + FR)(6CR + 3F + 3C(1)R^2 + F(1)R),$$

$$R_2 := 54(CR^2 + FR)(C + F(1) + C(1)R + F(1)R + C(2)R^2).$$

Clearly $\lim_{h \to 0} I_0(h) = \lim_{h \to 0} I_1(h) = \lim_{h \to 0} I_2(h) = 0$ if (H4) holds. L_1 actually depends on h

$$L_1 = L_1(h) := Ke^{M_2\sqrt{N(h)}\tau}, \tag{75}$$

where $M_2 := 3(C(1)R^2 + 2CR + F(1)R + F)$. Hence L_1 tends to the constant K by (H4), as h tends to 0.

Let us define the functions I_0', J, B, B' appearing in the statements of Theorems 2, 3, and 4 by the following equations:

$$I_0'(h, \alpha) := I_0(h) + \tau Ke^{(K-L_1)T} \cdot \left[\frac{K_\alpha}{2} + 2\left(1 + \min\left\{\frac{1}{\tau}, p\right\}\right)\|\alpha\|_\infty\right], \tag{76}$$

$$J(h, T) := 8T\left\{L_1^2\left[I_2(h)\tau^2 + I_1(h)^2(T + \tau)\right] + N(h)^2\left[2I_2(h) + \tau L_2^2\right]\right\}, \tag{77}$$

$$B(h, \delta) := \tau\|B\|_\infty + K_B\sqrt{2}I_1(h) + K_B(\delta + I_0(h)T)\left(e^{L_1 T} + \frac{e^{L_1 T} - 1}{L_1}\right) \tag{78}$$

$$+ \frac{3}{2^{\frac{1}{3}}} \left[e^{L_1 T} + \frac{e^{L_1 T} - 1 + \frac{\tau}{2}}{L_1} \right]^{\frac{2}{3}} \cdot K_B^{\frac{1}{3}} ||B||_\infty^{\frac{1}{3}} J(h, T)^{\frac{1}{3}} (T + 1)^{\frac{2}{3}} \quad (79)$$

and $B'(h, \delta)$ has the same expression as $B(h, \delta)$ replacing $I_0(h)$ by $I'_0(h, \alpha)$. From (H4) and (H6), follow that $\lim_{h \to 0} J(h, T) = \lim_{h \to 0} I'_0(h, \alpha) = 0$ and $\lim_{\substack{h \to 0 \\ \delta \to 0}} B'(h, \delta) = \lim_{\substack{h \to 0 \\ \delta \to 0}} B(h, \delta) = 0$.

3.5 Constructing an Optimal Policy

By means of Corollary 1, an optimal action function for the mean field limit is asymptotically optimal for the system of small players. This provides a way for constructing an asymptotically optimal policy.

We denote by $u(x, t)$ the optimal cost over horizon $[t, T]$ for the limiting system. Under our hypothesis, the following proposition holds:

Proposition 2 *The value function $u(x, t)$ is the unique, bounded, and uniformly continuous, viscosity solution in $S \times [0, T] \subset \ell^2 \times [0, T]$ of the Hamilton-Jacobi-Bellman equation*

$$-\frac{\partial u(x, t)}{\partial t} - \max_{b \in E} \{ \nabla u(x, t) \cdot f(x, b) + B(x, b) \} = 0 \quad (80)$$

which satisfies the terminal condition $u(x, T) = V_0(x)$.

Let us recall that the definition of viscosity solution in a Hilbert space, as ℓ^2, does not differ from the usual one. Further, under our assumptions, the Hamiltonian defined as $H(x, p) := \max_{b \in E} \{ p \cdot f(x, b) + B(x, b) \}$ for any $(x, p) \in S \times \ell^2$ is such that

$$|H(x, p) - H(y, p)| \leq C ||x - y||(1 + ||p||)$$

$$|H(x, p) - H(x, q)| \leq C ||p - q||,$$

where C is a constant. Therefore, existence and uniqueness of bounded and uniformly continuous viscosity solutions to (80) are implied by Theorem 5.1 in [6].

We state the algorithm presented in [8] for constructing an asymptotically optimal policy for the system of small players X^h via an optimal action function for the mean field limit X:

- Let $X^h(0)$ be the initial condition of the limiting system. Solve the Hamilton-Jacobi-Bellman equation (80) on $[0, \tau n(h)]$. Assume this provides an optimal control function α_*;
- Construct a policy π for the system of small players: the action to be taken under state $X^h(k\tau)$ at step k is

$$\pi_k(X^h(k\tau)) := \alpha_*(k\tau).$$

The asymptotic optimality of the related value is ensured by Corollary 1. The policy π constructed above is static in the sense that it does not depend on the state $X^h(k\tau)$ but only on the initial state $X^h(0)$. The deterministic estimation of $X^h(k\tau)$ is provided by the differential equation.

The algorithm described above uses an optimal action function for the limiting system which may not exist, as we observed above. A sufficient condition for its existence is that the set $f(x, E) \times B(x, E)$ is convex for all $x \in S$; a proof of this fact can be found in [1].

However, the existence of an optimal action function is actually not necessary in the algorithm described above. Indeed, if there is no optimal control of the HJB equation (80), one can replace the optimal α_* used in the algorithm by an action function which is h-optimal. This still provides an asymptotically optimal policy.

3.6 Example

We show here that in a class of applications, the limiting problem can be reduced to an optimization problem in one dimension. This provides a computationally much more effective scheme in order to obtain an asymptotically optimal policy for the system of small players, in view of the algorithm presented above.

Firstly, we study a particular case in which the mean field limit admits an explicit solution. So let us consider $E = [0, 1]$ and a particular shape of the functions C_{ij} and F_{ij}, in which there is no dependence on x:

$$C_{ij}(x, b) = b \tag{81}$$

$$F_{ij}(x, b) = \frac{1-b}{i-1}. \tag{82}$$

In particular if b is 1, then only merging is possible. While if it is 0, only splitting is possible.

Let $m(x) := \sum_i x_i$ be the norm of $x \in \ell^1$. With these rates, the limiting evolution f in (11) can be studied considering only the dynamics of the norm. Evaluating the limiting generator (16) on $G = m$, we obtain

$$\sum_i f_i(x, b) = -bm^2 + (1 - b)m. \tag{83}$$

Hence, the evolution of m is described by

$$\begin{cases} \dot{m} = -bm^2 + (1 - b)m \\ m(0) = m_0, \end{cases} \tag{84}$$

which is in fact an ODE on $\mathbb{R}_+$.

If b is constant, then the explicit solution to (84) is

$$m(t, m_0, b) = \frac{m_0(1-b)e^{(1-b)t}}{1 - b - m_0 b + m_0 e^{(1-b)t}}. \tag{85}$$

if $b \neq 1$. In particular, if $b = 0$, $m(t, m_0, 0) = m_0 e^t$. While if $b = 1$,

$$m(t, m_0, 1) = \frac{m_0}{1 + tm_0}. \tag{86}$$

Observe that $m(t, m_0, b)$ is a continuous and strictly decreasing function of $b \in [0, 1]$, for any value of $0 < t \leq T$ and $m_0 > 0$.

We assume that there is no instantaneous reward and that the final reward $V_0 = V_0(m)$ is a strictly concave function of m which has a unique maximum in m^*. The value function V in this case can be computed explicitly, that is,

$$V(t, m) = \begin{cases} V_0(me^{T-t}) & \text{if } m < m^* e^{-(T-t)} \\ V_0(m^*) & \text{if } m^* e^{-(T-t)} \leq m \leq \frac{m^*}{1-(T-t)m^*} \\ V_0\left(\frac{m}{1+(T-t)m}\right) & \text{if } m > \frac{m^*}{1-(T-t)m^*}. \end{cases} \tag{87}$$

If $m^* e^{-(T-t)} \leq m_0 \leq \frac{m^*}{1-(T-t)m^*}$, denote by $b^*(T, m_0)$ the unique value of $b \in [0, 1]$ such that $m(T, m_0, b) = m^*$. Therefore, if $m(0) = m_0$, an optimal action function for the mean field limiting problem is to choose the constant value

$$\alpha^*(t, m_0) = \alpha^*(m_0) := \begin{cases} 1 & \text{if } m_0 < m^* e^{-(T-t)} \\ b^*(T, m_0) & \text{if } m^* e^{-(T-t)} \leq m_0 \leq \frac{m^*}{1-(T-t)m^*} \\ 0 & \text{if } m_0 > \frac{m^*}{1-(T-t)m^*}. \end{cases} \tag{88}$$

This example can be generalized to the case in which the rates C_{ij} and F_{ij} depend also on the norm of x. Namely, let

$$C_{ij}(x, b) = bf_C(m(x)) \tag{89}$$

$$F_{ij}(x, b) = \frac{1-b}{i-1} f_B(m(x)), \tag{90}$$

where f_C and f_B are some Lipschitz continuous and nonnegative functions. The dynamics (84) for m becomes

$$\begin{cases} \dot{m} = -bf_C(m)m^2 + (1-b)f_B(m)m \\ m(0) = m_0. \end{cases} \tag{91}$$

Also in this case, the mean field limiting problem reduces to an optimization problem in one dimension. Thus, there are numerical schemes which provide an

$\frac{1}{N}$-optimal action function in feedback form (see, e.g., [1]). These are much more efficient than trying to solve the Bellman equation (26). In fact the prelimit problem is allowed to be tackled only if N is lower than a few tens: see [20].

4 Proofs of Convergence

In this section, we show that (A1)–(A3) hold in our fragmentation-coagulation model. So assume (H1)–(H6) and recall that the compact state space $S \subset \ell^1$ is stable for all the dynamics considered.

4.1 Lipschitz Continuity of the Limit

We verify that (66) and (64) are satisfied for the function f in (11), if (H2) holds. We actually do not need f to be in $\mathscr{C}^2(S)$, but only in $\mathscr{C}^1(S)$. We use the fact that

$$||x|| = ||x||_{\ell^2} \leq ||x||_{\ell^1} \leq ||x||_{\ell^1(L)} \leq R \tag{92}$$

for any $x \in S$. So equation (34) gives

$$||f(x)|| \leq 3CR^2 + 3FR$$

for all $x \in S$, which is (64) with bound

$$L_2 := 3CR^2 + 3FR. \tag{93}$$

By (92) and its definition (2), we get also

$$||Df(x)||_{\ell^2} \leq ||Df(x)||_{\ell^1} \quad \forall x \in S$$

which, together with equation (36), yields

$$||Df(x)||_{\ell^2} \leq 6CR + 3F + 3R[C(1)R + F(1)]. \tag{94}$$

Applying then the mean value theorem in the convex space S, for every x and y in S, there exist a point z in the segment $[x, y]$ such that

$$f(y) - f(x) = Df(z).(y - x).$$

Thus, taking the ℓ^2-norm and using the fact that Df is a bounded linear map, we get

$$||f(y) - f(x)|| \leq ||Df(z)||_{\ell^2}||y - x||$$

for all x and y in S. According to (94), we find that the limiting function is Lipschitz continuous in x, for the ℓ^2-norm, with

$$K = 6CR + 3F + 3R[C(1)R + F(1)] \tag{95}$$

as a Lipschitz constant.

Clearly (H3) implies (67), (68) and (69).

4.2 Convergence of the Drift

Here we prove (63). Recall that the drift of the model is

$$F^h(x, b) := E[X^h(\tau, x, b) - x].$$

The semigroup U^h_t of the Markov process X^h is defined by

$$U^h_t G(x, b) := E[G(X^h(t, x, b))]$$

for any $G \in \mathscr{C}(S, \mathbb{R})$ and $t \geq 0$. U_0 is the identity and U_t satisfies the Kolmogorov differential equation

$$\frac{\partial}{\partial t} U^h_t G(x, b) = \Lambda_{h,b} U^h_t G(x, b) = U^h_t \Lambda_{h,b} G(x, b), \tag{96}$$

where $\Lambda_{h,b}$ is the infinitesimal generator defined in (10).

We apply U^h_t to the projection on the k-th coordinate G_k. This function is in $\mathscr{C}(S, \mathbb{R})$ and

$$F^h_k(x, t) := G_k\left(E[X^h(\tau, x, b) - x]\right) = EG_k(X^h(\tau, x, b)) - x_k = U^h_\tau G_k(x, b) - G_k(x). \tag{97}$$

Let us denote $u_k(t, x, b) := U^h_t G_k(x, b)$. The Taylor formula applied to u_k in the variable t gives

$$u_k(\tau, x, b) = u_k(0, x, b) + \tau \frac{\partial u_k}{\partial t}(0, x, b) + \frac{\tau^2}{2} \frac{\partial^2 u_k}{\partial t^2}(s, x, b) \tag{98}$$

for a fixed $s \in]0, \tau[$ and for every x and b. We have

$$u_k(0, x, b) = G_k(x) = x_k \tag{99}$$

and according to (96)

$$\frac{\partial u_k}{\partial t}(0, x, b) = \Lambda_{h,b} u_k(0, x, b) = \Lambda_{h,b} G_k(x) \tag{100}$$

for every x and b.

The generator (10) calculated on the projection, using $G_k(e_i) = \delta_{ik}$, gives

$$\Lambda_{b,h}G_k(x) = \frac{1}{h}\sum_{i,j} C_{ij}(x,b)x_ix_j\left[G_k(x - he_i - he_j + he_{i+j}) - G_k(x)\right]$$

$$+ \frac{1}{h}\sum_i\sum_{i<j} F_{ij}(x,b)x_i\left[G_k(x - he_i + he_j + he_{i-j}) - G_k(x)\right]$$

$$= \sum_{i,j} C_{ij}(x,b)x_ix_j\left[\delta_{i+j,k} - \delta_{ik} - \delta_{jk}\right]$$

$$+ \sum_i\sum_{i<j} F_{ij}(x,b)x_i\left[\delta_{jk} + \delta_{i-j,k} - \delta_{ik}\right]$$

$$= \sum_{i<k} C_{i,k-i}(x,b)x_ix_{k-i} - \sum_j C_{kj}(x,b)x_kx_j - \sum_i C_{ik}(x,b)x_ix_k$$

$$+ \sum_{i>k} F_{ik}(x,b)x_i + \sum_{i>k} F_{i,i-k}(x,b)x_i - \sum_{j<k} F_{kj}(x,b)x_k$$

and, observing that $C_{ij} = C_{ji}$ and $F_{ij} = F_{i,i-j}$, the latter expression is exactly $f_k(x,b)$ defined in (11), whence

$$\Lambda_{b,h}G_k(x) = f_k(x,b) \tag{101}$$

for every x and b.

We need also an estimate of the latter term in (98). Equation (96) yields

$$\frac{\partial^2 u_k}{\partial t^2}(s,x,b) = \frac{\partial}{\partial t}U_s^h\Lambda_{b,h}G_k(x) = U_s^h\Lambda_{b,h}\Lambda_{b,h}G_k(x) = U_s^h\Lambda_{b,h}f_k(x,b).$$

We use then the fact that the semigroup U_s^h is, for any s, a contraction in the space $\mathscr{C}(S,\mathbb{R})$ equipped with the sup-norm $\|G\|_\infty := \sup_{x\in S}|G(x)|$. Thus, the latter term in (98) is bounded by

$$\left|\frac{\partial^2 u_k}{\partial t^2}(s,x,b)\right| \leq \left\|U_s^h\Lambda_{b,h}f_k\right\|_\infty \leq \|\Lambda_{b,h}f_k\|_\infty \tag{102}$$

for any $x \in S$, $b \in E$ and $s > 0$.

The estimate for the norm of $\Lambda_{b,h}f_k$ is found applying again the Taylor formula to f_k:

$$\Lambda_{b,h}f_k(x) = \frac{1}{h}\sum_{i,j} C_{ij}(x,b)x_ix_j\left[f_k(x - he_i - he_j + he_{i+j}) - f_k(x)\right]$$

$$+ \frac{1}{h}\sum_i\sum_{j<i} F_{ij}(x,b)x_i\left[f_k(x - he_i + he_j + he_{i-j}) - f_k(x)\right]$$

$$= \sum_{i,j} C_{ij}(x,b)x_i x_j \sum_l \frac{\partial f_k}{\delta x_l}(x) \left(\delta_{i+j,l} - \delta_{il} - \delta_{jl}\right)$$

$$+ h \sum_{i,j} C_{ij}(x,b)x_i x_j \sum_l \sum_m \frac{\partial}{\delta x_l} \frac{\partial}{\delta x_m} f_k(y) \left(\delta_{i+j,l} - \delta_{il} - \delta_{jl}\right)\left(\delta_{i+j,m} - \delta_{im} - \delta_{jm}\right)$$

$$+ \sum_i \sum_{j<i} F_{ij}(x,b)x_i \sum_l \frac{\partial f_k}{\delta x_l}(x) \left(\delta_{jl} + \delta_{i-j,l} - \delta_{il}\right)$$

$$+ h \sum_i \sum_{j<i} F_{ij}(x,b)x_i \sum_l \sum_m \frac{\partial}{\delta x_l} \frac{\partial}{\delta x_m} f_k(z) \left(\delta_{jl} + \delta_{i-j,l} - \delta_{il}\right)\left(\delta_{jm} + \delta_{i-j,m} - \delta_{im}\right)$$

for certain fixed points y and z in S. This gives, using estimates (31), (36) and (38) and the definitions of the norms of the derivatives Df and D^2f,

$$|\Lambda_{b,h}f_k(x)| \le C||x||_{\ell^1}^2 \left(3||Df(x)||_{\ell^1} + 9h||D^2f(y)||_{\ell^1}\right)$$

$$+ F||x||_{\ell^1} \left(3||Df(x)||_{\ell^1} + 9h||D^2f(z)||_{\ell^1}\right)$$

$$\le C||x||_{\ell^1}^2 [3\left(6C||x||_{\ell^1} + 3F + 3[C(1)||x||_{\ell^1} + F(1)]||x||_{\ell^1}\right)$$

$$+ 54h\left(C + F(1) + [C(1) + F(2)]||y||_{\ell^1} + C(2)||y||_{\ell^1}^2\right)]$$

$$+ F||x||_{\ell^1}[3\left(6C||x||_{\ell^1} + 3F + 3[C(1)||x||_{\ell^1} + F(1)]||x||_{\ell^1}\right)$$

$$+ 54h\left(C + F(1) + [C(1) + F(2)]||z||_{\ell^1} + C(2)||z||_{\ell^1}^2\right)]$$

$$\le (CR^2 + FR)[3(6CR + 3F + 3C(1)R^2 + F(1)R)$$

$$+ 54h(C + F(1) + C(1)R + F(1)R + C(2)R^2)]$$

for every $x \in S$ and $b \in E$ and $k = 1, \ldots, N(h)$.

Therefore,

$$\|\Lambda_{b,h}f_k\|_\infty \le R_1 + hR_2 \tag{103}$$

for every $x \in S$ and $b \in E$ and $k = 1, \ldots, N(h)$, where $R_1 := 3(CR^2 + FR)(6CR + 3F + 3C(1)R^2 + F(1)R)$ and $R_2 := 54(CR^2 + FR)(C + F(1) + C(1)R + F(1)R + C(2)R^2)$.

Equations (97) and (98), by means of (99), (100) and (101), lead to

$$F_k^h(x,b) - \tau f_k(x,b) = \frac{\tau^2}{2} \frac{\partial^2 u_k}{\partial t^2}(s,x,b)$$

and this applying (102) and (103) yields

$$\left| F_k^h(x, b) - \tau f_k(x, b) \right| \leq \frac{\tau^2}{2}(R_1 + hR_2)$$

for any $k = 1, \ldots, N(h)$. Considering the ℓ^2-norm, we have hence

$$\left\| \frac{F^h(x, b)}{\tau} - f(x, b) \right\| \leq \sqrt{N(h)} \frac{\tau}{2}(R_1 + hR_2)$$

which is (63), whereas I_0 is given by (72).

4.3 Lipschitz Continuity of the Drift

Here we verify that the drift F^h of the model is Lipschitz continuous, and we also find a constant for which it is bounded in ℓ^2. We use two tools. The first is the fact that the processes

$$M_G(t) := G(X(t)) - G(X(0)) - \int_0^t \Lambda(X(s))ds \qquad (104)$$

are martingales with respect to the filtration generated by the Markov process $X(t)_{t \geq 0}$, for every G in the domain of its generator. And the second tool is the notion of *coupling* for Markov chains.

We want to study the behavior of the two Markov chains $X^h(t, x, b)$ and $X^h(t, y, b)$ for $x \neq y$ and link them in some sense in the product space. So we could define a coupling of these stochastic processes in terms of their distributions in the space of paths $D([0, T], S)$, the space of cadlag functions, for fixed initial points. However, for given marginal Markov processes, the resulting coupled process might not be Markovian. So we introduce the following fundamental definition, which can be found, for instance, in [3].

Definition 4 Given two Markov processes with semigroups $U_j(t)$ and generators Λ_j, or transition probabilities $P_j(t, x_1, \cdot)$, on $(E_j, \mathcal{E}_j), j = 1, 2$, a *Markovian coupling* is a Markov process with semigroup $\tilde{U}(t)$ and generator $\tilde{\Lambda}$, or transition probability $\tilde{P}(t; x_1, x_2, \cdot)$, on the product space $(E_1 \times E_2, \mathcal{E}_1 \otimes \mathcal{E}_2)$ having the marginality:

1.

$$\tilde{P}(t; x_1, x_2, A_1 \times E_2) = P_1(t, x_1, A_1) \quad \forall t \geq 0, x_1 \in E_1, A_1 \in \mathscr{E}_1,$$

$$\tilde{P}(t; x_1, x_2, E_1 \times A_2) = P_2(t, x_2, A_2) \quad \forall t \geq 0, x_2 \in E_2, A_2 \in \mathscr{E}_2.$$

Or equivalently

$$\tilde{U}(t)G(x_1, x_2) = U_1(t)G(x_1), \quad \forall t \geq 0, x_1 \in E_1, G \in B(\mathscr{E}_1),$$

$$\tilde{U}(t)G(x_1, x_2) = U_2(t)G(x_2), \quad \forall t \geq 0, x_2 \in E_2, G \in B(\mathscr{E}_2),$$

where $B(\mathscr{E})$ is the set of all bounded $\mathscr{E}$-measurable functions. Here on the left-hand side, G is regarded as a bivariate function, although it depends on only one variable.

2.

$$\tilde{\Lambda}G(x_1, x_2) = G(x_1) \quad \forall x_1 \in E_1, G \in B(\mathscr{E}_1),$$

$$\tilde{\Lambda}G(x_1, x_2) = G(x_2) \quad \forall x_2 \in E_2, G \in B(\mathscr{E}_2),$$

where G on the left-hand side is regarded as above.

The main result concerning coupling of Markov chains, whose proof can be found in [3], is that the two parts given in the definition are equivalent under certain conditions, for instance, if the two Markov chains take value in a finite state space. A Markovian coupling always exists: the simplest is the *independent coupling*. Consider a finite set $E = E_1 = E_2$, in the notations of the above definition. The generator of the independent coupling is defined by

$$\tilde{\Lambda}G(x_1, x_2) := \Lambda_1 G(\cdot, x_2)(x_1) + \Lambda_1 G(x_1, \cdot)(x_2)$$

for any $G \in \mathscr{C}(E^2)$.

Hence, for fixed h and b (which will be omitted in the following writing), we now consider any coupling operator $\tilde{\Lambda}_h$ of the Markov chain $X^h(t)$ and itself, which gives the semigroup $\tilde{U}_t^h$. The coupled process will be called $(X^h(t), Y^h(t))$, although we consider the same process, to avoid misunderstandings. Then by (104), the process

$$M_g(t) := g(X^h(t), Y^h(t)) - g(X^h(0), Y^h(0)) - \int_0^t \tilde{\Lambda}g(X^h(s), Y^h(s))ds \quad (105)$$

is a martingale for any $g \in \mathscr{C}(S(h) \times S(h))$.

We let

$$E^x[G(X^h(t))] := E[G(X^h(t))|X(0) = x] = E[G(X^h(t, x))] = U_t^h G(x)$$

and

$$\tilde{E}^{x,y}[g(X^h(t), Y^h(t))] := \tilde{U}_t^h g(x, y) = \tilde{E}\left[g(X^h(t), Y^h(t))|(X^h(0), Y^h(0)) = (x, y)\right]$$

for any x and y in $S(h)$. Therefore, the martingale M_g defined in (105) leads to

$$\tilde{E}^{x,y}[g(X^h(t), Y^h(t))] - g(x, y) = \tilde{E}^{x,y}\left[\int_0^t \tilde{\Lambda}g(X^h(s), Y^h(s))ds\right] \tag{106}$$

for any $t \geq 0$ and $g \in \mathscr{C}(S(h) \times S(h))$.

Considering $g_k(x, y) := x_k - y_k$ and $G_k(x) = x$, by means of Definition 4 and identity (101), we have

$$\tilde{\Lambda}g_k(x, y) = \Lambda G_k(x) - \Lambda G_k(y) = f_k(x) - f_k(y)$$

and

$$\tilde{E}^{x,y}[g_k(X^h(\tau), Y^h(\tau))] - g_k(x, y) = E^x[X_k^h(\tau)] - x - E^y[X_k^h(\tau)] + y = F_k^h(x) - F_k^h(y)$$

by definition (60) of the drift. Hence, equation (106) yields

$$F^h(x) - F^h(y) = \tilde{E}^{x,y}\left[\int_0^\tau \left(f(X^h(s)) - f(Y^h(s))\right)ds\right]$$

which, taking the ℓ^2-norm and applying Fubini's theorem and the Lipschitz continuity of f (66), becomes

$$||F^h(x) - F^h(y)|| \leq K \int_0^\tau \tilde{E}^{x,y} \left\|X^h(s) - Y^h(s)\right\| ds \tag{107}$$

for any coupling operator $\tilde{\Lambda}_h$, where K is defined in (95).

Now we need the following lemma, which is stated, for instance, in [4]:

Lemma 3 *Let U_t be a strongly continuous semigroup on E with generator Λ whose domain is $\mathscr{D}$, $\alpha \in \mathbb{R}$ be a constant, and $f \in \mathscr{D}$. Then, $U_t f \leq e^{\alpha t} f$ if and only if $\Lambda f \leq \alpha f$.*

Proof ($\Leftarrow$) Let $\Lambda f \leq \alpha f$; then, by Kolmogorov's equation

$$\frac{d}{dt}U_t f = \Lambda U_t f \leq \alpha U_t f.$$

Thus, by Gronwall's lemma

$$U_t f \leq e^{\alpha t} U_0 f = e^{\alpha t} f.$$

($\Rightarrow$) Let $U_t f \leq e^{\alpha t} f$; then, by definition of the generator

$$\Lambda f = \lim_{t \to 0} \frac{U_t f - f}{t} \leq \lim_{t \to 0} \frac{e^{\alpha t} - 1}{t} f = \alpha f.$$

Hence we want to apply this lemma to the coupling operator $\tilde{\Lambda}_h$ and the function $\rho(x, y) = ||x - y||$, in order to obtain an upper bound of the right-hand side in (107). So we have to choose a particular coupling for which there exists $\alpha \in \mathbb{R}$ such that the condition

$$\tilde{\Lambda}_h \rho \leq \alpha \rho \tag{108}$$

is satisfied.

We use the so-called *coupling of marching soldiers*, introduced by Chen in 1986 and whose description can be found, for instance, in [3]. This gives a Markov chain in $S(h) \times S(h)$ such that if it is in (x, y), it jumps to

- $(x + z, y + z)$ at rate $\min\{q(x, x + z), q(y, y + z)\}$,
- $(x + z, y)$ at rate $[q(x, x + z) - q(y, y + z)]^+$,
- $(x, y + z)$ at rate $[q(x, x + z) - q(y, y + z)]^-$,

for any $z \in S(h)$, where $q(x, y)$ are the rates, i.e., the elements of the Q-matrix, of the Markov chain X^h. It is a Markov coupling, i.e., it satisfies Definition 4 part 2, since $\min\{a, b\} + (a - b)^+ = a$ for any $a, b \geq 0$.

Set $h_{ij} = -he_i - he_j + he_{i+j}$ and $h^{ij} = -he_i + he_j + he_{i-j}$. Thus, the generator of the marching coupling $\tilde{\Lambda}$ of the Markov chain X^h, whose generator is defined in (10), is given by

$$\tilde{\Lambda}_h g(x, y) = \frac{1}{h} \sum_{i,j} \min\{C_{ij}(x) x_i x_j, C_{ij}(y) y_i y_j\} \left[g(x + h_{ij}, y + h_{ij}) - g(x, y) \right]$$

$$\tag{109}$$

$$+ \frac{1}{h} \sum_{i,j} \left[C_{ij}(x) x_i x_j - C_{ij}(y) y_i y_j \right]^+ \left[g(x + h_{ij}, y) - g(x, y) \right]$$

$$+ \frac{1}{h} \sum_{i,j} \left[C_{ij}(x) x_i x_j - C_{ij}(y) y_i y_j \right]^- \left[g(x, y + h_{ij}) - g(x, y) \right]$$

$$+ \frac{1}{h} \sum_{i} \sum_{j<i} \min\{F_{ij}(x) x_i, F_{ij}(y) y_i\} \left[g(x + h^{ij}, y + h^{ij}) - g(x, y) \right]$$

$$+ \frac{1}{h} \sum_{i} \sum_{j<i} \left[F_{ij}(x) x_i - F_{ij}(y) y_i \right]^+ \left[g(x + h^{ij}, y) - g(x, y) \right]$$

$$+ \frac{1}{h} \sum_{i} \sum_{j<i} \left[F_{ij}(x) x_i - F_{ij}(y) y_i \right]^- \left[g(x, y + h^{ij}) - g(x, y) \right]$$

for any $g \in \mathscr{C}(S(h) \times S(h))$. Hence, calculating this generator (109) on the distance function $\rho(x, y) = ||x - y||$, we obtain

$$\tilde{\Lambda}_h \rho(x, y) \leq 3 \sum_{ij} \left| C_{ij}(x)x_i x_j - C_{ij}(y)y_i y_j \right| + 3 \sum_i \sum_{j<i} \left| F_{ij}(x)x_i - F_{ij}(y)y_i \right|,$$

(110)

using the identity $a^+ + a^- = |a|$ for any real number a and the upper bound $||x - y - z|| - ||x - y|| \leq ||z|| \leq 3h$, whereas z can be either $-he_i - he_j + he_{i+j}$ or $-he_i + he_j + he_{i-j}$.

In order to get an estimate of the above equation, we shall consider the two functions $u, v : S \to \mathbb{R}^{N \times N} \cong \mathbb{R}^{N^2}$, where $N = N(h)$, defined by

$$u_{ij}(x) := C_{ij}(x)x_i x_j \tag{111}$$

$$v_{ij}(x) := F_{ij}(x)x_i \mathbb{I}_{]0,+\infty[}(i - j). \tag{112}$$

So the right-hand side in (110) is equal to

$$3||u(x) - u(y)||_{\ell^1} + 3||v(x) - v(y)||_{\ell^1}.$$

The derivatives of u and v are given by

$$\frac{\partial u_{ij}}{\partial x_k}(x) = \frac{\partial C_{ij}(x)}{\partial x_k} x_i x_j + C_{ij}(x)\delta_{ik}x_j + C_{ij}(x)x_i \delta_{jk}$$

and

$$\frac{\partial v_{ij}}{\partial x_k}(x) = \left(\frac{\partial F_{ij}(x)}{\partial x_k} x_i + F_{ij}(x)\delta_{ik} \right) \mathbb{I}_{]0,+\infty[}(i - j).$$

Thus, we apply the mean value theorem to get

$$||u(x) - u(y)||_{\ell^1} + ||v(x) - v(y)||_{\ell^1}$$

$$= \left\| \frac{\partial u}{\partial x}(z).\xi \right\|_{\ell^1} + \left\| \frac{\partial v}{\partial x}(w).\xi \right\|_{\ell^1}$$

$$\leq \sum_{ijk} \left| \frac{\partial u_{ij}}{\partial x_k}(z) \right| |\xi_k| + \sum_{ik} \sum_{j<i} \left| \frac{\partial v_{ij}}{\partial x_k}(w) \right| |\xi_k|$$

$$\leq C(1)||z||_{\ell^1}^2 ||\xi||_{\ell^1} + 2C||z||_{\ell^1}||\xi||_{\ell^1} + F(1)||w||_{\ell^1}||\xi||_{\ell^1} + F||\xi||_{\ell^1}$$

for any $x, y \in S(h)$ and for certain z and w in S, where $\xi = x - y$. The latter inequality follows from (31) and (32).

Therefore, (110) becomes

$$\tilde{\Lambda}_h \rho(x, y) \leq 3(C(1)R^2 + 2CR + F(1)R + F)||x - y||_{\ell^1}$$

for any x and y in $S(h)$. If we use the estimate $||x - y||_{\ell^1} \leq \sqrt{N}||x - y||$ for any x and y in $S(h)$, then

$$\tilde{\Lambda}_h \rho(x, y) \leq M_2 \sqrt{N(h)}||x - y||, \tag{113}$$

where $M_2 := 3(C(1)R^2 + 2CR + F(1)R + F)$ is constant, which says that (108) holds with $\alpha := M_2 \sqrt{N(h)} > 0$.

Thus, we can apply Lemma 3 to the marching coupling, so that

$$\tilde{E}^{x,y} \left\| X^h(s) - Y^h(s) \right\| \leq e^{\alpha s}||x - y||$$

for any $s \in [0, \tau]$ and $x \neq y \in S(h)$. Hence (107) becomes

$$||F^h(x) - F^h(y)|| \leq K \int_0^\tau e^{\alpha s}||x - y||ds \leq K\tau e^{\alpha \tau}||x - y||,$$

which is (65) where L_1 is the function defined in (75):

$$||F^h(x) - F^h(y)|| \leq K\tau e^{M_2\tau\sqrt{N(h)}}||x - y||.$$

4.3.1 Boundedness of the Drift

Applying (104) simply to the process X^h, without considering couplings, we obtain

$$F^h(x, b) = E\left[\int_0^\tau f(X^h(s, x, b))ds\right]. \tag{114}$$

Hence from (114), by means of (64), we get also an upper bound for the drift:

$$||F^h(x, b)|| \leq L_2\tau \tag{115}$$

for every $x \in S(h)$ and $b \in E$, where L_2 is defined in (93).

4.4 Bounds for Δ

We find an estimate for $E(\Delta_\pi^h(k)|X_\pi^h = x)$ where $\Delta_\pi^h(k)$ is the number of coalitions that perform a transition between time step $k\tau$ and $(k + 1)\tau$. Because of Markovianity, the above expectation is independent of k, so we can suppose $k = 0$. Hence we consider $E(\Delta^h|X^h(0) = x_0)$, where Δ^h is the number of coalitions that change their state between 0 and τ.

If the system is in x_0 in $t = 0$, there is an exponential clock of parameter $s(x_0, b)$ such that, when it clicks, the system changes its state, say it goes in x_1. Now there is another exponential clock of parameter $s(x_1, b)$ such that, when it clicks, the system changes its state, say it goes in x_2. We repeat this procedure until we arrive at time τ. Note that Δ^h is then less or equal than the number of clicks that we get from 0 to τ.

Thus, to estimate Δ^h, we take an upper bound of $s(x, b)$ defined in (8)

$$s(x, b) = \sum_{i,j} n_i n_j h C_{ij}(x, b) + \sum_i n_i F_{ij}(x, b),$$

for any $x \in S(h)$. Using assumption (31) on C_{ij} and F_{ij}, we have

$$s(x, b) \leq Ch \sum_{i,j} n_i n_j + F \sum_i n_i = Ch \left(\sum_i n_i \right)^2 + F \sum_i n_i$$

for any x and b. According to (44), $\sum_i n_i \leq N(h) \leq \frac{R}{h}$, which gives

$$s(x, b) \leq \frac{1}{h}(CR^2 + FR).$$

Hence, for any $x \in S(h)$ and $b \in E$, $s(x, b)$ is bounded by a constant s^h that depends only on h:

$$s^h := \frac{1}{h}(CR^2 + F). \tag{116}$$

For a constant s, the number of occurrences of clicks is known to be a Poisson process of intensity s. Thus, for a constant s, the number of clicks from 0 to τ is a random variable X with Poisson distribution of parameter $s\tau$. This implies in particular that $E(X) = s\tau$ and $E(X^2) = s\tau(1 + s\tau)$.

Hence, the expectations for Δ^h are bounded by

$$E(\Delta^h | X^h(0) = x_0) \leq s^h \tau$$

and

$$E((\Delta^h)^2 | X^h(0) = x_0) \leq s^h \tau(1 + s^h \tau)$$

for any $x_0 \in S(h)$. Using (116), we find

$$E(\Delta^h | X^h(0) = x_0) \leq \frac{1}{h}\tau(CR^2 + F)$$

and

$$E((\Delta^h)^2 | X^h(0) = x_0) \leq \frac{1}{h}\tau(CR^2 + F)[1 + \frac{1}{h}(CR^2 + F)\tau]$$

$$= \frac{1}{h^2}\tau(CR^2 + F)[\tau(CR^2 + F) + h].$$

Therefore (61) and (62) hold with the bounds specified in (73) and (74).

Acknowledgements A. Cecchin was supported by the PhD programme in Mathematical Sciences, Dipartimento di Matematica, Università di Padova (Italy) and Progetto Dottorati - Fondazione Cassa di Risparmio di Padova e Rovigo. We thank the associate editor and the referees for the useful comments and remarks.

References

1. Bardi, M., Capuzzo Dolcetta, I.: Optimal control and viscosity solutions of Hamilton-Jacobi-Bellman equations. Birkhauser (1997)
2. Caines, P.E., Huang, M., Malhamé, R.P.: Large population stochastic dynamic games: Closed-loop McKean-Vlasov systems and the Nash certainty equivalence principle. Commun. Inf. Syst. 6(3), 221–252 (2006)
3. Chen, M.F.: From Markov Chains to non equilibrium particle systems. World scientific (1992)
4. Chen, M.F.: Eigenvalues, Inequalities and ergodic theory. Springer (2005)
5. Chen, W., Huang, D., Kulkarni, A.A, Unnikrishnan, J., Zhu, Q., Mehta, P., Meyn, S., Wierman, A.: Approximate dynamic programming using fluid and diffusion approximations with applications to power management. In: Proceedings of the 48th IEEE Conference on Decision and Control held jointly with 28th Chinese Control Conference, CDC/CCC 2009, Shangai, September 2009, pp. 3575–3580 (2009)
6. Crandall, M.G., Lions, P.L.: Hamilton-Jacobi equations in infinite dimensions. II. Existence of viscosity solutions. J. Funct. Anal. 65, 360–405 (1986)
7. Finus, M., Rundshagen, B., Eyckmans, J.: Simulating a sequential coalition formation process for the climate change problem: first come, but second served? Ann. Oper. Res. 220, 5–23 (2014)
8. Gast, N., Gaujal, B., Le Boudec, J.Y.: Mean Field for Markov Decision Processes: from discrete to continuous optimization. IEEE Trans. Autom. Control 57, 2266–2280 (2012)
9. Inal, H.: Core of coalition formation games and fixed-point methods. Soc. Choice Welfare 45(4), 745–763 (2015)
10. Jouida, J.B., Sihem, K., Krichen, S., Klibi, W.: Coalition-formation problem for sourcing contract design in supply networks. Eur. J. Oper. Res. 257(2), 539–558 (2017)
11. Kallenberg, O.: Foundations of Modern Probability. 2nd ed. Springer (2002)
12. Karos. D.: Coalition formation in general apex games under monotonic power indices. Games Econom. Behav. 87, 239–252 (2014)
13. Kolokoltsov, V.N.: Hydrodynamic limit of coagulation-fragmentation type models of k-nary interacting particles. J. Stat. Phys. 115(5/6), 1621–1653 (2004)
14. Kolokoltsov, V.N.: Kinetic equations for the pure jump models of k-nary interacting particle systems. Markov Process. Relat. 12, 95–138 (2006)

15. Kolokoltsov, V.N.: Nonlinear Markov games on a finite state space (mean-field and binary interactions). Int. J. Stat. Probab. 1(1), 77–91(2012). doi:10.5539/ijsp.v1n1p77
16. Kolokoltsov, V.N.: The evolutionary game of pressure (or interference), resistance and collaboration. Math. Oper. Res. ISSN 0364-765X (2016, in press). doi:10.1287/moor.2016.0838
17. Krapivsky, P.L., Redner, S.: Organization of growing random networks. Phys. Rev. E. 63, 066123 (2001)
18. Lasry, J.M., Lions, P.L.: Mean field games. Jpn. J. Math. 2(1), 229–260 (2007)
19. Norris, J.: Cluster Coagulation. Comm. Math. Phys. 209, 407–435 (2000)
20. Papadimitriou, C.H., Tsitsiklis, J.N.: The complexity of optimal queuing network control. Math. Oper. Res. 24(2), 292–305 (1999)
21. Pushkin, D.O., Aref, H.: Bank mergers as scale-free coagulation. Physica A 336, 571–584 (2004)
22. Saichev, A., Malvergne, Y., Sornette, D.: Theory of Zipf's Law and Beyond. Springer, Berlin (2010)
23. Tsitsiklis, J.N., Van Roy, B.: An analysis of temporal-difference learning with function approximation. IEEE Tran. Autom. Control 42(5), 674–690 (1997)
24. Weese, E.; Political mergers as coalition formation: an analysis of the Heisei municipal amalgamations. Quant. Econ. 6(2), 257–307 (2015)

The Execution Problem in Finance with Major and Minor Traders: A Mean Field Game Formulation

Dena Firoozi and Peter E. Caines

Abstract The theory of partially observed mean field games is extended to cover indefinite LQG problems and then is applied to an optimal execution problem in finance; following standard financial models, controlled linear system dynamics are postulated where an institutional investor (interpreted as a major agent) in the market aims to liquidate a specific amount of shares and has partial observations of its own state (which includes its inventory). Furthermore, the market is assumed to have two populations of high-frequency traders (interpreted as minor agents) who wish to liquidate or acquire a certain number of shares within a specific time, and each one of them has partial observations of its own state and the major agent's state (which include the corresponding inventories). The objective for each agent is to maximize its own wealth and to avoid the occurrence of large execution prices, large rates of trading, and large trading accelerations which are appropriately weighted in the agent's performance function. The existence of ϵ-Nash equilibria together with the individual agents' trading strategies yielding the equilibria was established. A simulation example is provided.

1 Introduction

Partially observed mean field game (PO MFG) theory was introduced and developed in [1–4, 14] where it is assumed the major agent's state is partially observed by each minor agent and the major agent completely observes its own state. Accordingly, each minor agent can recursively estimate the major agent's state, compute the system's mean field, and thence generate the feedback control which yields the ϵ-Nash property. This PO MFG theory was further extended in the work in [5] to major-minor (MM) LQG systems in which both the major agent and the minor agents partially observe the major agent's state. The existence of ϵ-Nash equilibria, together with the individual agents' control laws yielding the equilibria,

D. Firoozi (✉) • P.E. Caines
The Centre for Intelligent Machines (CIM) and the Department of Electrical and Computer Engineering (ECE), McGill University, Montreal, QC, Canada
e-mail: dena.firoozi@mail.mcgill.ca; peterc@cim.mcgill.ca

© Springer International Publishing AG 2017
J. Apaloo, B. Viscolani (eds.), *Advances in Dynamic and Mean Field Games*,
Annals of the International Society of Dynamic Games 15,
https://doi.org/10.1007/978-3-319-70619-1_5

was established wherein each minor agent recursively generates (i) an estimate of the major agent's state and (ii) an estimate of the major agent's estimate of its own state (in order to estimate the major agent's control feedback) and hence generates a version of the system's mean field. It is to be noted that the case where each agent has only partial observations on its own state was addressed in the LQG case in [6] and in the nonlinear case in [7, 8, 14].

Optimal execution problems have been addressed in the literature (see, e.g., [11, 16, 18, 19]) where an agent must liquidate or acquire a certain amount of shares over a prespecified time horizon at a trading speed to balance the price impact (from trading quickly) and the price uncertainty (from trading slowly) while it maximizes its final wealth. Further, in [20] the partially observed setting where the market liquidity variable is not observed was studied. This problem with the linear models in [11] was formulated as for the nonlinear major-minor (MM) MFG model in [22]. The PO MM LQG MFG theory was first applied to an optimal execution problem with the linear models of [11] in [9] where an institutional investor, interpreted as a major agent, aims to liquidate a specific amount of shares and it has only partial observations of its own state (which includes its inventory). Furthermore, there is a large population of high-frequency traders (HFTs), interpreted as minor agents, who wish to liquidate their shares, and each of them has partial observations of its own state and the major agent's state (which include the corresponding inventories). In the current paper, this work is refined in the formulation of the market dynamics in the MFG framework and also is extended to consider two populations of HFTs with liquidation or acquisition objectives who wish to, respectively, liquidate or acquire a certain number of shares within a specific duration of time. The initial results of this extension have been presented in [10]. Moreover, to address this problem, the PO LQG MFG theory is extended in the current work to cover indefinite LQ problems. The theory is then utilized to establish the existence of ϵ-Nash equilibria together with the best response trading strategies such that each agent attempts to maximize its own wealth and avoid the occurrence of large execution prices and large trading accelerations which are appropriately weighted in the agent's performance function.

We note that the terms major trader (respectively, minor trader) and institutional trader (respectively, HFT) are used interchangeably in this paper.

The paper is organized as follows. Section 2 is devoted to the description of trading dynamics in the market and the execution problem. In Section 3, the optimal execution problem is formulated in the mean field game framework. Completely observed and partially observed optimal execution problems are then addressed in Sections 4 and 5, respectively. Section 6 presents the simulation results.

2 Trading Dynamics of Agents in the Market

As stated in the Introduction, the institutional investor is considered as a major agent in the mean field model of the market which liquidates its shares, and the HFTs are considered as minor agents, where two types of them are considered: liquidators and acquirers. Employing the trading model in [11], the trading dynamics of the

major agent and any generic minor agent in the market are described by the linear
time evolution of the inventories, trading rates, and prices, while the bilinear cash
process appears in the quadratic performance function for each agent.

2.1 Inventory Dynamics

It is assumed that the institutional investor liquidates its inventory of shares, $Q_0(t)$,
by trading at a rate $v_0(t)$ during the trading period $[0, T]$. Hence the major agent's
inventory dynamics is given by

$$dQ_0(t) = v_0(t)dt + \sigma_0^Q dw_0^Q(t), \quad 0 \le t \le T,$$

where w_0^Q is a Wiener process modeling the noise in the inventory information that
the institutional trader collects from its branches in different locations and σ_0^Q is
a positive scalar, and we assume that $Q_0(0) \gg 1$. The same dynamical model is
adopted for the trading dynamics of a generic HFT

$$dQ_i(t) = v_i(t)dt + \sigma_i^Q dw_i^Q(t), \quad 1 \le i \le N_a + N_l, \ 0 \le t \le T$$

where N_a and N_l are, respectively, liquidator and acquirer populations of N minor
traders, i.e., $N = N_a + N_l$, w_i^Q is a Wiener process that models the HFT's information
noise, σ_i^Q is a positive scalar, $v_i(t)$ is the agent's rate of trading which can be positive
or negative depending on whether the agent is acquirer or liquidator, respectively,
and $Q_i(t)$ is the minor liquidator agent's remaining shares at time t or the shares the
minor acquirer agent has bought until time t. However, the initial inventories of the
HFTs, $\{Q_i(0), 1 \le i \le N_a + N_l\}$, are not considered to be large.

We assume that the trading rate of the major agent is controlled via $u_0(t)$ as

$$dv_0(t) = u_0(t)dt, \quad 0 \le t \le T,$$

where the trading strategy $u_0(t)$ can be seen to be the trading acceleration of the
major trader. Correspondingly, $u_i(t)$ controls the trading rate of minor agent, $\mathscr{A}_i$, by

$$dv_i(t) = u_i(t)dt, \quad 1 \le i \le N_a + N_l, \ 0 \le t \le T.$$

2.2 Price Dynamics

The trading rate of the major agent and the average trading rate of the minor agents
give rise to the asset midprice which models the permanent effect of agents' trading
rates on the market price. Further, each agent has a temporary effect on the asset

price which only persists during the action of the trade and which determines the execution price, that is to say the price at which each agent can trade.

2.2.1 Asset Midprice

We model the dynamics of the asset midprice, as seen from the major agent's viewpoint, by

$$dF_0(t) = \left(\lambda_0 v_0(t) + \frac{\lambda}{N} \sum_{i=1}^{N} v_i(t)\right) dt + \sigma dw_0^F(t), \quad 0 \le t \le T,$$

where the Wiener process $w_0^F(t)$ models the aggregate effect of all traders in the market which – unlike the major and minor agents $\mathscr{A}_0, \mathscr{A}_i$, – have no partial observations on any of the state variables appearing in the dynamical market model (these are termed uninformed traders). Further, σ denotes the intensity of the market volatility, and $\lambda_0, \lambda \ge 0$ denote the strength of the linear permanent impact of the major and minor agents' tradings on the asset midprice, respectively. Similarly, we model the asset midprice dynamics, as seen by a minor agent $\mathscr{A}_i$, by

$$dF_i(t) = \left(\lambda_0 v_0(t) + \frac{\lambda}{N} \sum_{i=1}^{N} v_i(t)\right) dt + \sigma dw_i^F(t), \quad 1 \le i \le N_a + N_l, \ 0 \le t \le T,$$

where the Wiener process, $w_i^F(t)$, represents the mass effect of all uninformed traders in the market. The time differences between agents in getting data from fast-changing limit order book make the Wiener processes, $w_i^F, \ 0 \le i \le N_a + N_l$, independent.

2.2.2 Execution Price

The major agent's execution price $S_0(t)$ evolution is assumed to be given by

$$dS_0(t) = dF_0(t) + a_0 dv_0(t), \quad 0 \le t \le T, \tag{1}$$

where $a_0 \ge 0$ is the temporary impact strength of the major agent on the asset midprice. Likewise, a minor agent's execution price, $S_i(t)$, is assumed to evolve as

$$dS_i(t) = dF_i(t) + a dv_i(t), \quad 1 \le i \le N_a + N_l, \ 0 \le t \le T, \tag{2}$$

where a models the temporary impact of a minor agent's trading on its execution price.

2.3   Cash Process

The cash processes for the major agent and a generic minor agent, $Z_0(t)$, $Z_i(t)$, are given by

$$dZ_0(t) = -S_0(t)dQ_0(t), \quad 0 \leq t \leq T, \tag{3}$$

$$dZ_i(t) = -S_i(t)dQ_i(t), \quad 1 \leq i \leq N_a + N_l, \ 0 \leq t \leq T, \tag{4}$$

where $Z_0(t)$, $Z_i(t)$, $1 \leq i \leq N_l$, are the cash obtained through liquidation of shares and $Z_i(t), 0 \leq i \leq N_a$, is the cash paid for acquisition of shares up to time t. We note that the value of $dQ_0(t)$ in a stock sale is negative and hence for positive $S_0(t)$, $Z_0(t)$ increases.

2.4   Cost Function

2.4.1   Major Liquidator Trader

The objective for the major trader is to liquidate $\mathcal{N}_0$ shares and maximize the cash it holds at the end of the trading horizon, i.e., maximize $Z_0(T)$, and if the remaining inventory at the final time T is $Q_0(T)$, it can liquidate it at a lower price than the market asset price, reflected in the cost function by $Q_0(T)(F_0(T) - \alpha_0 Q_0(T))$. Further, the major trader's utility in minimizing the inventory over the period $[0, T]$ is modeled by including the penalty $\int_0^T Q_0^2(s)ds$ in its objective function and the utility of avoiding very high execution prices, large trading intensities, and large trading accelerations by including the terms $S_0^2(T)$, $\int_0^T S_0^2(s)ds$, $v_0^2(T)$, $\int_0^T v_0^2(s)ds$, and $\int_0^T R_0 u_0^2(s)ds$ in the objective function. Therefore, its cost function to be minimized is given by

$$J_0(u_0, u_{-0}) = \mathbb{E}\Big[-\psi_0 Z_0(T) - \mu_0 Q_0(T)\big(F_0(T) - \alpha_0 Q_0(T)\big) + \xi_0 S_0^2(T) + \gamma_0 v_0^2(T)$$

$$+ \int_0^T \big(\phi_0 Q_0^2(s) + \delta_0 S_0^2(s) + \theta_0 v_0^2(s) + R_0 u_0^2(s)\big)ds \Big], \tag{5}$$

where ψ_0, μ_0, α_0, ξ_0, γ_0, ϕ_0, δ_0, θ_0, and R_0 are positive scalars and $u_{-0} := (u_1, \ldots, u_{N_a+N_l})$ are trading strategies of the minor traders. Note that for larger values of ϕ_0, the trader attempts to liquidate its inventory more quickly.

2.4.2   Minor Liquidator Trader

In a similar way, the objective function to be minimized for a liquidator HFT who wants to liquidate $\mathcal{N}_l$ shares during the time interval $[0, T]$ is given by

$$J_i(u_i, u_{-i}) = \mathbb{E}\Big[-\psi_l Z_i(T) - \mu_l Q_i(T)\big(F_i(T) - \alpha_l Q_i(T)\big) + \xi_l S_i^2(T) + \gamma_l v_i^2(T)$$

$$+ \int_0^T \big(\phi_l Q_i^2(s) + \delta_l S_i^2(s) + \theta_l v_i^2(s) + R_l u_i^2(s)\big)ds\Big], \ 1 \le i \le N_l, \quad (6)$$

where ψ_l, μ_l, α_l, ξ_l, γ_l, ϕ_l, δ_l, θ_l, and R_l are positive scalars and $u_{-i} :=$ $(u_0, u_1, \ldots, u_{i-1}, u_{i+1}, \ldots, u_{N_a+N_l})$. Note that $\mathcal{N}_l \ll \mathcal{N}_0$.

2.4.3 Minor Acquirer Trader

The objective for a minor acquirer trader is to buy $\mathcal{N}_a$ shares over the trading horizon $[0, T]$ while it minimizes the execution cost including the cash $Z_i(T)$ paid up to time T, and the cash must be paid at time T to buy the remaining shares at once at a higher price than the market's asset price, i.e., $(\mathcal{N}_a - Q_i(T))\big(F_i(T) + \alpha_a(\mathcal{N} - Q_i(T))\big)$. It also intends to avoid high execution prices, large trading intensities, and large trading accelerations modeled by including $\xi_a S_i^2(T) + \gamma_a v_i^2(T) + \int_0^T \big(\delta_a S_i^2(s) + \theta_a v_i^2(s) + R_a u_i^2(s)\big)ds$ in its objective function

$$J_i(u_i, u_{-i}) = \mathbb{E}\Big[\psi_a Z_i(T) + \mu_a(\mathcal{N}_a - Q_i(T))\big(F_i(T) + \alpha_a(\mathcal{N}_a - Q_i(T))\big) + \xi_a S_i^2(T) +$$

$$\gamma_a v_i^2(T) + \int_0^T \big(\phi_a(\mathcal{N}_a - Q_i(s))^2 + \delta_a S_i^2(s) + \theta_a v_i^2(s) + R_a u_i^2(s)\big)ds\Big], \ 1 \le i \le N_a, \tag{7}$$

where $\int_0^T \phi_a(\mathcal{N}_a - Q_i(s))^2 ds$ is to penalize the agent for the remaining shares to be bought up to T and to expedite the acquisition. The parameters ψ_a, μ_a, α_a, ξ_a, γ_a, ϕ_a, δ_a, θ_a, and R_a are positive scalars, and $u_{-i} := (u_0, u_1, \ldots, u_{i-1}, u_{i+1}, \ldots, u_{N_a+N_l})$.

3 MFG Formulation of the Optimal Execution Problem

In this section we formulate the optimal execution problem into the major minor LQG MFG framework.

3.1 Finite Populations

3.1.1 Major Agent

The stochastic optimal control problem for the major trader is modeled as

$$dv_0(t) = u_0(t)dt, \tag{8}$$

$$dQ_0(t) = v_0(t)dt + \sigma_0^Q dw_0^Q(t), \tag{9}$$

$$dS_0(t) = \Big(\lambda_0 v_0(t) + \frac{\lambda}{N}\sum_{i=1}^{N} v_i(t)\Big)dt + a_0 u_0(t)dt + \sigma dw_0^F(t), \tag{10}$$

with the cost function

$$J_0(u_0, u_{-0}) = \mathbb{E}\Big[-\mu_0 Q_0(T)\big(S_0(T) - a_0 v_0(T) - \alpha_0 Q_0(T)\big) + \xi_0 S_0(T)^2 + \gamma_0 v_0^2(T) +$$
$$\int_0^T \Big(\phi_0 Q_0^2(s) + \psi_0 S_0(s)v_0(s) + \delta_0 S_0^2(s) + \theta_0 v_0^2(s) + R_0 u_0^2(s)\Big)ds\Big],$$

wherein the final cash process in (5) was replaced by $\mathbb{E}[Z_0(T)] = -\mathbb{E}[\int_0^T S_0(s) v_0(s)ds]$ and the asset midprice $F_0(T)$ was replaced using (1).

As can be seen, the major agent is coupled with the minor agents by the average term $\frac{\lambda}{N}\sum_{i=1}^N v_i$ in the execution price dynamics (10).

Now let the major agent's state be denoted by

$$x_0 = \begin{bmatrix} v_0 \\ Q_0 \\ S_0 \end{bmatrix}.$$

Then the major agent's cost function will be written in the standard quadratic form

$$J_0(u_0) = \mathbb{E}\Big[\|x_0(T)\|_{\bar{P}_0}^2 + \int_0^T \big(\|x_0(s)\|_{P_0}^2 + \|u_0(s)\|_{R_0}^2\big)ds\Big], \tag{11}$$

with

$$\bar{P}_0 = \begin{bmatrix} \gamma_0 & \frac{1}{2}\mu_0 a_0 & 0 \\ \frac{1}{2}\mu_0 a_0 & \mu_0 \alpha_0 & -\frac{1}{2}\mu_0 \\ 0 & -\frac{1}{2}\mu_0 & \xi_0 \end{bmatrix}, \quad P_0 = \begin{bmatrix} \theta_0 & 0 & \frac{1}{2}\psi_0 \\ 0 & \phi_0 & 0 \\ \frac{1}{2}\psi_0 & 0 & \delta_0 \end{bmatrix}, \quad R_0 > 0.$$

3.1.2 Minor Liquidator Agent

Similarly, the stochastic optimal control problem for a minor liquidator trader $\mathscr{A}_i$, $1 \le i \le N_l$, is given by the set of dynamical equations

$$dv_i(t) = u_i(t)dt, \tag{12}$$

$$dQ_i(t) = v_i(t)dt + \sigma^Q dw_i^Q(t), \tag{13}$$

$$dS_i(t) = \Big(\lambda_0 v_0(t) + \frac{\lambda}{N}\sum_{i=1}^{N} v_i(t)\Big)dt + a u_i(t)dt + \sigma dw_i^F(t), \tag{14}$$

The equations above show that a minor agent is coupled with the major agent and other minor agents through the execution price dynamics (14).

Similar to the major trader, we define a generic minor trader's state vector as

$$x_i = \begin{bmatrix} v_i \\ Q_i \\ S_i \end{bmatrix},$$

and its quadratic cost function where the final cash process in (6) has been replaced by $\mathbb{E}[Z_i(T)] = -\mathbb{E}[\int_0^T S_i(s)v_i(s)ds]$ using (4), and the asset midprice $F_i(T)$ was replaced using (2) is given by

$$J_i(u_i, u_{-i}) = \mathbb{E}\left[\|x_i(T)\|_{\bar{P}_l}^2 + \int_0^T \left(\|x_i(s)\|_{P_l}^2 + \|u_i(s)\|_{R_l}^2 \right) ds \right], \tag{15}$$

where

$$\bar{P}_l = \begin{bmatrix} \gamma_l & \frac{1}{2}\mu_l a & 0 \\ \frac{1}{2}\mu_l a & \mu_l \alpha_l & -\frac{1}{2}\mu_l \\ 0 & -\frac{1}{2}\mu_l & \delta_l \end{bmatrix}, \quad P_l = \begin{bmatrix} \theta_l & 0 & \frac{1}{2}\psi_l \\ 0 & \phi_l & 0 \\ \frac{1}{2}\psi_l & 0 & \delta_l \end{bmatrix}, \quad R_l > 0.$$

3.1.3 Minor Acquirer Agent

The stochastic optimal control problem for a minor acquirer trader $\mathscr{A}_i$, $1 \le i \le N_a$, is given by the set of dynamical equations

$$dv_i(t) = u_i(t)dt, \tag{16}$$

$$dY_i(t) = -v_i(t)dt + \sigma_i^Q dw_i^Q(t), \tag{17}$$

$$dS_i(t) = \left(\lambda_0 v_0(t) + \frac{\lambda}{N} \sum_{i=1}^N v_i(t) \right) dt + au_i(t)dt + \sigma dw_i^F(t), \tag{18}$$

where $Y_i(t) = \mathscr{N}_a - Q_i(t)$ is the remaining shares at t to be acquired until the end of the trading horizon. Accordingly, the cost function for acquisition is given by

$$J_i(u_i, u_{-i}) = \mathbb{E}\left[\psi_a Z_i(T) + \mu_a Y_i(T)\left(F_i(T) + \alpha_a Y_i(T)\right) + \xi_a S_i^2(T) + \gamma_a v_i^2(T) + \right.$$

$$\left. \int_0^T \left(\phi_a Y_i(s)^2 + \delta_a S_i^2(s) + \theta_a v_i^2(s) + R_a u_i^2(s) \right) ds \right], \quad 1 \le i \le N_a.$$

We define a generic minor acquirer trader's state vector as

$$x_i = \begin{bmatrix} v_i \\ Y_i \\ S_i \end{bmatrix},$$

and its quadratic cost function is given by

$$J_i(u_i, u_{-i}) = \mathbb{E}\left[\|x_i(T)\|_{\bar{P}_a}^2 + \int_0^T \left(\|x_i(s)\|_{P_a}^2 + \|u_i(s)\|_{R_a}^2 \right) ds \right], \tag{19}$$

where

$$\bar{P}_a = \begin{bmatrix} \gamma_a & -\frac{1}{2}\mu_a a & 0 \\ -\frac{1}{2}\mu_a a & \mu_a \alpha_a & \frac{1}{2}\mu_a \\ 0 & \frac{1}{2}\mu_a & \xi_a \end{bmatrix}, \quad P_a = \begin{bmatrix} \theta_a & 0 & -\frac{1}{2}\psi_a \\ 0 & \phi_a & 0 \\ -\frac{1}{2}\psi_a & 0 & \delta_a \end{bmatrix}, \quad R_a > 0.$$

3.2 Mean Field Evolution

Following the LQG MFG methodology [17], the mean field, $\bar{x}$, is defined as the L^2 limit, when it exists, of the average of minor agents' states when the population size goes to infinity

$$\bar{x}(t) = \lim_{N \to \infty} x^N(t) = \lim_{N \to \infty} \frac{1}{N} \sum_{i=1}^N x_i(t), \quad q \cdot m.$$

Now, if the control strategy for each minor agent is considered to have the general feedback form

$$u_i = L_1 x_i + L_2 x_0 + \sum_{j \neq i, j=1}^N L_4 x_j + L_3, \quad 1 \leq i \leq N, \tag{20}$$

then the mean field dynamics can be obtained by substituting (20) in the minor agents' dynamics (12)–(14), (16)–(18) and taking the average and then its L^2 limit as $N \to \infty$.

However, the only element of the mean field directly active in the dynamics in our setup is the L^2 limit

$$\bar{v} = \lim_{N \to \infty} \frac{1}{N} \sum_{i=1}^N v_i, \quad q \cdot m. \tag{21}$$

To derive the evolution equation of $\bar{v}$, we substitute (20) in the trading dynamics (12), (16) which yields

$$dv_i = \left[L_1 x_i + L_2 x_0 + L_3 \right] dt + \sum_{j \neq i, j=1}^N L_4 x_j dt, \quad 1 \leq i \leq N. \tag{22}$$

Then expanding the vector products in (22) and summing them over $\{i : 1 \le i \le N\}$, we get

$$Ndv^N = N[\ell_{11}v^N + \ell_{12}Q^N + \ell_{13}S^N + \ell_{21}v_0 + \ell_{22}Q_0 + \ell_{23}S_0 + \ell_3]dt$$

$$+ \Big[\sum_{i=1}^{N}\sum_{j \ne i, j=1}^{N} \ell_{41}v_j + \sum_{i=1}^{N}\sum_{j \ne i, j=1}^{N} \ell_{42}Q_j dt + \sum_{i=1}^{N}\sum_{j \ne i, j=1}^{N} \ell_{43}S_j\Big]dt,$$

where $\{\ell_{km}; \; k = 1, \ldots, 4, \; m = 1, 2, 3\}$ is the mth element of the vector L_k. Then with $N \to \infty$ (as in [1, 17]), the mean field equation for $\bar{v}$ is given by

$$d\bar{v} = [\bar{\ell}_{11}\bar{v} + \bar{\ell}_{12}\bar{Q} + \bar{\ell}_{13}\bar{S}]dt + [\bar{\ell}_{21}v_0 + \bar{\ell}_{22}Q_0 + \bar{\ell}_{23}S_0]dt + \bar{\ell}_3 dt, \qquad (23)$$

where $\bar{\ell}_{11}, \bar{\ell}_{12}, \bar{\ell}_{13}, \bar{\ell}_{21}, \bar{\ell}_{22} \bar{\ell}_{23}$, and $\bar{\ell}_3$ are the parameters which can be calculated from the consistency conditions and, by (21)

$$d\bar{Q} = \lim_{N \to \infty} \frac{1}{N}\sum_{i=1}^{N} dQ_i = \lim_{N \to \infty}\Big[\frac{1}{N}\sum_{i=1}^{N} v_i dt + \frac{\sigma^Q}{N}\sum_{i=1}^{N} dw_i^Q\Big] = \bar{v}dt, \; q \cdot m$$

$$d\bar{S} = \lim_{N \to \infty} \frac{1}{N}\sum_{i=1}^{N} dS_i = \lim_{N \to \infty}\Big[(\lambda_0 v_0 + \frac{\lambda}{N}\sum_{i=1}^{N} v_i)dt + \frac{a}{N}\sum_{i=1}^{N} u_i(t)dt + \frac{\sigma}{N}\sum_{i=1}^{N} dw_i\Big]$$

$$= [(\lambda + a\bar{\ell}_{11})\bar{v} + a\bar{\ell}_{12}\bar{Q} + a\bar{\ell}_{13}\bar{S}]dt + [(\lambda_0 + a\bar{\ell}_{21})v_0 + a\bar{\ell}_{22}Q_0 + a\bar{\ell}_{23}S_0]dt + a\bar{\ell}_3 dt, \, q \cdot m.$$

The dynamical equation of the mean field $\bar{x} = [\bar{v}, \bar{Q}, \bar{S}]^T$ for the optimal execution problem can be written as

$$d\bar{x} = \bar{A}\bar{x}dt + \bar{G}x_0 dt + \bar{m}dt, \qquad (24)$$

with the matrices to be defined as

$$\bar{A} = \begin{bmatrix} \bar{\ell}_{11} & \bar{\ell}_{12} & \bar{\ell}_{13} \\ 1 & 0 & 0 \\ (\lambda + a\bar{\ell}_{11}) & a\bar{\ell}_{12} & a\bar{\ell}_{13} \end{bmatrix}, \quad \bar{m} = \begin{bmatrix} \bar{\ell}_3 \\ 0 \\ a\bar{\ell}_3 \end{bmatrix}, \quad \bar{G} = \begin{bmatrix} \bar{\ell}_{21} & \bar{\ell}_{22} & \bar{\ell}_{23} \\ 0 & 0 & 0 \\ (\lambda_0 + a\bar{\ell}_{21}) & a\bar{\ell}_{22} & a\bar{\ell}_{23} \end{bmatrix},$$

which can be determined from the consistency equations.

3.3 Infinite Populations

Following the mean field game methodology with a major agent [21], the optimal execution problem is first solved in the infinite population case where the average term in the finite population dynamics and cost function of each agent is replaced with its infinite population limit, i.e., the mean field. Then specializing to MFG linear systems [17], the major agent's state is extended with the mean field, while the minor agent's state is extended with the mean field and the major agent's state; this yields LQG problems for each trader linked only through the mean field and the major agent's state. Finally the infinite population best response strategies are applied to the finite population system which yields ϵ-Nash equilibria (see Theorem 1).

In this paper we address the optimal execution problem in the MFG framework when the traders have, first, complete observations and, second, partial observations of their state and the major trader's state in Sections 4 and 5, respectively.

The stochastic optimal control problem for each agent in the infinite population case is given below.

3.3.1 Major Liquidator Agent

The major trader's stochastic optimal control problem in the infinite population case is given by

$$dx_0 = A_0 x_0 dt + B_0 u_0 dt + E_0 \bar{x} dt + D_0 dw_0, \tag{25}$$

where

$$A_0 = \begin{bmatrix} 0 & 0 & 0 \\ 1 & 0 & 0 \\ \lambda_0 & 0 & 0 \end{bmatrix}, \quad B_0 = \begin{bmatrix} 1 \\ 0 \\ a_0 \end{bmatrix}, \quad E_0 = \begin{bmatrix} 0 & 0 & 0 \\ 0 & 0 & 0 \\ \lambda & 0 & 0 \end{bmatrix}, \quad D_0 = \begin{bmatrix} 0 & 0 \\ \sigma_0^Q & 0 \\ 0 & \sigma \end{bmatrix}, \quad w_0 = \begin{bmatrix} w_0^Q \\ w_0^F \end{bmatrix},$$

together with the cost function (11).

3.3.2 Minor Liquidator Agent

The stochastic optimal control problem for a minor liquidator agent in the infinite population case is given by

$$dx_i = A_l x_i dt + E_l \bar{x} dt + B_l u_i dt + G_l x_0 dt + D_l dw_i, \tag{26}$$

$1 \leq i \leq N_l$, with the matrices

$$A_l = \begin{bmatrix} 0 & 0 & 0 \\ 1 & 0 & 0 \\ 0 & 0 & 0 \end{bmatrix}, \quad E_l = \begin{bmatrix} 0 & 0 & 0 \\ 0 & 0 & 0 \\ \lambda & 0 & 0 \end{bmatrix}, \quad B_l = \begin{bmatrix} 1 \\ 0 \\ a \end{bmatrix}, \quad G_l = \begin{bmatrix} 0 & 0 & 0 \\ 0 & 0 & 0 \\ \lambda_0 & 0 & 0 \end{bmatrix},$$

$$D_l = \begin{bmatrix} 0 & 0 \\ \sigma^Q & 0 \\ 0 & \sigma \end{bmatrix}, \quad w_i = \begin{bmatrix} w_i^Q \\ w_i^F \end{bmatrix},$$

together with the cost function (15).

3.3.3 Minor Acquirer Agent

The stochastic optimal control problem for an acquirer agent in the infinite population case is given by

$$dx_i = A_a x_i dt + E_a \bar{x} dt + B_a u_i dt + G_a x_0 dt + D_a dw_i, \tag{27}$$

$1 \leq i \leq N_a$, where

$$A_a = \begin{bmatrix} 0 & 0 & 0 \\ -1 & 0 & 0 \\ 0 & 0 & 0 \end{bmatrix}, \quad E_a = \begin{bmatrix} 0 & 0 & 0 \\ 0 & 0 & 0 \\ \lambda & 0 & 0 \end{bmatrix}, \quad B_a = \begin{bmatrix} 1 \\ 0 \\ a \end{bmatrix},$$

$$G_a = \begin{bmatrix} 0 & 0 & 0 \\ 0 & 0 & 0 \\ \lambda_0 & 0 & 0 \end{bmatrix}, \quad D_a = \begin{bmatrix} 0 & 0 \\ \sigma^Q & 0 \\ 0 & \sigma \end{bmatrix}, \quad w_i = \begin{bmatrix} w_i^Q \\ w_i^F \end{bmatrix},$$

together with the cost function (19).

4 Completely Observed Optimal Execution Problem

In the completely observed (CO) optimal execution problem, it is assumed that the major trader completely observes its own state and each generic minor trader completely observes its own state and the major trader's state. The best response MFG trading strategies which are obtained later in this section yield ϵ-Nash equilibria for the market by the following theorem.

Theorem 1 (ϵ-Nash Equilibria for CO MM-MF Systems). *Subject to reasonable technical assumptions (see [17]), the system equations (8)–(19) together with the mean field equations (39) generate the set of control laws $\mathscr{U}_{MF}^N \triangleq \{u_i^\circ; 0 \leq i \leq N\}, 1 \leq N < \infty$, given by (34) and (37) such that*

(i) *All agent systems $\mathscr{A}_i$, $0 \leq i \leq N$, are second order stable.*
(ii) *$\{\mathscr{U}_{MF}^N; 1 \leq N < \infty\}$ yields an ϵ-Nash equilibrium for all ϵ, i.e. for all $\epsilon > 0$, there exists $N(\epsilon)$ such that for all $N \geq N(\epsilon)$;*

$$J_i^{s,N}(u_i^\circ, u_{-i}^\circ) - \epsilon \leq \inf_{u_i \in \mathscr{U}_{i,y}^N} J_i^{s,N}(u_i, u_{-i}^\circ) \leq J_i^{s,N}(u_i^\circ, u_{-i}^\circ).$$

After applying the mean field methodology to decouple the agents, the problem of obtaining the best response trading strategy is transformed to a stochastic indefinite LQ problem that is solved for using the following theorem which is a restriction to the constant matrix parameter case of the general result in [12].

Theorem 2 (Stochastic Indefinite LQ Problem). *Let $T > 0$ be given. For any $(s, y) \in [0, T) \times \mathbb{R}^n$, consider the following linear system*

$$dx = \left[Ax + Bu + b\right]dt + \left[Cx + Du + \sigma\right]dw, \tag{28}$$

where $t \in [s, T]$, $x(s) = y$ and A, B, C, D, b, σ are matrix valued functions of suitable sizes and $w(.)$ is a standard Wiener process. In addition, a quadratic cost function is given

$$J(s, y, u(.)) = \mathbb{E}\{\frac{1}{2} \int_0^T \left[\langle Px(t), x(t) \rangle + \langle Nx(t), u(t) \rangle \right.$$

$$\left. + \langle Ru(t), u(t) \rangle \right]dt + \frac{1}{2} \langle \bar{P}x(T), x(T) \rangle\}, \tag{29}$$

where P, N, and R are $\mathscr{S}^n$, $\mathbb{R}^{m \times n}$, and $\mathscr{S}^m$-valued functions, respectively, and $G \in \mathscr{S}^n$. Let $\Pi(.) \in C([s, T]; \mathscr{S}^n)$ be the solution of the Riccati equation

$$\dot{\Pi} + \Pi A + A^T \Pi + C^T PC + P - (B^T \Pi + N + D^T \Pi C)^T$$

$$\times (R + D^T \Pi D)^{-1}(B^T \Pi + N + D^T \Pi C) = 0, \ a.e. \ t \in [s, t]$$

$$\Pi(T) = \bar{P}, \tag{30}$$

where $R + D^T PD > 0$, $a.e.t \in [s, T]$, and $s(.) \in C([s, T]; \mathbb{R}^n)$ be the solution of the offset equation

$$\dot{s} + [A - B(R + D^T \Pi D)^{-1}(B^T P + s + D^T PC)]^T s$$

$$+ [C - D(R + D^T \Pi D)^{-1}(B^T \Pi + N + D^T \Pi C)]^T \Pi \sigma$$

$$+ \Pi b = 0, \ a.e.t \in [s, T], \quad s(T) = 0. \tag{31}$$

Let's define $\Psi \triangleq (R + D^T \Pi D)^{-1}[B^T \Pi + N + D^T \Pi C]$, and $\psi \triangleq (R + D^T \Pi D)^{-1}[B^T s + D^T \Pi \sigma]$. Then the stochastic LQ problem (28)–(29) is solvable at s with the optimal control $u^\circ(.)$ being of a state feedback form

$$u^\circ(t) = -\Psi(t)x(t) - \psi(t), \quad t \in [s, T]. \tag{32}$$

□

Henceforth we discuss the stochastic optimal control problem for the major trader and a generic minor trader.

4.1 Major Liquidator Agent

The dynamics for the major agent's extended state $x_0^{ex} = [x_0^T, \bar{x}^T]^T$ in the infinite population is given by

$$\begin{bmatrix} dx_0 \\ d\bar{x} \end{bmatrix} = \begin{bmatrix} A_0 & E_0 \\ \bar{G} & \bar{A} \end{bmatrix} \begin{bmatrix} x_0 \\ \bar{x} \end{bmatrix} dt + \begin{bmatrix} 0_{3\times1} \\ \bar{m} \end{bmatrix} dt + \begin{bmatrix} B_0 \\ 0_{3\times1} \end{bmatrix} u_0(t)dt + \begin{bmatrix} D_0 & 0 \\ 0 & 0 \end{bmatrix} \begin{bmatrix} dw_0 \\ 0 \end{bmatrix}. \tag{33}$$

Accordingly, the following matrices are defined

$$\mathbb{A}_0 = \begin{bmatrix} A_0 & E_0 \\ \bar{G} & \bar{A} \end{bmatrix}, \quad \mathbb{M}_0 = \begin{bmatrix} 0_{3\times1} \\ \bar{m} \end{bmatrix}, \quad \mathbb{B}_0 = \begin{bmatrix} B_0 \\ 0_{3\times1} \end{bmatrix}, \quad \mathbb{D}_0 = \begin{bmatrix} D_0 & 0 \\ 0 & 0 \end{bmatrix}.$$

Consequently, using *Theorem 2*, the infinite population best response control is given by

$$u_0^\circ(t) = -R_0^{-1}\mathbb{B}_0^T \Pi_0 \left(x_0^T, \bar{x}^T\right)^T, \tag{34}$$

$$-\frac{d\Pi_0}{dt} = \Pi_0 \mathbb{A}_0 + \mathbb{A}_0^T \Pi_0 - \Pi_0 \mathbb{B}_0 R_0^{-1} \mathbb{B}_0^T \Pi_0 + \mathbb{P}_0, \quad \Pi_0(T) = \bar{\mathbb{P}}_0,$$

where in the above Riccati equation

$$\mathbb{P}_0 = [I_{3\times3}, 0_{3\times3}]^T P_0 [I_{3\times3}, 0_{3\times3}],$$

$$\bar{\mathbb{P}}_0 = [I_{3\times3}, 0_{3\times3}]^T \bar{P}_0 [I_{3\times3}, 0_{3\times3}].$$

4.2 Minor Acquirer/Liquidator Agent

For brevity, the notation $(.)_{a/l}$ is used in the rest of this paper to denote the matrices and parameters corresponding to a generic acquirer or a liquidator agent, respectively. Accordingly, a generic minor (acquirer/liquidator) agent $\mathscr{A}_i$'s extended dynamics with the extended state $x_i^{ex} = [x_i^T, x_0^T, \bar{x}^T]^T$ is

$$\begin{bmatrix} dx_i \\ dx_0 \\ d\bar{x} \end{bmatrix} = \begin{bmatrix} A_{a/l} & [G_{a/l} \ E_{a/l}] \\ 0_{6\times3} & \mathbb{A}_0 \end{bmatrix} \begin{bmatrix} x_i \\ x_0 \\ \bar{x} \end{bmatrix} dt + \begin{bmatrix} 0_{3\times1} \\ \mathbb{M}_0 \end{bmatrix} dt + \begin{bmatrix} 0_{3\times1} \\ \mathbb{B}_0 \end{bmatrix} u_0(t)dt$$

$$+ \begin{bmatrix} B_{a/l} \\ 0_{6\times1} \end{bmatrix} u_i(t)dt + \begin{bmatrix} D_{a/l} \ 0_{3\times6} \\ 0_{6\times3} \ \mathbb{D}_0 \end{bmatrix} \begin{bmatrix} dw_i \\ dw_0 \\ 0 \end{bmatrix}. \tag{35}$$

Substituting the major agent's control action (34) into (35) yields

$$dx_i^{ex} = \mathbb{A}_{a/l} x_i^{ex} dt + \mathbb{M}_{a/l} dt + \mathbb{B}_{a/l} dt + \mathbb{D}_{a/l} dW_i, \tag{36}$$

where

$$\mathbb{A}_{a/l} = \begin{bmatrix} A_{a/l} & [G_{a/l} \ E_{a/l}] \\ 0_{6\times3} & \mathbb{A}_0 - \mathbb{B}_0 R_0^{-1} \mathbb{B}_0^T \Pi_0 \end{bmatrix}, \quad \mathbb{M}_{a/l} = \begin{bmatrix} 0_{3\times1}, \\ \mathbb{M}_0 \end{bmatrix},$$

$$\mathbb{B}_{a/l} = \begin{bmatrix} B_{a/l} \\ 0_{6\times1} \end{bmatrix}, \quad \mathbb{D}_{a/l} = \begin{bmatrix} D_{a/l} \ 0_{3\times6} \\ 0_{6\times3} \ \mathbb{D}_0 \end{bmatrix}, \quad W_i = \begin{bmatrix} w_i \\ w_0 \\ 0 \end{bmatrix}.$$

We utilize *Theorem 2* again to obtain the best response control for a generic minor agent as

$$u_i^\circ(t) = -R_{a/l}^{-1} \mathbb{B}_{a/l}^T \Pi_{a/l} (x_i^T, x_0^T, \bar{x}^T)^T, \tag{37}$$

where $\Pi_{a/l}$ is calculated by the following Riccati equation

$$-\frac{d\Pi_{a/l}}{dt} = \Pi_l \mathbb{A}_{a/l} + \mathbb{A}_{a/l}^T \Pi_{a/l} - \Pi_{a/l} \mathbb{B}_{a/l} R_{a/l}^{-1} \mathbb{B}_{a/l}^T \Pi_{a/l} + \mathbb{P}_{a/l}, \quad \Pi_{a/l}(T) = \bar{\mathbb{P}}_{a/l},$$

with the matrices

$$\mathbb{P}_{a/l} = [I_{3\times3}, 0_{3\times6}]^T P_{a/l} [I_{3\times3}, 0_{3\times6}],$$

$$\bar{\mathbb{P}}_{a/l} = [I_{3\times3}, 0_{3\times6}]^T \bar{P}_{a/l} [I_{3\times3}, 0_{3\times6}].$$

4.3   Consistency Conditions

The closed loop trading dynamics of a generic minor agent $\mathscr{A}_i$, $1 \leq i \leq N$ applying (37) is given by

$$dv_i = -R_{a/l}^{-1}\mathbb{B}_{a/l}^T \Pi_{a/l}\big(x_i^T, x_0^T, \bar{x}^T\big)^T dt.$$

Accordingly, the average of closed loop trading dynamics over minor traders' population is obtained as

$$\sum_{i=1}^N dv_i = -\pi_a^N \sum_{i=1}^{N_a} R_a^{-1}\mathbb{B}_a^T \Pi_a\big(x_i^T, x_0^T, \bar{x}^T\big)^T dt - \pi_l^N \sum_{i=1}^{N_l} R_l^{-1}\mathbb{B}_l^T \Pi_l\big(x_i^T, x_0^T, \bar{x}^T\big)^T dt, \tag{38}$$

where $\pi_a^N = N_a/N$ and $\pi_l^N = \frac{N_l}{N}$ are the empirical distributions of acquirers and the liquidators.

Then taking the L^2 limit of (38) as the population size N goes to infinity

$$\lim_{N\to\infty} dv^N = -\Pi_a R_a^{-1}\mathbb{B}_a^T \Pi_a \lim_{N_a\to\infty} \big((x^N)^T, x_0^T, \bar{x}^T\big)^T dt$$

$$-\Pi_l R_l^{-1}\mathbb{B}_l^T \Pi_l \lim_{N_l\to\infty} \big((x^N)^T, x_0^T, \bar{x}^T\big)^T dt, q \cdot m.$$

yields the trading rate mean field dynamics

$$d\bar{v} = -\Pi_a R_a^{-1}\mathbb{B}_a^T \Pi_a (\bar{x}^T, x_0^T, \bar{x}^T)^T dt - \Pi_l R_l^{-1}\mathbb{B}_l^T \Pi_l(\bar{x}^T, x_0^T, \bar{x}^T)^T dt$$

$$= -\Pi_a R_a^{-1}\big(\Pi_{a,11}\bar{v} + \Pi_{a,12}\bar{Q} + \Pi_{a,13}\bar{F} + \Pi_{a,14}v_0 + \Pi_{a,15}Q_0$$

$$+ \Pi_{a,16}F_0 + \Pi_{a,17}\bar{v} + \Pi_{a,18}\bar{Q} + \Pi_{a,19}\bar{F}\big)dt$$

$$-\Pi_a a R_a^{-1}\big(\Pi_{a,31}\bar{v} + \Pi_{a,32}\bar{Q} + \Pi_{a,33}\bar{F} + \Pi_{a,34}v_0 + \Pi_{a,35}Q_0$$

$$+ \Pi_{a,36}F_0 + \Pi_{a,37}\bar{v} + \Pi_{a,38}\bar{Q} + \Pi_{a,39}\bar{F}\big)dt$$

$$-\Pi_l R_l^{-1}\big(\Pi_{l,11}\bar{v} + \Pi_{l,12}\bar{Q} + \Pi_{l,13}\bar{F} + \Pi_{l,14}v_0 + \Pi_{l,15}Q_0$$

$$+ \Pi_{l,16}F_0 + \Pi_{l,17}\bar{v} + \Pi_{l,18}\bar{Q} + \Pi_{l,19}\bar{F}\big)dt$$

$$-\Pi_l a R_l^{-1}\big(\Pi_{l,31}\bar{v} + \Pi_{l,32}\bar{Q} + \Pi_{l,33}\bar{F} + \Pi_{l,34}v_0 + \Pi_{l,35}Q_0$$

$$+ \Pi_{l,36}F_0 + \Pi_{l,37}\bar{v} + \Pi_{l,38}\bar{Q} + \Pi_{l,39}\bar{F}\big)dt, \quad q \cdot m.$$

where $\Pi_a = \lim_{N\to\infty} \Pi_a^N$, $\Pi_l = \lim_{N\to\infty} \Pi_l^N$.

Accordingly, the consistency equations ([18]) between (24) and the equations above are

$$\bar{\ell}_{11} = -\Pi_a R_a^{-1}\big[(\Pi_{a,11} + \Pi_{a,17}) - a(\Pi_{a,31} + \Pi_{a,37})\big] - \Pi_l R_l^{-1}\big[(\Pi_{l,11} + \Pi_{l,17}) - a(\Pi_{l,31} + \Pi_{l,37})\big],$$

$$\bar{\ell}_{12} = -\Pi_a R_a^{-1}\big[(\Pi_{a,12} + \Pi_{a,18}) - a(\Pi_{a,32} + \Pi_{a,38})\big] - \Pi_l R_l^{-1}\big[(\Pi_{l,12} + \Pi_{l,18}) - a(\Pi_{l,32} + \Pi_{l,38})\big],$$

$$\bar{\ell}_{13} = -\Pi_a R_a^{-1}\left[(\Pi_{a,13} + \Pi_{a,19}) - a(\Pi_{a,33} + \Pi_{a,39})\right] - \Pi_l R_l^{-1}\left[(\Pi_{l,13} + \Pi_{l,19}) - a(\Pi_{l,33} + \Pi_{l,39})\right],$$

$$\bar{\ell}_{21} = -\Pi_a R_a^{-1}(\Pi_{a,14} + a\Pi_{a,34}) - \Pi_l R_l^{-1}(\Pi_{l,14} + a\Pi_{l,34}),$$

$$\bar{\ell}_{22} = -\Pi_a R_a^{-1}(\Pi_{a,15} + a\Pi_{a,35}) - \Pi_l R_l^{-1}(\Pi_{l,15} + a\Pi_{l,35}),$$

$$\bar{\ell}_{23} = -\Pi_a R_a^{-1}(\Pi_{a,16} + a\Pi_{a,36}) - \Pi_l R_l^{-1}(\Pi_{l,16} + a\Pi_{l,36}),$$

$$\bar{\ell}_3 = 0, \tag{39}$$

where the scalars $\bar{\ell}_{(\cdot)}$ were defined in (23) and $\Pi_{a/l,ij} = \Pi_{a/l}(i,j)$ for $i = 1,3$; $j = 1,2,\ldots,9$.

5 Partially Observed Optimal Execution Problem

In this section it is assumed that the major trader has partial observations of its own state. This can happen, for example, in the foreign exchange (Forex) market, where an electronic communication network (ECN) Forex broker as a major agent trades on behalf of banks, high net worth (HNW) traders, and other brokers and hence needs to estimate the trades, amount of exchanges, and prices of each agent regularly. Figure 1 depicts this scenario.

It is also assumed that each minor trader has partial observations of its own state and the major trader's state. A justification for the partial observation assumption on the minor agents' own state is similar to that of the major trader but at a smaller scale. However one may also argue that the minor agents have complete

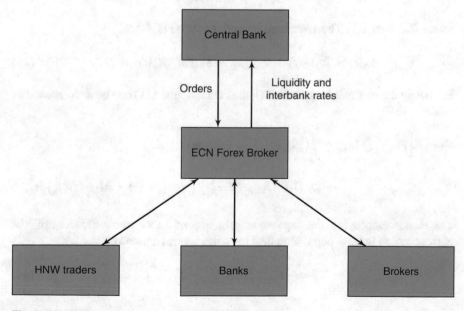

Fig. 1 ECN broker (major trader) in the Forex market

observations of their states because they carry out smaller trades which they manage individually. We note that the latter special case may be obtained from the former more general one by setting the corresponding part of the filter equations to zero; this will not cause any singularities because the observability and noise controllability conditions will still hold.

We now follow the general development in [5, 11] for PO MM LQG MFG systems to address the partially observed optimal execution problem for the major trader and a generic minor trader.

5.1 Major Liquidator Agent

Let the major agent's observation process be

$$dy_0 = \mathbb{H}_0[x_0^T, \bar{x}^T]^T dt + \sigma_{v_0} dv_0, \tag{40}$$

where $\mathbb{H}_0$ is a constant matrix with appropriate dimension. Then the corresponding Kalman filter equation to generate the estimates of the major agent's states is given by

$$d\hat{x}^{ex}_{0|\mathscr{F}_0^y} = \mathbb{A}_0 \hat{x}^{ex}_{0|\mathscr{F}_0^y} dt + \mathbb{M}_0 dt + \mathbb{B}_0 \hat{u}_{0|\mathscr{F}_0^y} dt + K_0(t)[dy_0 - \mathbb{H}_0 \hat{X}_{0|\mathscr{F}_0^y} dt], \tag{41}$$

where the filter gain is given by

$$K_0(t) = V_0(t) \mathbb{H}_0^T R_{v_0}^{-1}, \tag{42}$$

where $R_{v_0} = \sigma_{v_0} \sigma_{v_0}^T$. The associated Riccatti equation is

$$\dot{V}_0(t) = \mathbb{A}_0 V_0(t) + V_0(t) \mathbb{A}_0^T - K_0(t) R_{v_0} K_0(t)^T + Q_{w_0}. \tag{43}$$

Following the methodology in [5, 11], the cost function (11) can be decomposed as

$$J_0 = \mathbb{E}\Big[\|\hat{x}_{0|\mathscr{F}_0^y}(T)\|^2_{P_0} + \int_0^T \big(\|\hat{x}_{0|\mathscr{F}_0^y}(s)\|^2_{P_0} + \|u_0(s)\|^2_{R_0} \big) ds$$

$$+ \|x_0(T) - \hat{x}_{0|\mathscr{F}_i^y}(T)\|^2_{P_0} + \int_0^T \big(\|x_0(s) - \hat{x}_{0|\mathscr{F}_0^y}(s)\|^2_{P_0} \big) ds \Big],$$

and thence employing the separation principle of LQG stochastic control, the corresponding infinite population best response control action is given by

$$\hat{u}_0^\circ = -R_0^{-1} \mathbb{B}_0^T \big[\Pi_0 \big(\hat{x}^T_{0|\mathscr{F}_0^y}, \hat{\bar{x}}^T_{|\mathscr{F}_0^y} \big)^T \big]. \tag{44}$$

5.2 Minor (Acquirer/Liquidator) Agent

The extended state shall be denoted by

$$X_i = [x_i^T, x_0^T, \bar{x}^T, \hat{x}_{0|\mathscr{F}_0^y}^T, \hat{\bar{x}}_{|\mathscr{F}_0^y}^T]^T. \tag{45}$$

Let the minor agent's observation process be given by

$$dy_i(t) = \mathbb{H}_{a/l}[x_i^T, x_0^T, \bar{x}^T, \hat{x}_{0|\mathscr{F}_0^y}^T, \hat{\bar{x}}_{0|\mathscr{F}_0^y}^T]^T dt + \sigma_{v_i} dv_i, \tag{46}$$

with the constant matrix $\mathbb{H}_{a/l}$. Then the extended dynamics of the minor agent is given by

$$
\begin{bmatrix} dx_i \\ dx_0 \\ d\bar{x} \\ d\hat{x}_{0|\mathscr{F}_0^y} \\ d\hat{\bar{x}}_{|\mathscr{F}_0^y} \end{bmatrix} =
\begin{bmatrix} A_{a/l} & [G_{a/l}, E_{a/l}] & & 0_{3\times 6} \\ 0_{6\times 3} & \mathbb{A}_0 & & -\mathbb{B}_0 R_0^{-1} \mathbb{B}_0^T \Pi_0 \\ 0_{6\times 3} & K_0 \mathbb{H}_0 & & \mathbb{A}_0 - \mathbb{B}_0 R_0^{-1} \mathbb{B}_0^T \Pi_0 - K_0 \mathbb{H}_0 \end{bmatrix}
\begin{bmatrix} x_i \\ x_0 \\ \bar{x} \\ \hat{x}_{0|\mathscr{F}_0^y} \\ \hat{\bar{x}}_{|\mathscr{F}_0^y} \end{bmatrix} dt +
\begin{bmatrix} 0_{3\times 1} \\ M_0 \\ M_0 \end{bmatrix} dt
$$

$$
+ \begin{bmatrix} \mathbb{B}_{a/l} \\ 0_{6\times 1} \end{bmatrix} u_i(t) dt +
\begin{bmatrix} \mathbb{D}_{a/l} & 0 \\ 0 & K_0 \end{bmatrix}
\begin{bmatrix} dW_i \\ dW_0 \\ 0 \\ dv_0 \end{bmatrix}, \tag{47}
$$

or equivalently

$$dX_i = \mathscr{A}_{a/l} X_i dt + \mathscr{M}_{a/l} dt + \mathscr{B}_{a/l} u_i dt + \Sigma_{a/l} [dW_i^T, dW_0^T, 0, dv_0]^T.$$

The Kalman filter which generates the estimates of the minor (liquidator/acquirer) agent's states is

$$d\hat{X}_{i|\mathscr{F}_i^y} = \mathscr{A}_{a/l} \hat{X}_{i|\mathscr{F}_i^y} dt + \mathscr{M}_{a/l} dt + \mathscr{B}_{a/l} \hat{u}_{i|\mathscr{F}_i^y} dt + K_{a/l}(t)[dy_i - \mathbb{H}_{a/l} \hat{X}_{i|\mathscr{F}_i^y} dt], \tag{48}$$

where the filter gain is given as

$$K_{a/l}(t) = V_{a/l}(t) \mathbb{H}_{a/l}^T R_{v_i}^{-1}, \tag{49}$$

with $R_{v_i} = \sigma_{v_i} \sigma_{v_i}^T$. The corresponding Riccati equation is

$$\dot{V}_{a/l}(t) = \mathscr{A}_{a/l} V_{a/l}(t) + V_{a/l}(t) \mathscr{A}_{a/l}^T - K_{a/l}(t) R_v K_{a/l}(t)^T + Q_w. \tag{50}$$

The same procedure as in [5, 11] can be used to decompose the cost function (15) or (19) as

$$J_i = \mathbb{E}\Big[\|\hat{x}_{i|\mathscr{F}_i^y}(T)\|^2_{P_{a/l}} + \int_0^T \big(\|\hat{x}_{i|\mathscr{F}_i^y}(s)\|^2_{P_{a/l}} + \|u_i(s)\|^2_{R_{a/l}}\big)ds$$

$$+ \|x_i(T) - \hat{x}_{i|\mathscr{F}_i^y}(T)\|^2_{P_{a/l}} + \int_0^T \|x_i(s) - \hat{x}_{i|\mathscr{F}_i^y}(s)\|^2_{P_{a/l}}ds\Big].$$

So employing the separation principle, the corresponding infinite population best response control for a generic minor trader is seen to be

$$\hat{u}_i^\circ = -R_{a/l}^{-1}\mathbb{B}_{a/l}^T\big[\Pi_{a/l}\big(\hat{x}_{i|\mathscr{F}_i^y}^T, \hat{x}_{0|\mathscr{F}_i^y}^T, \hat{\bar{x}}_{|\mathscr{F}_i^y}^T\big)^T\big]. \tag{51}$$

Moreover, one may show (see [5, 11]) that the infinite population best response control laws applied to a finite population system yield the following ϵ-Nash equilibrium.

Theorem 3. *ϵ-Nash Equilibria for PO MM-MF Systems: The KF-MF state estimation schemes (41)–(43) and (48)–(50) together with the MM-MFG equation scheme (39) generate the set of control laws $\hat{\mathscr{U}}_{MF}^N \triangleq \{\hat{u}_i^\circ; 0 \leq i \leq N\}$, $1 \leq N < \infty$, given by*

$$\hat{u}_0^\circ = -R_0^{-1}\mathbb{B}_0^T\Pi_0(\hat{x}_{0|\mathscr{F}_0^y}^T, \hat{\bar{x}}_{|\mathscr{F}_0^y}^T)^T,$$

$$\hat{u}_i^\circ = -R^{-1}\mathbb{B}^T\Pi(\hat{x}_{i|\mathscr{F}_i^y}^T, \hat{x}_{0|\mathscr{F}_i^y}^T, \hat{\bar{x}}_{|\mathscr{F}_i^y}^T)^T, \ 1 \leq i \leq N$$

such that

(i) All agent systems $\mathscr{A}_i$, $0 \leq i \leq N$, are second order stable.
(ii) $\{\hat{\mathscr{U}}_{MF}^N; 1 \leq N < \infty\}$ yields an ϵ-Nash equilibrium for all ϵ, i.e., for all $\epsilon > 0$, there exists $N(\epsilon)$ such that for all $N \geq N(\epsilon)$,

$$J_i^{s,N}(\hat{u}_i^\circ, \hat{u}_{-i}^\circ) - \epsilon \leq \inf_{u_i \in \mathscr{U}_{i,y}^N} J_i^{s,N}(u_i, \hat{u}_{-i}^\circ) \leq J_i^{s,N}(\hat{u}_i^\circ, \hat{u}_{-i}^\circ).$$

6 Simulation

In the numerical experiments it is assumed that the trading action takes place within $T = 1$. The temporary impact strength of the major agent's trading and a generic minor agent's trading on the market are $a_0 = a = 5.43 \times 10^{-6}$, while their permanent impact strengths are taken to be $\lambda_0 = \lambda = 2 \times 10^{-8}$. The diffusion coefficients in the trading dynamics are selected as $\sigma_0^Q = 0.05$ and $\sigma_i^Q = 0.02$. The weights in the cost function for the major trader are $\psi_0 = 100$, $\mu_0 = 100$,

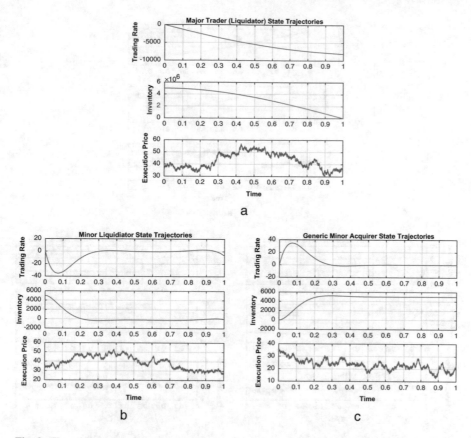

Fig. 2 The trading rate, inventory, and execution price trajectories of (a) the major liquidator trader, (b) a generic minor liquidator, and (c) a generic minor acquirer trader in the market

$\alpha_0 = 5a_0 \times 10^5$, $\phi_0 = 10^{-6}a_0$, $\delta_0 = 1/(2a0)$, $\xi_0 = 1/(2\alpha_0)$, $\theta_0 = 1/(2\delta_0)$, and $\gamma_0 = 10$ and those of a generic minor (liquidator/acquirer) trader are $\psi_l = \psi_a = 1$, $\mu_l = \mu_a = 1000$, $\alpha_l = \alpha_a = 5a \times 10^5$, $\phi_l = \phi_a = 10^{-1}a$, $\xi_l = 1/(2\alpha_l)$, $\xi_a = 1/(2\alpha_a)$, $\delta_l = 1/(2a_l)$, $\delta_a = 1/(2a_a)$, $\theta_l = 1/(2\delta_l)$, $\theta_a = 1/(2\delta_a)$, and $\gamma_l = \gamma_a = 10$. Furthermore, the market volatility is $\sigma = 0.6565$, the initial asset price is taken to be $F_0(0) = F_i(0) = \$35$, and the initial inventory stock of the major trader to be liquidated is set to $Q_0(0) = 5 \times 10^6$, while the minor liquidator HFT aims to sell $Q_i(0) = 5000$ shares and the acquirer HFT wishes to buy $Q_i(0) = 5000$ shares. In the estimation part, the measurement noise standard deviation for the major trader is $\sigma_0 = 0.05$, and for the HFT is $\sigma = 0.05$.

The resulting ϵ-Nash equilibrium trajectories of the major agent and generic acquirer/liquidator HFTs for the complete observation case are displayed in Figure 2, and the corresponding estimated trajectories in the partial observation case are depicted in Figure 3. As can be seen in Figure 2, the major trader liquidates its shares gradually during the trading interval and comes up with 28520 shares at the end of

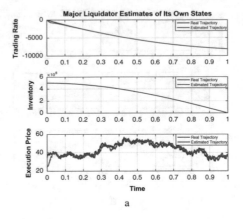

a

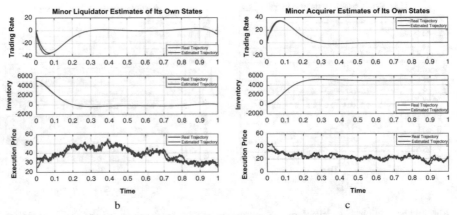

b   c

Fig. 3 The trading rate, inventory, and execution price trajectories and the corresponding estimated trajectories based on its own observations of (a) the major liquidator trader, (b) a generic minor liquidator, and (c) a generic minor acquirer trader in the market

trading horizon. The minor acquirer buys 5004 shares and the minor liquidator sells 4974 shares during the trading horizon T. Moreover, in the partial observation case shown in Figure 3, the estimated trajectories generated by the Kalman filter closely follow the real ones.

7 Conclusion

In this paper, an execution problem in finance with major and minor traders having liquidation or acquisition objectives was formulated and addressed in the mean field game framework by application of the separation principle of stochastic optimal control theory extended to indefinite partially observed LQG problems. Our future work will include parameter estimation of dynamic models of real market data employing methodologies including those in [15].

Acknowledgements We gratefully acknowledge conversations with Sebastian Jaimungal and Tamer Başar concerning this work.

References

1. Caines P. E. , Kizilkale A. C. (2013) Recursive estimation of common partially observed disturbances in MFG systems with application to large scale power markets. In: Proceedings of the 52nd IEEE Conference on Decision and Control (CDC): 2505–2512
2. Caines P. E. , Kizilkale A. C. (2014) Mean field estimation for partially observed LQG systems with major and minor agents. In: Proceedings of the 19th World Congress of the International Federation of Automatic Control (IFAC): 8705–8709
3. Şen N. , Caines P. E. (2014) Mean field games with partially observed major player and stochastic mean field. In: Proceedings of 53rd IEEE Conference on Decision and Control (CDC): 2709–2715
4. Şen N. , Caines P. E. (2015) ϵ-Nash equilibria for a partially observed mean field game with major player. In: Proceedings of 2015 American Control Conference (ACC): 4791–4797
5. Firoozi D. , Caines P. E. (2015) ϵ-Nash equilibria for partially observed LQG mean field games with major agent: Partial observations by all agents. In: Proceedings of 54th IEEE Conference on Decision and Control (CDC): 4430–4437
6. Huang M. , Caines P. E. , Malhamé R. P. (2006) Distributed multi-agent decision-making with partial observations: asymptotic Nash equilibria. In: Proceedings of the 17th International Symposium on Mathematical Theory of Networks and Systems (MTNS): 2725–2730
7. Şen N. , Caines P. E. (2016) On mean field games and nonlinear filtering for agents with individual-state partial observations. In: Proceedings of the 2016 American Control Conference (ACC): 4681–4686
8. Şen N. , Caines P. E. (2016) Mean field game theory for agents with individual-state partial observations. In: Proceedings of the 55th IEEE Conference on Decision and Control (CDC): 6105–6110
9. Firoozi D. , Caines P. E. (2016) Mean field game ϵ-Nash equilibria for partially observed optimal execution problems in finance. In: Proceedings of the 55th IEEE Conference on Decision and Control (CDC): 268–275
10. Firoozi D. , Caines P. E. (2017) An Optimal Execution Problem in Finance with Acquisition and Liquidation Objectives: an MFG Formulation. In: Proceedings of the 20th World Congress of the International Federation of Automatic Control (IFAC): 4960–4967
11. Cartea Á, Sebastian J. , Penalva J (2015) Algorithmic and high-frequency trading, 1st edn. Cambridge University Press, Cambridge, United Kingdom
12. Yong J. , Zhou X. Y. (1999) Stochastic Controls: Hamiltonian Systems and HJB Equations, Springer-Verlag, New York
13. Şen N. , Caines P. E. (2016) Mean field game theory with a partially observed major agent. SIAM Journal on Control and Optimization 54(6):3174–3224
14. Caines P. E. , Kizilkale A. C. (2016) ϵ-Nash equilibria for partially observed LQG mean field games with a major player. IEEE Transaction on Automatic Control: 3225–3234
15. Jaimungal S. , Kinzebulatov D. (2014) Optimal execution with a price limiter. Journal of Risk:49–54
16. Almgren R. , Chriss N. (2001) Optimal execution of portfolio transactions. Journal of Risk: 5–39

17. Huang M. (2010) Large-population LQG games involving a major player: The Nash certainty equivalence principle. SIAM Journal on Control and Optimization 48(5):3318–3353
18. Jaimungal S. , Kinzebulatov D. (2014) Optimal execution with a price limiter. Journal of Risk 27(7):49–54
19. Alfonsi A. , Fruth A. , Schied A. (2010) Optimal execution strategies in limit order books with general shape functions. Quantitative Finance 10(2): 143–157
20. Bayraktar E. , Ludkovski M. (2011) Optimal trade execution in illiquid markets. Mathematical Finance 21(4): 681–701
21. Nourian M. , Caines P. E. (2013) ϵ-Nash mean field game theory for nonlinear stochastic dynamical systems with major and minor agents. SIAM Journal on Control and Optimization 51(4): 3302–3331
22. Huang X. , Jaimungal S. , Nourian M. (2015) Mean-field game strategies for optimal execution. Available via SSRN. https://ssrn.com/abstract=2578733

Mean Field Limits Through Local Interactions

Suhail M. Shah and Vivek S. Borkar

Abstract We consider a controlled stochastic dynamics on a connected graph with gossip-like nearest neighbor affine interactions on a faster time scale. In the limit as the time scale separation diverges, followed by a limit as the graph grows to an infinite graph, we recover a mean field dynamics.

Keywords Mean field limits • Local interaction • Graph limits

1 Introduction

In [11], Tsitsiklis et al introduced a novel paradigm for distributed stochastic approximation which interfaced a gossip-like [9] averaging scheme across processors with a stochastic approximation like regular perturbation at each processor. What this ensured was a distributed stochastic approximation algorithm that achieved consensus across processors with only local averaging. After a talk on a related work by one of us (VSB), Prof. Roland Malhame suggested that this might provide a mechanism for recovering mean field game models from local interaction. This was attempted in [4] with rather limited success. Here we go a step further by adding to the averaging mechanism of [4] a time scale separation that allows us to carry out this program for an important special case, viz., one in which the averaging variables enter the dynamics in an affine manner. In particular this covers linear control systems with mean field interactions. We carry out this program here and also sketch some variants which give an indication as to what one might expect in more general models.

Specifically, we consider agents associated with the nodes of a graph, each with a coupled dynamics with two time scales. The fast component does pure local averaging, and the slow one is a diffusion process coupled to the corresponding diffusions of its neighbors on the graph via the local average obtained on the fast

S.M. Shah • V.S. Borkar (✉)
Department of Electrical Engineering, Indian Institute of Technology Bombay, Powai, Mumbai 400076, India
e-mail: suhailshah@iitb.ac.in; borkar.vs@gmail.com

© Springer International Publishing AG 2017
J. Apaloo, B. Viscolani (eds.), *Advances in Dynamic and Mean Field Games*,
Annals of the International Society of Dynamic Games 15,
https://doi.org/10.1007/978-3-319-70619-1_6

time scale. We first let the time scale separation, parametrized by a parameter $\epsilon > 0$, diverge as $\epsilon \downarrow 0$, and then let the graphs increase to infinite graphs, in order to obtain the limiting mean field dynamics.

The two limits are not interchangeable in principle because if one takes the graph limit first, then the stochastic matrix Q used for local averaging is an infinite matrix and one has to consider its recurrence/transience/ergodicity behavior explicitly. This is avoided if we first take the $\epsilon \downarrow 0$ limit. There is also a third limit, the limit as time $t \uparrow \infty$. If we take this limit last, since the mean field limit turns out to be a deterministic dynamics, we get its ω-limit set (possibly an equilibrium) that can depend on the initial condition. If, however, the limit is taken first, then the Freidlin–Wentzell phenomenon ([5], Section 6.4) kicks in, and the choice of limiting behaviors may narrow down significantly. For mean field limits in interacting, uncontrolled Markov chains, this program was carried out in [3], but the controlled situation appears to be much harder.

It is worth noting that a remarkable outcome of the present exercise is that the limit we obtain is quite robust to modeling details. Most importantly, it does not depend on specifics of the graphs except for their irreducibility, or of the stochastic matrix that dictates the probabilities with which a node polls its neighbors, again except for its irreducibility and a technical condition (3) below on the corresponding stationary distributions.

The limiting argument closely follows [4] and is given in the next section for the $\epsilon \downarrow 0$ limit and in Section 3 for the $N \uparrow \infty$ limit. Section 4 describes some possible variations.

Mean field games were introduced in [6–8]. For prior work on the uncontrolled mean field (McKean-Vlasov) equation, see [10].

We use the following notation throughout. For a Polish space $\mathcal{X}$, let $\mathcal{P}(\mathcal{X}) :=$ the Polish space of probability measures on $\mathcal{X}$ with Prohorov topology and $C(I; \mathcal{X}) :=$ the space of continuous functions $I \mapsto \mathcal{X}$ for some interval $I \subset \mathcal{R}^+$, with the topology of uniform convergence on compacts. Denote by $C_b(\mathcal{R}^d), C_b^2(\mathcal{R}^d)$, resp., the spaces of bounded continuous $f : \mathcal{R}^d \mapsto \mathcal{R}$ and twice continuously differentiable functions $\mathcal{R}^d \mapsto \mathcal{R}$ that are bounded along with their first- and second-order partial derivatives. As usual, ∇ and Δ will denote, resp., the gradient and the Laplacian.

2 The $\epsilon \downarrow 0$ Limit

Consider a sequence of connected directed graphs $\mathcal{G}_N = (\mathcal{V}_N, \mathcal{E}_N)$, $N \geq 1$, with node and edge sets $\mathcal{V}_N, \mathcal{E}_N$, resp., such that each $\mathcal{G}_N$ is a subgraph of $\mathcal{G}_{N+1}$ and $\tilde{N} := |\mathcal{V}_N| \uparrow \infty$. Each graph is assumed to be irreducible, i.e., there is a directed path from any node to any other node in the graph. Let $\mathcal{N}(i) := \{j \in \mathcal{V}_N : (i,j) \in \mathcal{E}_N\}$ denote the set of successors of i in $\mathcal{G}_N$. We also assume that with each node i is associated a copy of the "action space" $\mathcal{A}$ which is compact metric. Without loss of generality, we set $\mathcal{V}_N = \{1, 2, \cdots, \tilde{N}\}$.

With each node i in the graph $\mathcal{G}_N$, we associate $\mathcal{R}^d$-valued processes $Y_N^{i,\epsilon}(t), X_N^{i,\epsilon}(t), t \geq 0$, indexed by i, N and a parameter $\epsilon > 0$ which will dictate the time scale separation of fast and slow components of the dynamics. Their evolution is described by the coupled stochastic differential equations

$$dY_N^{i,\epsilon}(t) = \frac{1}{\epsilon} \left(\sum_{j \in \mathcal{N}(i)} q_{ij}^N Y_N^{j,\epsilon}(t) - Y_N^{i,\epsilon}(t) \right) dt + dX_N^{i,\epsilon}(t), \tag{1}$$

$$dX_N^{i,\epsilon}(t) = m(X_N^{i,\epsilon}(t), Y_N^{i,\epsilon}(t), u_i(t))dt + dW^i(t), \tag{2}$$

where:

- $\epsilon > 0$,
- $Q_N = [[q_{ij}^N]]_{i,j \in \mathcal{V}_N}$ is an irreducible stochastic matrix compatible with $\mathcal{G}_N$ with unique stationary distribution π_N satisfying

$$\lim_{N \uparrow \infty} \sum_{i \in \mathcal{V}_N} \pi_N(i)^2 = 0. \tag{3}$$

(As an example, consider the Q_N corresponding to a random walk on $\mathcal{G}_N$. Thus $\pi_N(i) = \frac{d(i)}{2|\mathcal{E}_N|}$ where $d(i) :=$ the degree of i. Then the above holds if, e.g., the degrees are bounded by a common constant independent of N.)
- the map $(x, y, u) \in \mathcal{R}^d \times \mathcal{R}^d \times \mathcal{A} \mapsto m(x, y, u) \in \mathcal{R}^d$ is of the form

$$m(x, y, u) = m_1(x, u)y + m_2(x, u) \tag{4}$$

where $m_1 : \mathcal{R}^d \times \mathcal{A} \mapsto \mathcal{R}^{d \times d}, m_2 : \mathcal{R}^d \times \mathcal{A} \mapsto \mathcal{R}^d$ are Lipschitz in the first argument uniformly w.r.t. the second,
- $Y_N^{i,\epsilon}(0) = X_N^{i,\epsilon}(0)$ are i.i.d. across $i \geq 1$ with bounded moments, with a common law ν_0,
- $W^i(\cdot), i \geq 1$, are independent standard Brownian motions in $\mathcal{R}^d$,
- $u_i(\cdot)$ is an $\mathcal{A}$-valued measurable process satisfying: for $t > s$, $W^k(t) - W^k(s)$ is independent of $X_N^{i,\epsilon}(y), W^i(y), u_i(y), y \leq s \,\forall\, t > s \geq 0, \forall k$. (This will be referred to as an *admissible control*.)

We shall further assume the *relaxed control* framework, i.e., we take $\mathcal{A}$ to be of the form $\mathcal{P}(A)$ for a compact metric A and $m_i(x, u)$ to be of the form $\int m_i'(x, y)u(dy)$ for an $m_i' : \mathcal{R}^d \times A \mapsto \mathcal{R}^{d \times d}$ (resp., $\mathcal{R}^d$) which is Lipschitz in its first argument uniformly w.r.t. the second, for $i = 1, 2$. Depending on the problem specifics, one may consider further restrictions on the controls, e.g., that the ith control depends only on the past state-control processes of the ith node and its neighbors. This leads to further issues concerning "information structures" that we do not address here. The present level of generality is appropriate for the kind of results we are aiming for.

These assumptions in particular are adequate to ensure the well posedness of (1)–(2) for prescribed $\{u_i(\cdot), W^i(\cdot), i \geq 1\}$ on some probability space. The parameter $\epsilon > 0$ is taken to be ≈ 0, which renders the dynamics (1) "fast" in the time scale of its evolution as compared to (2), with the second term on the right-hand side of (1) a "slow" perturbation of the fast dynamics. We shall first analyze the pair dynamics above in the $\epsilon \downarrow 0$ limit, i.e., as the time scale separation diverges, followed by the $N \uparrow \infty$ limit in a suitable sense. We keep the processes $u_i(\cdot), W^i(\cdot), i \geq 1$, fixed on a given probability space throughout this limiting operation. The fast dynamics (1) is nothing but a continuous time analog of the scheme of [11], and our first lemma will reflect this. Let $Q^* :=$ the rank one matrix with all rows equal to π_N.

Lemma 1. *As* $\epsilon \downarrow 0$,

$$(Y_N^{i,\epsilon}(\cdot), X_N^{i,\epsilon}(\cdot), i \geq 1) \to (\bar{X}_N^i(\cdot), X_N^i(\cdot), i \geq 1)$$

in law, where

$$\bar{X}_N^i(\cdot) = \bar{X}_N(\cdot) := \sum_{i=1}^{\tilde{N}} \pi_N(i) X_N^i(\cdot).$$

and $X_N^i(\cdot)$ *satisfy the dynamics*

$$dX_N^i(t) = m(X_N^i(t), \bar{X}_N(t), u_i(t))dt + dW^i(t), \ i \geq 1. \tag{5}$$

Proof. Using standard arguments based on Ito formula and the Gronwall inequality, one can establish the moment bound

$$E\left[\|X_N^{i,\epsilon}(t) - X_N^{i,\epsilon}(s)\|^4\right] \leq C|t - s|^2$$

for $t > s$ in $[0, T]$, $T > 0$, with $C > 0$ a constant that may depend on T but is independent of ϵ, i. A similar inequality then follows for $Y_N^{i,\epsilon}(\cdot)$. By (12.51), p. 95, [2], the laws of $(Y_N^{i,\epsilon}(\cdot), X_N^{i,\epsilon}(\cdot))$ are tight, hence relatively sequentially compact in $\mathcal{P}((C(\mathcal{R}^+; \mathcal{R}^d)^2)^\infty)$. By the variation of constants formula,

$$Y_N^{i,\epsilon}(t) = e^{\frac{1}{\epsilon}(Q-I)t} Y_N^{i,\epsilon}(0) + \int_0^t e^{\frac{1}{\epsilon}(Q-I)(t-s)} dX_N^{i,\epsilon}(s)$$

$$\overset{\epsilon \downarrow 0}{\to} Q^*(Y_N^i(0) - X_N^i(0)) + Q^* X_N^i(t)$$

$$= \bar{X}_N(t)$$

in mean square by standard arguments. From this and the affine dependence of m on its second argument (cf. (4)), it follows that the limiting dynamics along any subsequential limit in law is given by (5). The claim follows.

Define the $\mathcal{P}(\mathcal{R}^d)$-valued process $v_{.,N}$ by

$$\int f dv_{t,N} := \sum_{i \in V_N} \pi_N(i) f(X_N^i(t)), \ t \geq 0,$$

for $f \in C_b(\mathcal{R}^d)$. This is the natural counterpart in our framework of time t empirical measure for the coupled processes indexed by V_N that arises in classical mean field analysis. Similarly, define the $\mathcal{P}(\mathcal{R}^d \times A)$ valued process $\mu_{.,N}$ by

$$\int g d\mu_{t,N} := \sum_{i \in V_N} \pi_N(i) \int g(X_N^i(t), \cdot) du_i(t), \ t \geq 0,$$

for $g \in C_b(\mathcal{R}^d \times A)$. We shall use the notation $\kappa(f)$ for $\int f d\kappa$ for a function f and a measure κ. Then $v_{.,N}$ satisfies the stochastic p.d.e.

$$v_{t,N}(f) = v_{0,N}(f) + \int_0^t \mu_{s,N}(\mathcal{L}_{v_{s,N}}f) ds +$$

$$\int_0^t \langle v_{s,N}(\nabla f), dW^i(s) \rangle \tag{6}$$

where for $\kappa \in \mathcal{P}(\mathcal{R}^d)$,

$$\mathcal{L}_\kappa f(x,z) := \frac{1}{2}\Delta f(x) + \langle \int m'(x,y,z)\kappa(dy), \nabla f(x) \rangle, f \in C_b^2(\mathcal{R}^d). \tag{7}$$

We view $v_{.,N}$ as a process with path space $C(\mathcal{R}^+, \mathcal{P}(\mathcal{R}^d))$ and $\mu_{.,N}$ as a process with path space $C(\mathcal{R}^+, \mathcal{P}(\mathcal{R}^d \times A))$. These processes play a key role in the $N \uparrow \infty$ limit we take up next.

3 The $N \uparrow \infty$ Limit

For passage to the infinite graph limit, it will be convenient to view $v_t, \mu_t, t \geq 0$, as processes of marginals associated with probability measures on path spaces. The path space for $X_N^i(\cdot)$ is clearly $C(\mathcal{R}^+; \mathcal{R}^d)$. For the control process $u_i(\cdot)$, the path space is $\mathcal{U} :=$ the space of measurable maps $\mathcal{R}^+ \mapsto A$ topologized as follows. Give it the coarsest topology that renders continuous the maps

$$u(\cdot) \in \mathcal{U} \mapsto \int_s^t g(s) \int_A f du(s) ds \ \forall f \in C(A), g \in L_2[s,t], t > s \text{ in } \mathcal{R}^+.$$

Then $\mathcal{U}$ is compact metrizable as proved in, e.g., Theorem 2.3.2, p. 50, [1]. The idea of proof is as follows: Let $\{f_i\}$ be a countable dense set in the unit ball of $C(A)$ whence it is a convergence determining class for $\mathcal{A}$. Then the map $\kappa \in \mathcal{A} \mapsto [\int f_1 d\kappa, \int f_2 d\kappa, \cdots] \in [-1, 1]^\infty$ is a homeomorphism onto its range. Let Ψ denote the map that maps the $\mathcal{A}$-valued measurable function $t \in [0, \infty) \to v_t \in \mathcal{A}$, viewed as an element of $\mathcal{U}$, to the $[-1, 1]^\infty$-valued function $t \in [0, \infty) \mapsto [\int f_1 dv_t, \int f_2 dv_t, \cdots]$. We topologize the range of Ψ as follows.

Let $L_2^*[0, T]$:= the space $L_2[0, T]$ with weak* topology. View the space of measurable maps $[0, T] \mapsto [-1, 1]$ as a closed bounded subset of $L_2^*[0, T]$ which makes it compact by the Banach-Alaoglu theorem. It is also metrizable by the metric

$$d_T(h, g) := \sum_n 2^{-n} \left| \int_0^T e_n^T(t) h(t) dt - \int_0^T e_n^T(t) g(t) dt \right| \wedge 1,$$

where $\{e_n^T(\cdot)\}$ is a complete orthonormal basis for $L_2[0, T], T = 1, 2, \cdots$. The space χ_1 of measurable maps $[0, \infty) \mapsto [-1, 1]^\infty$ can be viewed as a subset of the space χ_2 := the space of measurable maps $[0, \infty) \mapsto \mathcal{R}^\infty$ whose restriction to $[0, T]$ is componentwise square-integrable for each $T \in [0, \infty)$. χ_2 is given the inductive topology obtained from $(L_2^*[0, T])^\infty$, $T > 0$. In particular, χ_1 is a compact subset of χ_2 metrizable with the metric

$$d(h, g) := \sum_N 2^{-N} d_N \left(h|_{[0, N]}, g|_{[0, N]} \right).$$

In turn, the range $\Psi(\mathcal{U})$ of Ψ is a closed subset of χ_1, which makes it compact and metrizable by the above metric. Ψ is also a homeomorphism $\mathcal{U} \leftrightarrow \Psi(\mathcal{U})$. Thus $\mathcal{U}$ is compact and can be identified with $\Psi(\mathcal{U})$ through Ψ. The metric on $\Psi(\mathcal{U})$ is then pulled back to a metric on $\mathcal{U}$ via Ψ. Being compact and metrizable, $\mathcal{U}, \Psi(\mathcal{U})$ are also Polish.

Let $\Phi_N^i \in \mathcal{P}(C(\mathcal{R}^+; \mathcal{R}^d) \times \mathcal{U})$ denote the joint law of $(X_N^i(\cdot), u_i(\cdot))$. By Lemma 2.3.8, p. 55, [1], $\{\Phi_N^i, i \geq 1, N \geq 1\}$ are tight: Since $\mathcal{U}$ is compact, one needs to establish only the tightness of $X_N^i(\cdot)$, which follows from the sufficient condition

$$\sup_{i,N} E \left[\|X_N^i(t) - X_N^i(s)\|^4 \right] \leq K(t - s)^2$$

for $t > s$ in $[0, T]$ and a constant $K > 0$ that depends on T ((12.51), p. 95, [2]). This is easily verified using the Ito formula and Gronwall inequality; see [1] for details. Let $\Phi_N \in \mathcal{P}((C(\mathcal{R}^+; \mathcal{R}^d) \times \mathcal{U})^N)$ denote the joint laws of $\{(X_N^i(\cdot), u_i(\cdot)), 1 \leq i \leq \tilde{N}\}$. Then these are tight as well because their marginals are. For $N > M \geq 1$, the marginal of Φ_N restricted to $\mathcal{G}_M$, given by $\Phi_{N,M}$:= the law of $\{(X_N^i(\cdot), u_i(\cdot)), 1 \leq i \leq M\}$, is well defined because of $\mathcal{G}_M \subset \mathcal{G}_N$. Let Φ be a subsequential limit of Φ_N as $N \uparrow \infty$ in the sense that its marginal on $\mathcal{G}_M$ is given by a subsequential limit of $\Phi_{N,M}$ as $M \leq N \uparrow \infty$ with the subsequence being chosen to be independent of M by a diagonal argument. We characterize this Φ next.

Let $\bar{\mathcal{R}}^d$ denote the one-point compactification of $\mathcal{R}^d$. View $\mathcal{R}^d$, resp., $\mathcal{P}(\mathcal{R}^d \times A)$ as subsets of $\bar{\mathcal{R}}^d$, resp., $\mathcal{P}(\bar{\mathcal{R}}^d \times A)$ with their natural embedding. View the functions $t \mapsto \mu_{t,N}$ as elements of the space $\mathcal{U}'$ of measurable maps $t \in [0, \infty) \mapsto \mathcal{P}(\bar{\mathcal{R}}^d \times A)$ with a compact metrizable topology defined analogously to that of $\mathcal{U}$. Then along a further subsequential limit, we can have $\mu_{\cdot,N} \to \mu_{\cdot}$ in law for some $\mu_{\cdot}$. By a standard Gronwall inequality based argument, we have

$$E\left[\int_0^T \|X_N^i(t)\|^2 dt\right] \leq C_T$$

for some constant $C_T < \infty$ that may depend on $T \in [0, \infty)$, but is independent of i, N. It follows that for $T > 0$,

$$E\left[\int_0^T \int \|x\|^2 \mu_{t,N}(dx \times A)dt\right] \leq C_T \ \forall N.$$

By Fatou's lemma, this also holds for any weak limit $\mu_{\cdot}$ of $\mu_{\cdot,N}$ as $N \uparrow \infty$. That is, $\mu_t(\mathcal{R}^d \times A) = 1$ a.e. t, a.s., where we may drop the qualification "a.e. t" by choosing an appropriate version. Hence we can view $\mu_{\cdot}$ as a process in $\mathcal{U}'' :=$ the space of measurable maps $[0, \infty) \mapsto \mathcal{P}(\mathcal{R}^d \times A)$, viewed as a subset of $\mathcal{U}'$ with relative topology and with the metric inherited from $\mathcal{U}'$. Likewise, its marginal $\nu_t(\cdot) := \mu_t(\cdot \times A)$ belongs to the space $\mathcal{U}^0$ of measurable maps $[0, \infty) \mapsto \mathcal{P}(\mathcal{R}^d)$ with the corresponding topology and metric.

Let $(\check{X}^i(\cdot), u_i(\cdot))$ denote the canonically realized random processes with joint law Φ and without loss of generality, let μ_t, ν_t be as above, defined on the same probability space. Then for compactly supported $f \in C_b(\mathcal{R}^d)$,

$$\mathcal{L}_{\nu_{t,N}} f \to \mathcal{L}_{\nu_t} f$$

pointwise. By the Stone-Weirstrass theorem, linear combinations of functions of the form $(t, x) \mapsto g(t)h(x)$ for $g \in C_b(\mathcal{R}^+), h \in C_b(\mathcal{R}^d)$ uniformly approximate any $f \in C^{0,2}(\mathcal{R}^+ \times \mathcal{R}^d)$ on compacts. Hence it follows that

$$\int_s^t \mu_{y,N}(\mathcal{L}_{\nu_{y,N}} f)dy \to \int_s^t \mu_y(\mathcal{L}_{\nu_y} f)dy$$

for $t > s \geq 0$, f as above. This allows us to pass to the $N \uparrow \infty$ limit in (6) to obtain:

Lemma 2. *The process* ν_t, $t \geq 0$, *satisfies*

$$\nu_t(f) = \nu_0(f) + \int_0^t \mu_y(\mathcal{L}_{\nu_y} f)dy. \tag{8}$$

Proof. Note that the quadratic variation of the stochastic integral term in (6) is

$$\left(\sum_i \pi_N(i)^2\right) O(t) = o(N).$$

Also, by the strong law of large numbers, $\int f dv_{0,N} \to \int f dv_0$ a.s. for $f \in$ a suitable countable convergence determining class $\subset C_b(\mathcal{R}^d)$. Thus $v_{0,N} \to v_0$ in $\mathcal{P}(\mathcal{R}^d)$, a.s. The claim now follows by letting $N \uparrow \infty$ in (6) in view of the preceding observations and our condition (3).

One therefore has the classic mean field limit [6–8]:

Theorem 1. *The limiting processes* $\check{X}^i(\cdot), i \geq 1$, *satisfy the dynamics*

$$d\check{X}^i(t) = v_t(m(\check{X}^i(t), \cdot, u_i(t)))dt + dW^i(t), \ t \geq 0, \tag{9}$$

where $\mu_t, t \geq 0$, *satisfies the deterministic controlled McKean-Vlasov equation: for* $v_t(dx) = \mu_t(dx \times A)$,

$$v_t(f) = v_0(f) + \int_0^t \mu_s(\mathcal{L}_{v_s}f)ds, \ t \geq 0, f \in C_b^2(\mathcal{R}^d). \tag{10}$$

Observe that (10) is a deterministic dynamics. Thus any random solution to it amounts to a probability measure on its deterministic trajectories. If we pick a deterministic solution to (10), then since the initial conditions and driving Brownian motions of (10) are independent, the processes $\{X^i(\cdot)\}$ will be independent, recovering the "propagation of chaos" result of the classical McKean-Vlasov theory [10]. However, we do not have any reasons to claim that the limit obtained by our limiting procedures will be deterministic. The minimal conditions we have imposed on $u_i(\cdot)$ are inadequate to ensure any such result.

Also, $\check{X}^i(\cdot), i \geq 1$, need not be identically distributed. The latter will be the case in, e.g., a symmetric Nash equilibrium in the ensuing mean field game. We do not touch these issues here as they have been extensively studied in [6–8] and the many works that followed them. Instead we take up in the next section a few variations that raise interesting issues.

4 Variations

We look at a few variations that result by tweaking the above model a bit, in what might appear as "natural" extensions. In each case we first write the $\epsilon \downarrow 0$ limit, followed by the modification needed in the drift term both in (9) and in the definition (7) of $\mathcal{L}_\kappa$ that appears in (10).

1. If we drop the affine structure (4), what we get in place of (5) is

$$dX_N^i(t) = m(X_N^i(t), \sum_{j \in V_N} \pi_N(j)X_N^j(t), u_i(t))dt + dW^i(t).$$

This is distinct from the classical mean field limit

$$dX_N^i(t) = \sum_{j \in V_N} \pi_N(j)m(X_N^i(t), X_N^j(t), u_i(t))dt + dW^i(t).$$

Furthermore, the drift term $v_t(m(x, \cdot, u))$ in (9), resp.,

$$\int \mu_t(dx, dz) \int v_t(dy)m'(x, y, z)$$

in (10) (as a part of $\mu_t(\mathcal{L}_{v_t}f)$) gets replaced by $m(x, \int yv_t(dy), u)$, resp., $\int \mu_t(dx, dz)m'(x, \int yv_t(dy), z)$.

2. If we change (1) to

$$dY_N^{i,\epsilon}(t) = \frac{1}{\epsilon}(\sum_{j \in \mathcal{N}(i)} q_{ij}^N Y_N^{j,\epsilon}(t) - Y_N^{i,\epsilon}(t))dt +$$

$$\sum_{j \in \mathcal{N}(i)} q_{ij}^N m(X_N^{i,\epsilon}(t), X_N^{j,\epsilon}(t), u_i(t))dt,$$

and (2) to

$$dX_N^{i,\epsilon}(t) = Y_N^{i,\epsilon}(t)dt + dW^i(t),$$

then (9) is replaced by

$$dX_N^i(t) = \sum_k \sum_{j \in \mathcal{N}(k)} \pi_N(k)q_{kj}^N m(X_N^k(t), X_N^j(t), u_k(t))dt$$

$$+ dW^i(t).$$

Furthermore, the drift term $v_t(m(x, \cdot, u))$ in (9) and

$$\int \mu_t(dx, dz) \int v_t(dy)m'(x, y, z)$$

in (10) (appearing in $\mu_t(\mathcal{L}_{v_t}f)$) gets replaced by

$$\int \zeta_t(dx, dy, dz)m'(x, y, z),$$

where $\zeta_\cdot$ is a weak limit (along appropriate subsequence) of $\zeta_{\cdot,N}$ defined by

$$\int f(x, y, z)\zeta_{t,N}(dx, dy, dz) :=$$

$$\sum_k \sum_{j \in \mathcal{N}(k)} \pi_N(k)q_{kj}^N f(X_N^k(t), X_N^j(t), u_k(t)).$$

3. The limit obtained for the simpler dynamics

$$dX_N^{i,\epsilon}(t) = \frac{1}{\epsilon}(\sum_{j \in \mathcal{N}(i)} q_{ij}^N X_N^{j,\epsilon}(t) - X_N^{i,\epsilon}(t))dt +$$

$$\sum_{j \in \mathcal{N}(i)} q_{ij}^N m(X_N^{i,\epsilon}(t), X_N^{j,\epsilon}(t), u_i(t))dt + dW^i(t)$$

is

$$dX_N^i(t) = \sum_k \sum_{j \in \mathcal{N}(k)} \pi_N(k)(q_{kj}^N m(X_N^k(t), X_N^j(t), u_k(t))dt$$

$$+ dW^k(t)).$$

Furthermore, the drift term $v_t(m(x, \cdot, u))$ (9), resp.,

$$\int \mu_t(dx, dz) \int v_t(dy)m'(x, y, z)$$

in (10) (appearing in $\mu_t(\mathcal{L}_{v_t}f)$) gets replaced as in the preceding bullet.

This suggests that the program suggested by Prof. Malhame is perhaps unfeasible in the absence of (4). On the positive side, it suggests alternative models of mean field interaction that may be of independent interest.

Acknowledgements The work of VSB was supported in part by a grant for "Approximation of High Dimensional Optimization and Control Problems" from the Department of Science and Technology, Government of India.

References

1. Arapostathis, A., Borkar, V. S. and Ghosh, M. K. *Ergodic Control of Diffusion Processes*. Cambridge University Press, Cambridge, UK, 2012.
2. Billingsley, P. *Convergence of Probability Measures*, John Wiley and Sons, New York, 1968.
3. Borkar, V. S. and Sundaresan, R. 'Asymptotics of the invariant measure in mean field models with jumps' *Stochastic Systems* **2(2)** (2012), 322–380.
4. Borkar, V. S. and Suresh Kumar, K. 'Coordination in networked control processes through gossip-like local interactions.' In *Modern Trends in Controlled Stochastic Processes* (A. B. Piunovskiy, ed.), Luniver Press, Frome, UK, 2015.
5. Freidlin, M. I. and Wentzell, A. D. *Random Perturbations of Dynamical Systems* (3rd ed.) Springer Verlag, New York, 2011.
6. Huang, M., Malhamé, R. P. and Caines, P. E. 'Large population stochastic dynamic games: closed-loop McKean-Vlasov systems and the Nash certainty equivalence principle' *Communications in Information and Systems* **6(3)** (2006), 221–252.

7. Huang, M., Caines, P. E. and Malhamé, R. P. 'Large-population cost-coupled LQG problems with nonuniform agents: individual-mass behavior and decentralized ε-Nash equilibria' *IEEE Transactions on Automatic Control* **52(9)** (2007), 1560–1571.
8. Lasry, J.-M. and Lions, P.-L. 'Mean field games' *Japanese Journal of Mathematics* **2(1)** (2007), 229–260.
9. Shah, D.: Gossip algorithms. *Foundations and Trends in Networking*, **3** (2009) 1–125.
10. Sznitman, A.-S. 'Topics in propagation of chaos.' In *Ecole d'Ete de Probabilites de Saint-Flour XIX - 1989*, (P.-L. Hennequin, ed.), Lecture Notes in Math. No. 1464, Springer Verlag, Berlin-Heidelberg, 1991, 164–251.
11. Tsitsiklis, J. N., Bertsekas, D. P. and Athans M. 'Distributed asynchronous deterministic and stochastic gradient optimization algorithms.' *IEEE Transactions on Automatic Control*, **31(9)** (1986) 803–12.

Part II
Dynamic Games and Applications

Differential Games in Health-Care Markets: Models of Quality Competition with Fixed Prices

Kurt R. Brekke, Roberto Cellini, Luigi Siciliani, and Odd Rune Straume

Abstract We present a review of models that investigate quality competition in health-care markets under price regulation, taking a differential game approach. We analyse and discuss three different variants of a unified modelling framework, each of which incorporates specific institutional and behavioural characteristics of health-care markets. In each case we derive and compare equilibrium strategies under open-loop and feedback closed-loop information structures. We also address potential policy implications from these analyses.

Keywords Health-care • Competition • Quality • Differential games

This Chapter represents the reference of the plenary talk of R. Cellini at the 17th ISDG (Urbino, July, 2016); the content is based on Brekke et al. [3, 4] and Siciliani et al. [46]. We thank the participants at the ISDG, and the colleagues that provided us with comments and criticism on this work and the articles upon which it bases. In particular, we would like to mention T. Basar, G. Feichtinger, L. Grilli, F. Lamantia, G. Zaccour for helpful comments. The responsibility remains on the authors alone.

K.R. Brekke
Department of Economics, Norwegian School of Economics, Helleveien 30, N-5045 Bergen, Norway
e-mail: kurt.brekke@nhh.no

R. Cellini (✉)
Department of Economics, University of Catania, Corso Italia 55, I95129 Catania, Italy
e-mail: cellini@unict.it

L. Siciliani
Department of Economics and Related Studies, University of York, Heslington, York YO10 5DD, UK
e-mail: luigi.siciliani@york.ac.uk

O.R. Straume
Department of Economics/NIPE, EEG, University of Minho, Braga, Portugal; Department of Economics, University of Bergen, Bergen, Norway
e-mail: o.r.straume@eeg.uminho.pt

© Springer International Publishing AG 2017
J. Apaloo, B. Viscolani (eds.), *Advances in Dynamic and Mean Field Games*,
Annals of the International Society of Dynamic Games 15,
https://doi.org/10.1007/978-3-319-70619-1_7

1 Introduction

In this paper we present a critical review of some differential game models that investigate quality competition between health-care providers. Our aim is to highlight the usefulness of differential games for understanding and analysing specific characteristics of health-care markets. This might, in turn, help policymakers to improve the institutional setting of these markets and thereby to achieve a more favourable outcome in terms of the quality of the services offered.

The importance of quality competition in health-care markets is underlined by the fact that, in recent years, several countries have implemented various types of reforms that introduce some aspects of competition between health-care providers. Among the intended aims of these reforms is the promotion of increased availability and quality of health-care services. In particular, the combination of prospective payment systems and free patient choice aims at giving health-care providers incentives to attract patients (and thus payments) by improving the quality of the services offered. We will show that differential games are able to capture some relevant and important features of health-care providers' behaviour over time, with potentially important implications for the effect of competition on quality provision and for the optimal regulation of such markets.

Health systems affect our everyday lives and constitute an important aspect of our societies. The consumption of health services represents a source of great concern, for individuals (on the demand side) as well as for providers and policymakers (on the supply side). According to OECD [42], health-care services and products (including preventive, curative and rehabilitative treatments, home and hospital services, long-term care; in sum, medical treatments and pharmaceutical products) account for 9–15% of GDP in OECD countries, peaking at 17% in the USA. Thus, developing a better understanding of the key mechanisms that determine the quantity and quality of health-care provision is crucial for the design and regulation of health-care institutions. To this end, differential games as an analytical tool are less diffused in the field of health economics than it perhaps ought to be. This is surprising but, at the same time, promising for future advances in the field. It is surprising because individual and social choices concerning both the demand and the supply for health goods and services take place in a dynamic context. Hence, differential games provide a natural framework to analyse individual choices and social dynamics in this field. The promise lies in the ability of differential game arguments to explain and analyse important features of existing health-care systems and thereby provide new and relevant policy recommendations.

In the present paper, we concentrate on the supply side of health-care service provision, taking an 'industrial organisation' approach with strategic interaction between health-care providers (hospitals) in a context of regulated markets. In Section 2 we discuss the relation between competition and quality in health-care markets, with specific reference to hospital competition, before introducing the basic elements of a unified analytical framework in Section 3. This framework is common to three specific model variants, analysed in Section 4. These variants correspond to the models presented in Brekke et al. [3, 4] and Siciliani et al. [46]. For each of the three model variants, we describe the individual optimal choice of providers under different solution concepts (open-loop and feedback closed-loop information structures) and compare the outcomes with respect to quality provision. We also discuss the efficiency properties of the market outcome with respect to the socially optimal allocation. In Section 5 we mention some possible extensions to our main framework, before offering some concluding remarks in Section 6.

2   Quality and Competition in Health-Care Markets

In health-care markets, consumers typically do not pay the full cost of goods and services. Since payments are mostly made by the public sector or insurance companies, a patient's choice of health-care provider is usually made on the basis of non-price criteria such as *quality*, which is arguably the most important aspect of the goods and services under evaluation. In addition, travel distance, habits and trust in specific providers are also likely to be important aspects. Prices are typically regulated (at least in most Western countries) and fixed.[1] These characteristics affect the way providers compete. If consumers make their purchasing decisions mainly based on quality, a provider can attract more consumers by increasing the quality of the goods and services offered.

The body of existing theoretical literature on quality competition in health-care markets with fixed prices is generally based on static models in a setting of spatial competition. These analyses are either based on a Hotelling model with two providers (e.g. Calem and Rizzo [12]; Lyon [36]; Gravelle and Masiero [28]; Beitia [1]; Montefiori [39]; Brekke, Nuscheler, and Straume [6]; Karlsson [31]) or a Salop model with n providers (e.g. Gravelle [27]; Nuscheler [41]; Brekke, Siciliani, and Straume [8]). In the former type of model, the effect of competition on equilibrium quality provision is typically modelled as a reduction in patients' transportation costs, which makes each provider's demand more quality-elastic. The latter type of model also enables the use of the number of providers as an additional competition measure. Regardless of whether stronger competition comes along with a reduction in transportation costs or with an increase in the number of providers, the typical result emanating from the theoretical literature is that more competition leads to higher-quality provision when prices are fixed (regulated). A transportation cost reduction makes patient choices more sensitive to quality changes and therefore stimulates each provider to increase quality in order to attract

---

[1]Countries where price is not fixed for segments of the health-care market include the USA, Switzerland, France, and the Netherlands. Many issues need to be taken into account for the question of whether prices should be regulated or not in the health-care markets. One important issue is how endogenous pricing affects quality provision and whether competition along both dimensions (price and quality) will lead to a socially optimal quality provision or not. See Brekke, Siciliani, and Straume [10] for a discussion.

more patients. The same is true for a higher number of providers (which implies a smaller distance between each provider) if transportation costs are convex in distance.[2] The positive relationship between competition and quality is also in line with more general theoretical studies of quality competition in regulated markets.[3] One notable exception to this result is provided by Brekke, Siciliani, and Straume [8] who show that the positive relationship between competition and quality might be reversed (regardless of which competition measure is used) if health-care providers have semi-altruistic preferences since providers may be working at a negative profit margin for the marginal patient, therefore competing to avoid rather than attract patients.[4]

It is worth stressing that the static framework overlooks potentially important dynamic issues related to quality. For instance, static models assume that demand responds immediately to quality changes. This assumption is unrealistic, as long as demand tends to be sluggish, which implies that, if a provider increases its quality, it will take some time before the potential demand increase is fully realised. Such demand sluggishness typically arises for two different reasons. The first is imperfect information on the demand side, which is particularly relevant in markets where quality is the main competition variable: while prices usually are easily and immediately observable, quality is often less readily observable and much more difficult to measure. Health goods and services typically involve asymmetric information, and they can be interpreted as experience goods, or credence goods, or even merit goods. The second is sticky behaviour of consumers, motivated by personal or familiar habits or by trust or confidence in one specific provider. Consider, for example, the case of people who choose a dentist simply because relatives went to her/him. Moreover, the relational content in the service exchange between provider and consumer in health-care markets plays an important role in making demand sluggish.

The above discussion indicates the need for specific dynamic models of competition that incorporate the institutional and behavioural idiosyncrasies of health-care markets. In the next section, we present a unified framework for a theoretical analysis of quality competition in health-care markets under regulated prices, taking a differential game approach. Subsequently, there are three variants of the model differing in key assumptions. In the first model, we assume that quality is a stock

---

[2]Even with linear transportation costs, a higher number of providers can lead to higher-quality provision if treatment costs are convex. In this case, a higher number of providers, which leads to lower demand for each provider, will increase the price-cost margin and therefore give each provider a stronger incentive to attract patients by raising quality.

[3]See, e.g. Ma and Burgess [37], Wolinsky (1997) and Brekke, Nuscheler, and Straume [5].

[4]The empirical evidence, however, is ambiguous: Kessler and McClellan [32] and Tay [47] find a positive effect, Gowrisankaran and Town [25] find a negative effect and Shen [44] finds mixed effects, while Shortell and Hughes [45] find no significant effects of competition on quality under regulated prices. Gaynor [22] offers a survey of theoretical and empirical literature on the relationship between competition and quality in the health sector. Recent evidence from England suggests that competition increases quality if prices are fixed ([15, 23]; Bloom et al. 2016).

variable which increases if investment is higher than quality depreciation and the marginal treatment cost is weakly increasing. In the second model, we assume that demand is sluggish and marginal treatment cost is again weakly increasing. In the third model, we assume that the provider of health services is altruistic but the marginal treatment cost is constant. It is worth underlining that the differential game approach permits stronger competition to occur not only through lower transportation costs or a larger number of providers (like in static models) but also through different channels, such as less demand sluggishness or stronger interaction between the providers via feedback solutions.

3 The Basics of the Model

Consider a market with two health-care providers located at either end of the unit line $S = [0, 1]$. On this line segment, there is a uniform distribution of consumers (i.e. patients), with total mass normalised to 1. Assuming unit demand of each consumer, the utility of a consumer who is located at $x \in S$ and buys from Provider i, located at $z_i \in \{0, 1\}$, is given by

$$U(x, z_i) = v + kq_i - \tau |x - z_i|, \tag{1}$$

where v is a positive parameter, q_i is the quality of the product (service) offered by Provider i, k is a parameter measuring the marginal willingness to pay for quality and τ is the marginal transportation cost. Notice that in the case of competition among hospitals, it makes sense to assume that physical locations are fixed. Since the distance between providers is equal to one, the consumer who is indifferent between i and j is located at x_i^D, implicitly given by

$$v - \tau x_i^D + kq_i = v - \tau \left(1 - x_i^D\right) + kq_j, \tag{2}$$

and explicitly given by

$$x_i^D = \frac{1}{2} + \frac{k\left(q_i - q_j\right)}{2\tau}. \tag{3}$$

This is also the notional (or 'optimal') demand for Provider i, $\widehat{D}_i$, given the assumptions of uniform consumer distribution (with mass 1), exogenous locations of providers and full market coverage. In what follows we set $k = 1$ without loss of generality.

Different variants of the model may assume that quality can be set by providers instantaneously at any instant of time (this is the case in Brekke et al. [4], and in Siciliani et al. [46]) or that quality changes over time due to specific investments by the providers (as in Brekke et al. [3]). In the latter case, the quality is a state variable, whose dynamics represent a state equation in the model. In the former case, where quality can be set instantaneously, the state variable is the demanded quantity (the demand), which is assumed to move sluggishly over time. More formally, we can have as the state dynamics of the problem of Provider i, *either* as

$$\frac{dq_i(t)}{dt} := \dot{q}_i(t) = I_i(t) - \delta q_i(t), \qquad (4)$$

where $I_i(t)$ is the investment in quality at time t, $\delta > 0$ is the depreciation rate of the quality stock and $q_i(0) = q_{i0} > 0$ is the initial condition, *or*

$$\frac{dD_i(t)}{dt} := \dot{D}_i(t) = \gamma(\widehat{D}_i(t) - D_i(t)), \qquad (5)$$

together with the initial condition $D_i(0) = D_{i0} > 0$, where D is the *actual demand*, as opposed to the *notional demand* $\widehat{D}$, and the parameter $\gamma > 0$ captures the demand sluggishness.[5]

Each provider has a cost function $C(\cdot)$, which, at each point in time, depends on the quantity, the investment and possibly the quality level itself. For analytical tractability, the cost function is assumed to be separable and linear or quadratic in any argument. It is worth emphasising that the linear *vs.* quadratic form in quantity has a correspondence to the institutional context: in more regulated systems (such as the UK, Spain and Italy), there is typically excess demand for hospital services, and capacity constraints do matter: in this case, capacity constraints are properly captured by convex costs.[6] In less regulated systems (such as the USA, Germany and France), there is often excess supply and hence relatively constant marginal cost of treatment, which corresponds to the linearity assumption.

The objective function to maximise is the present value of current and future profit, over an infinite time horizon,[7] and actualisation is exponential with a constant rate $\rho > 0$. However, we also allow for the possibility that providers are to some extent *motivated*, which means that they are not purely profit-oriented and exhibit semi-altruistic preferences. For example, physicians are typically portrayed as 'imperfect agents' for their patients, trading off patient benefits against lower profits (see, e.g. McGuire [38]). The assumption of 'mission-oriented' providers (doctors, nurses) implies that the agents (hospitals) to some extent share the objectives of the

---

[5]The lower is γ, the more sluggish is demand: the parameter γ is therefore an inverse measure of the degree of demand sluggishness in the market. When time is discrete, it has to be assumed that $0 < \gamma < 1$ to avoid that the adjustment is in excess to the disequilibrium; in continuous time, it is sufficient to assume $\gamma > 0$.

[6]Rigid constraints, corresponding to assumptions, e.g. $D \leq \bar{D}$ and $\lim_{D \to \bar{D}} C(D, \ldots) = +\infty$ are rare to observe in the real world. Hospitals can expand capacity by, for instance, hiring more doctors or nurses or inducing existing staff to work overtime.

[7]In reality, players may not have an infinite time horizon but reasonably long finite horizons. However, in general, the optimal path does not differ significantly from the solution with a very large but finite horizon, and the convenience of working with an infinite horizon model may be worth the loss of realism (see Léonard and van Long [35], p. 285).

principal (consumers or government).[8] Hence, in the objective function of providers, other components may appear beyond profit. Motivation is usually modelled by assuming that the objective function is a weighted average of profit and consumer surplus. As we will show later, the presence of provider motivation might lead to a reversal of the established positive relationship between competition and quality.

In this framework, two reasonable measures of *competition intensity* are easily available: travelling cost and/or the degree of demand sluggishness. Indeed, the degree of competition increases if travelling costs become lower (a decrease in τ) and/or if demand becomes less sluggish (an increase in γ). Notice that both these measures can be influenced by policy. For example, by making publicly available quality indicators measuring the performance of hospitals, the government can increase consumer awareness about quality and thereby reduce demand sluggishness.[9] Furthermore, full or partial reimbursement of travel and stay expenses related to health-care should make travel distance less important (relative to quality) and thereby increase the intensity of competition among different providers. Clearly, an increase in the number of providers could be an alternative way of measuring the effect of stronger competition. However, in the case of hospitals, the number of providers is a flexible policy variable only in the long run. Moreover, from an analytical point of view, this would require to move from model *à la* Hotelling to model *à la* Salop [43], which goes beyond the scope of this paper.[10]

Within the framework at hand, the next section illustrates the results from three specific variants of the model. In the first variant, providers are purely profit-oriented, and the interest is in capturing the effect of dynamic strategic interaction on quality provision. We will show that the solution under the feedback closed-loop information structure (or, simply, the feedback solution) entails dynamic effects that are overlooked in a static game and also by the solution under the open-loop information assumption. The dynamic interaction entails smaller advantages from competition with respect to increases in quality provision. The second variant of the model focuses on the case of sluggish demand. The interest is in studying how the quantity of services offered by a provider affects the quality of its service. We will show that a truly dynamic perspective of analysis leads to an opposite result

---

[8]For an introduction to the literature on motivated agents in the broader public sector, see, e.g. Francois [21] and Glazer [24]; for a specific focus on workers' motivation in the health-care sector, see, inter alia, Ellis and McGuire [20], Eggleston [19] and Kaarbøe and Siciliani [30]. Brekke, Siciliani and Straume [9] offer an extensive discussion of the assumption of motivated providers, with further references to relevant literature, including experimental evidence.

[9]In the USA, the State of New York was among the first to introduce 'cards for hospital quality' and for this reason has been intensively investigated in the empirical literature. There is evidence that market shares may be influenced by report cards with providers with better reports having larger market shares. Cutler et al. [16] show that hospitals with high mortality rates experience a 10% reduction in coronary bypasses, but this is not the case for hospitals with low mortality rates. Mukamel et al. [40] also find that higher mortality rates reduce market shares. Dranove and Sfekas [18] find that hospitals with bad reports have a smaller market share but only after accounting for the prior beliefs of the patients.

[10]See Section 4.2. of Siciliani et al. [46] for a model under the open-loop solution.

with respect to static model, as to the relation between quality and quantity. The third variant shows how conclusions change when semi-altruistic motivations of providers are taken into account. In particular, we will see that the presence of an altruistic motivation in the objective function of providers impacts on the steady-state levels of relevant variables and also on the transition to steady state and hence on the optimal policy of price regulation.

For each variant of the model, we will consider: (a) the *open-loop* solution, where players know the initial state of the world, set the optimal plan at $t = 0$ and then stick to it forever. Hence, the plan of the control variable only depends on time and initial conditions; (b) the *feedback* solution, where players are assumed to be able to observe the evolution of the state(s) and where the control variable(s) depend(s) on the current state (and *only* on the current state under the Markovian assumption).[11] We will focus on steady states, compare the outcome from these different assumptions and discuss their characteristics.

4   Three Specific Models

4.1   Purely Profit-Oriented Providers

We examine here the case analysed in Brekke et al. [3], where providers are purely profit-oriented, so that the objective function is the maximisation of the discounted value of current and future profits, $\int_0^{+\infty} \pi_i(t) e^{-\rho t} dt$, and instantaneous profit at any instant t is $\pi_i = T + pD_i\left(q_i(t), q_j(t)\right) - C\left(I_i(t), D_i\left(q_i(t), q_j(t)\right), q_i(t)\right) - F$, where p is the regulated price per treatment/patient, T is a potential lump-sum transfer received from a third-party payer and F denotes possible fixed cost. Quality moves over time according to (4), and investment in quality $I_i(t)$ is the choice variable in the dynamic problem of provider/player i. Hence, the problem of player i is

$$\underset{I_i(t)}{\text{Maximise}} \int_0^{+\infty} [T + px_i^D\left(q_i(t), q_j(t)\right) - C\left(I_i(t), x_i^D\left(q_i(t), q_j(t)\right), q_i(t)\right) - F]e^{-\rho t} dt,$$

$$(6)$$

$$\text{s.t. } \dot{q}_i(t) = I_i(t) - \delta q_i(t); \quad \dot{q}_j(t) = I_j(t) - \delta q_j(t),$$

$$q_i(0) = q_{i0} > 0; \quad q_j(0) = q_{j0} > 0$$

where x_i^D is given by (3). The open-loop solution of this problem is standard. We denote by $\mu_i(t)$ and $\mu_j(t)$ the current value co-state variables associated

---

[11]We refer to Dockner et al. [17] for a general discussion about pros and cons of different solution concepts; in Brekke et al. [3, 4] and Siciliani et al. [46], there is a short discussion applied to the specific cases at hand.

with the two state equations and formulate the current-value Hamiltonian H. The solution is given by (a) $\partial H_i/\partial I_i = 0$, (b) $\dot{\mu}_i = \rho\mu_i - \partial H_i/\partial q_i$, (c) $\dot{q}_i = \partial H_i/\partial \mu_i$, (d) $\dot{\mu}_j = \rho\mu_j - \partial H_i/\partial q_j$, to be considered along with the transversality condition $\lim_{t\to+\infty} \mu_i(t)q_i(t) = 0$. The second-order conditions are satisfied if the Hamiltonian is concave in the control and state variables [35].

Assuming a symmetric equilibrium (justified by the fact that providers are identical and symmetrically located), totally differentiating FOCs and manipulating them, we obtain the dynamics of investment and quality in equilibrium. The linearised form around the steady state can be represented in matrix form as follows:

$$
\begin{bmatrix} \dot{I}(t) \\ \dot{q}(t) \end{bmatrix} = \begin{bmatrix} (\delta + \rho) - \frac{C_{I_i I_i q_i}}{C_{I_i I_i}}(I_i - \delta q_i) & \dfrac{\left(\left(\frac{1}{2\tau}\right)^2 (C_{x_i x_i}) + (2\delta + \rho)C_{I_i q_i} \atop +C_{q_i q_i} - C_{I_i q_i q_i}(I_i - \delta q_i)\right)}{C_{I_i I_i}} \\ 1 & -\delta \end{bmatrix} \begin{bmatrix} I(t) \\ q(t) \end{bmatrix}
$$
$$
+ \begin{bmatrix} -(p - c)/(2\gamma\tau) \\ 0 \end{bmatrix}, \tag{7}
$$

where the subscripts of C denote the partial derivative arguments. The 2x2 matrix is the Jacobian J of the dynamic system.

Suppose that the cost function is given by

$$
C(I_i, x_i^D, q_i) = \frac{c}{2}\left(x_i^D\right)^2 + \frac{\theta}{2}I_i^2 + \frac{\beta}{2}q_i^2 + \varphi q_i I_i. \tag{8}
$$

We therefore assume that both quality and *investment* in quality increase costs directly. Investments in quality involve buying some new, more expensive and technologically advanced machineries (e.g. MRI). But once a higher level of quality is achieved, the cost of running such machines on a regular basis may also be higher if they involve specialised products and maintenance; they may also require more specialised staff (e.g. technicians) who can make them work. The latter assumption is however not critical since all our results hold in the case when quality does not directly affect costs, i.e. $\partial C/\partial q_i = 0$ (or $\beta = \varphi = 0$ in the linear-quadratic specification).

In this case, the steady state is

$$
I_{OL}^s = \frac{\left(p - \frac{c}{2}\right)\delta}{2\tau\left[\beta + (\delta + \rho)\theta\delta + \varphi(2\delta + \rho)\right]}; \quad q_{OL}^s = I_{OL}^s/\delta. \tag{9}
$$

Clearly (with $p > c/2$), lower travel costs (τ) or higher price (p) increases steady-state quality and investment. As to the dynamics, this equilibrium is a saddle point if the $\dot{I}_i = 0$ locus is negatively sloped in the (q, I) space and also if $\dot{I}_i = 0$

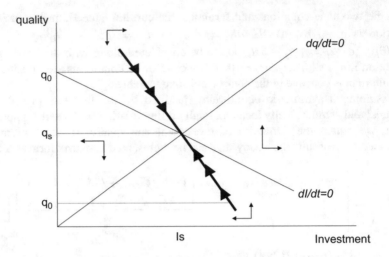

Fig. 1 Dynamics of quality and investment

is positively sloped, as long as it is not too positively sloped.[12] The saddle path towards the steady state is negatively sloped in the same space. Suppose we start off equilibrium, at level $q_0 < q_{OL}^s$. The solution is then characterised by a period of increasing quality and decreasing investment in quality. Suppose instead that $q_0 > q_{OL}^s$: in this case, a period of decreasing quality and increasing investment is observed (see Figure 1). Hence, off steady-state investment and quality move over time in opposite directions, while the steady-state investment is proportional to quality ($I_i = \delta q_i$).

Now let us move to the *feedback* solution. The problem has a linear-quadratic structure (irrespective of the linear or quadratic form of the cost function), so that the standard technique resorting to the Hamilton-Bellman-Jacobi equation can be used, positing a linear solution rule for any player. To ease computations, we assume that quality does not affect the marginal cost of investment, i.e. $\varphi = 0$, and we also set $\theta = 1$, which is a normalisation without loss of generality. With this specification, the optimal investment rule for Provider i is

$$I_i(t) = \phi_i^{FB}(q_i(t), q_j(t)) = \sigma_1 + \sigma_3 q_i(t) + \sigma_5 q_j(t), \tag{10}$$

where

$$\sigma_1 = \frac{2p - c}{4\tau} \frac{1}{\delta + \rho - \sigma_3} > 0, \tag{11}$$

---

[12]This case arises, for example, when the marginal cost of quality and the marginal cost of provision are constant but quality and investment are complements, which implies $C_{I_i q_i} < 0$. The equilibrium is still a saddle point if $\det(J) = -\delta(\delta + \rho) - (2\delta + \rho)C_{I_i q_i}/C_{I_i I_i} < 0$. If $\det(J) > 0$, then we have an unstable node.

$$\sigma_3 = \left(\frac{\rho}{2}+\delta\right) + \frac{18\tau^2}{c}\sigma_5^3 - \frac{\tau^2\left(5c + 16\beta\tau^2 + 16\tau^2\left(\frac{\rho}{2}+\delta\right)^2\right)}{2c}\sigma_5 < 0, \tag{12}$$

$$\sigma_5 = \frac{1}{6}\sqrt{\frac{3c}{\tau^2} + 8\left(\beta+\left(\frac{\rho}{2}+\delta\right)^2\right) - 4\sqrt{\left(\beta+\left(\frac{\rho}{2}+\delta\right)^2\right)\left[\frac{3c}{\tau^2}+4\left(\beta+\left(\frac{\rho}{2}+\delta\right)^2\right)\right]}} > 0. \tag{13}$$

The only fully stable steady-state solution[13] under the feedback rule is given by

$$I_{FB}^s = \frac{\delta(2p-c)}{4\tau\left(\delta+\rho-\sigma_3\right)\left(\delta-\sigma_3-\sigma_5\right)} > 0; \quad q_{FB}^s = I_{FB}^s/\delta > 0. \tag{14}$$

Also in this case, steady-state investment and quality levels increase with price and decrease with travel cost. Moreover, it is easy to check that the steady-state investment and quality in the feedback solution are lower than the corresponding investment and quality in the open-loop solution, under the *same* cost structure, i.e. $I_{OL}^s > I_{FB}^s$ and $q_{OL}^s > q_{FB}^s$, provided that $\varphi = 0$ and $\theta = 1$. With convex treatment costs, marginal revenues depend on rivals' quality levels, since quality changes will affect demand and, in turn, the marginal treatment cost faced by each provider. This introduces a strategic interaction in the sense that the optimal investment rule for Provider i depends, at each point in time, on the quality stock of Provider j. This dynamic interaction is reflected in the parameter $\sigma_5 > 0$, meaning that quality investments are strategic complements. More precisely, there exists an intertemporal strategic complementarity among the variables, according to the definition provided by Jun and Vives [29], as long as the control of a player responds positively to a change in the state pertaining to the opponent. From a substantial point of view, a quality increase by Provider i will shift demand away from j, implying that the marginal cost of Provider j decreases. Since the price is constant, this increases the profit margin of j, making quality investments more profitable on the margin for this Provider. Conversely, a decrease in quality investment by Provider i will invoke an investment-reducing response by j. From the viewpoint of Provider i, the instantaneous loss in market share by reducing the supply of quality is weighed against the future gain of a quality reduction by j. As long as the players value future profits, the dynamic strategic interaction will drive the supply of quality in the market to a lower level in steady state, compared with competition based on open-loop rules, where this form of dynamic strategic interaction is absent.

Not surprisingly, the outcome is different if treatment costs are linear, that is, if the cost function is given by $C(I_i, x_i^D, q_i) = cx_i^D + \frac{\theta}{2}I_i^2 + \frac{\beta}{2}q_i^2 + \varphi q_i I_i$. In such case, it is easy to check that the fully stable steady-state equilibrium under the feedback solution coincides with the steady state of the open-loop solution. This coincidence result emerges because there is a constant marginal instantaneous revenue gain of quality investments, implying that the optimal dynamic investment rule for each

---

[13]There are four possible solutions of which three are unstable. We focus on the stable solution.

player is independent of the quality level provided by the other player. In other words, the optimal investment path for Provider i does not depend on the quality stock of Provider j, implying an absence of strategic interaction in this respect.

4.2   Sluggish Demand

In this variant of the model (which is borrowed from Brekke et al. [4]), we assume that quality levels can be set instantaneously by providers, without ad hoc investment, so that $q_i(t)$ and $q_j(t)$ are control variables rather than states. However, demand $D_i(t)$ and $D_j(t)$ are sluggish and represent the states. Notice that only one state variable is relevant, $D := D_i$, since $D_i + D_j = 1$, and hence $D_j = 1 - D$. In particular, D is the *actual demand* of provider i at time t, and $\widehat{D}(t)$ is the *notional demand*. Actual demand evolves over time according to (5), to be considered along with the initial condition $D(0) = D_0 \in [0, 1]$. The objective function remains (6), apart from the fact that there are no investments to increase quality levels. We assume that the cost structure corresponds to function (8), but with $\theta = \varphi = 0$ to reflect that quality can be adjusted instantaneously. In this case, the direct dependence of costs on the quality level is straightforward: higher quality, which is the key control variable, increases fixed costs and through increases in demand also increases the variable costs. The dynamic problem of Provider i is then given by

$$\underset{q_i(t)}{\text{Maximise}} \int_0^{+\infty} [T + pD(t) - \frac{c}{2}(D(t))^2 - \frac{\beta}{2}q_i^2 - F](t)\, e^{-\rho t} dt, \tag{15}$$

$$\text{subject to } \dot{D}(t) = \gamma(\widehat{D}(t) - D(t)) = \left(\frac{1}{2} + \frac{q_i(t) - q_j(t)}{2\tau} - D(t) \right),$$

$$D(0) = D_0.$$

Let $\mu_i(t)$ be the current value co-state variable associated with the state equation and the transversality condition $\lim_{t \to +\infty} e^{-\rho t} \mu_i(t) D(t) = 0$. The open-loop solution is standard, and the equilibrium dynamics, as evaluated around the steady state, for the problem of Provider i, are given by

$$\begin{bmatrix} \dot{q}_i(t) \\ \dot{D}(t) \end{bmatrix} = \begin{bmatrix} (\rho + \gamma) & \frac{\gamma c}{2\tau\beta} \\ 0 & -\gamma \end{bmatrix} \begin{bmatrix} q_i(t) \\ D(t) \end{bmatrix} + \begin{bmatrix} -\frac{\gamma p}{2\tau\beta} \\ \frac{\gamma}{2} \end{bmatrix}. \tag{16}$$

It is immediate to see that the Jacobian matrix has a positive trace and a negative determinant, implying that the equilibrium is stable in the saddle sense. The steady-state levels of relevant variables are

$$D_{OL}^s = \frac{1}{2}; \quad q_{OL}^s = \frac{\gamma}{\rho + \gamma} \cdot \frac{p - (c/2)}{2\tau\beta}. \tag{17}$$

If the price is above the marginal cost, then lower transportation costs (τ) or a higher price (p) increases quality in steady state, as we would expect. Moreover, a higher marginal cost of quality (β), a higher marginal cost of provision (c) or a higher time preference discount rate (ρ) reduces quality. Steady-state quality is also decreasing in the degree of demand sluggishness (measured by γ^{-1}).

As to the saddle path leading to the steady state, it is negatively sloped in the (D, q) space. On this equilibrium path, demand and quality move in opposite directions over time as long as the marginal cost of provision is increasing (i.e. $c > 0$). Suppose we start off steady state at a level where the initial demand is low: $D(0) < D^s_{OL}$ (this could correspond to the case of a provider who at time 0 enters a previously monopolistic market). The solution is then characterised by a period of increasing demand and decreasing quality. The relation observed over the time of transition to the steady state is the opposite of what emerges from static models, where the comparative statics typically establish a positive association between quality and demand.

Now let us consider the feedback solution. The structure of the problem is still linear-quadratic, and the standard solution employing the Hamilton-Bellman-Jacobi technique leads to the following linear Markovian strategies (assuming symmetry and confining the attention to the fully stable solution):

$$q_i = \phi_i(D) = \frac{\gamma}{2\tau c}(\sigma_1 + \sigma_2 D), \tag{18}$$

$$q_j = \phi_j(D) = \frac{\gamma}{2\tau c}(\sigma_1 + \sigma_2(1 - D)), \tag{19}$$

where

$$\sigma_1 = \frac{p + \frac{\gamma\sigma_2}{2}\left(1 - \frac{\gamma\sigma_2}{2\beta\tau^2}\right)}{\gamma + \rho - \frac{\gamma^2}{4\beta\tau^2}\sigma_2} > 0, \tag{20}$$

$$\sigma_2 = -\frac{2\beta\tau^2}{3\gamma^2}\left(\sqrt{(2\gamma + \rho)^2 + \frac{3c\gamma^2}{\beta\tau^2}} - 2\gamma - \rho\right) < 0. \tag{21}$$

Note that $\sigma_2 < 0$ implies a negative relationship between demand and quality over time along the equilibrium dynamic path, like in the open-loop solution. Note also that the quality difference at each point in time is $q_i - q_j = \gamma\sigma_2(D - 1/2)/\beta\tau$, implying that the provider with larger demand offers the higher quality. Differences vanish at the steady state, which is characterised by

$$D^s_{FB} = \frac{1}{2}; \quad q^s_{FB} = \left(\frac{\gamma + \rho}{\gamma} - \frac{\gamma\sigma_2}{4\beta\tau^2}\right)^{-1} \cdot \frac{p - (c/2)}{2\tau\beta}. \tag{22}$$

The comparative statics properties of this solution are qualitatively similar to the open-loop case: in particular, more competition – measured either by less demand sluggishness or lower transportation costs – will increase the steady-state level of quality.

It can be shown that steady-state quality is equal under the two solution concepts (open-loop and feedback) if costs do not depend on quality, i.e. $\beta = 0$. Otherwise, with $\beta > 0$ implying $\sigma_2 < 0$, steady-state quality is lower under the feedback solution than under the open-loop solution. The intuition is simple: under the feedback behaviour, a quality increase by Provider i will decrease the demand for Provider j and hence increase Provider j's profit margin. Qualities are strategic complements à la Jun and Vives [29]: the control of a player responds positively to a change in the state pertaining to the opponent ($\partial q_j / \partial D > 0$, $\partial q_i / \partial (1 - D) > 0$). Thus, under feedback behaviour, there is a dynamic incentive for each provider to reduce quality in order to induce future quality reductions – as a strategic response – from the competing provider.

4.3   Sluggish Demand and Providers' Motivation

In this third and final variant of the model, we focus on the effect of providers' motivation. This is a direct extension of the second model variant, where the instantaneous objective function of Provider i is assumed to be

$$\Omega_i(t) = T + pD(t) - C(D(t), q_i(t)) - F + \alpha B_i(q_i(t), D(t)). \tag{23}$$

The altruistic motivation is captured by the last term in (23). The degree to which providers are motivated is measured by $\alpha > 0$, while the function B denotes the surplus of consumers attending Provider i[14]:

$$B_i(q_i, D) := \int_0^D (v + q_i - \tau x)\, dx = (v + q_i)D - \frac{\tau}{2}D^2. \tag{24}$$

Note that, for a given level of demand, the marginal benefit of quality is $\partial B_i / \partial q_i = D$ and is thus increasing in demand.

In order to facilitate analytical tractability, we now assume that the cost structure is $C(I_i, x_i^D, q_i) = cx_i^D + \frac{\beta}{2}q_i^2$ (i.e. we assume linear treatment cost, joint with $\theta = \varphi = 0$). A set of parameter conditions is needed to avoid corner solutions and to meet the second-order conditions.[15]

---

[14]However, alternative choices are possible as far as the measurement of the altruistic component is concerned, for instance, by considering the average or the marginal utility of patients instead of the aggregate surplus. Different modelling choices may affect the linear-quadratic structure of the problem.

[15]In order to ensure that the second-order conditions are met, we assume that the parameters β and τ take values such that $\beta\tau > 1$. In order to ensure that steady-state quality is always non-negative, we have to assume that the price is above a certain threshold level; specifically, we assume that $p \geq \bar{p} := c - \alpha(v + \tau(1/2 + \rho/\gamma))$. A further parameter condition requiring α smaller than a threshold is sufficient to ensure stability of the open-loop solution (in the saddle sense) and the full stability of the feedback solution.

The solution under the open-loop information structure can be found, from the usual FOCs, adjoint equations and the transversality condition. Like in the previous variants, the symmetric equilibrium is unique and stable in the saddle sense, under meaningful parameter conditions. Steady-state demand and quality are given by

$$D^s_{OL} = \frac{1}{2}; \quad q^s_{OL} = \frac{2(p-c)\gamma + \alpha(\gamma(2v+\tau) + 2\tau\rho)}{4\beta\tau(\gamma+\rho) - 2\alpha\gamma}. \tag{25}$$

An interesting point that is easily derivable from this model is the relationship between competition intensity and quality in steady state:

$$\frac{\partial q^s_{OL}}{\partial \gamma} = \tau\rho \frac{2\beta(p-c) + \alpha(\alpha + \beta(2v - \tau))}{(\gamma(2\beta\tau - \alpha) + 2\beta\tau\rho)^2} > (<)0 \quad if \quad p > (<)p^{OL}_\gamma \tag{26}$$

$$\frac{\partial q^s_{OL}}{\partial(-\tau)} = \gamma \frac{4\beta(\gamma+\rho)(p-c+\alpha v) + \alpha^2(\gamma+2\rho)}{2(2\beta\tau(\gamma+\rho) - \alpha\gamma)^2} > (<)0 \quad if \quad p > (<)p^{OL}_\tau \tag{27}$$

where

$$p^{OL}_\gamma := c - \alpha\left(v - \frac{1}{2}\tau + \frac{\alpha}{2\beta}\right) \tag{28}$$

$$p^{OL}_\tau := c - \alpha\left(v + \frac{\alpha(\gamma+2\rho)}{4\beta(\gamma+\rho)}\right). \tag{29}$$

Thus, if providers face a positive price-cost margin ($p > c$), there is an *unambiguously positive relationship* between competition intensity and steady-state quality. However, this relationship is reversed if the providers face a sufficiently low price (lower than marginal production costs). If the price-cost margin is negative (and below a certain threshold level), the providers' optimal quality choices result from two counteracting incentives. On the one hand, the providers have an incentive to increase quality for altruistic reasons ($\alpha > 0$), but, on the other hand, they also have an incentive to reduce quality for profit-oriented reasons (since $p < c$). More quality-elastic demand, due to lower travelling costs or less sluggish demand, will strengthen both these incentives, but the profit incentive will increase more if the price is sufficiently low, implying that steady-state quality will decrease. Of course, if $p < c$ the activity is not profitable and even with motivated providers costs should be covered, to avoid the bankruptcy of the firm. This situation is not unusual in the real world, and lump-sum transfers from the Government are the rule in most countries.

Regarding the transition to steady state, let us define as $Q := q_i - q_j$ the difference in quality between the two providers. The dynamics of this quality difference are described by

$$\dot{Q} = \frac{1}{\beta}\left[\alpha(3\gamma + 2\rho)\left(\frac{1}{2} - D\right) + \left(\beta(\gamma+\rho) + \frac{\alpha}{2\tau}\gamma\right)Q\right]. \tag{30}$$

If the initial demand for Provider i is above one half ($D_0 > 1/2$), then the quality difference Q is strictly positive and converges towards zero as D converges towards the steady-state level. Intuitively, if the initial demand is above one half, the marginal benefit from quality is higher for Provider i as quality affects a larger number of consumers. Thus, Provider i has a stronger incentive than j to provide quality in the initial period of the game, implying a positive initial quality difference. However, on the equilibrium dynamic path, the quality difference may be sufficiently small such that $\widehat{D}(Q) < D_0$, implying that Provider i's potential demand is lower than its actual demand. As demand for Provider i reduces over time, this provider's incentive to invest in quality reduces correspondingly, while the opposite is true for the rival provider. This process continues until the steady state, where quality and demand differences vanish.

When solving the model under the feedback information structure, the following Markovian decision rules obtain

$$q_i = \phi_i(D) = \frac{\alpha}{\beta}D + (\sigma_1 + \sigma_2 D)\frac{\gamma}{2\tau\beta}, \tag{31}$$

$$q_j = \phi_j(D) = \frac{\alpha}{\beta}(1 - D) + (\sigma_1 + \sigma_2(1 - D))\frac{\gamma}{2\tau\beta}. \tag{32}$$

where

$$\sigma_1 = \frac{4\beta\tau^2(p - c + \alpha v) + 2\tau\gamma\sigma_2(\beta\tau - \alpha) - \gamma^2\sigma_2^2}{4\tau(\beta\tau(\gamma + \rho) - \alpha\gamma) - \gamma^2\sigma_2} \tag{33}$$

$$\sigma_2 = \frac{\tau}{3\gamma^2}\left((2\beta\tau(2\gamma + \rho) - 4\alpha\gamma) - \sqrt{\Delta}\right) < 0 \tag{34}$$

$$\Delta = (2\beta\tau(2\gamma + \rho) - 4\alpha\gamma)^2 + 12\alpha\gamma^2(\beta\tau - \alpha). \tag{35}$$

Note that $\partial q_i/\partial(1 - D) < 0$ and $\partial q_j/\partial D < 0$. Thus, also in this version of the model, qualities are *intertemporal strategic substitutes* [29], meaning that the control (quality) of each player responds negatively to a positive change in the state (demand) of the other player.[16]

The symmetric and fully stable steady-state solution in the feedback equilibrium is unique and given by

$$D_{FB}^s = \frac{1}{2}; \quad q_{FB}^s = \frac{12\beta\gamma(p - c + \alpha v) + 10\beta\alpha\tau\rho + 2\alpha\gamma(\beta\tau - \alpha) + \alpha\sqrt{\Delta}}{2\theta\left(8\gamma(\beta\tau - \alpha) + 10\beta\tau\rho + \sqrt{\Delta}\right)}. \tag{36}$$

As in the open-loop solution, steady-state quality is increasing in p and α, and a qualitatively similar relationship between competition intensity and steady-state

---

[16]It is easy to show that $\frac{\partial q_i}{\partial(1-D)} < 0$ and $\frac{\partial q_j}{\partial D} < 0$ if $\alpha < \frac{\beta\tau(2\rho + 5\gamma)}{2\gamma}$.

quality provision also obtains. That is, increased competition will increase (reduce) quality if the regulated price is above (below) a certain threshold level (lower than marginal treatment costs).

A comparison of steady-state levels of variables between the open-loop and feedback equilibria yields the following results: (i) if $\alpha = 0$ or $p = \widetilde{p}$, then $q_{FB}^s = q_{OL}^s$; (ii) if $\alpha > 0$ and $p > \widetilde{p}$, then $q_{FB}^s > q_{OL}^s$; and (iii) if $\alpha > 0$ and $p < \widetilde{p}$, then $q_{FB}^s < q_{OL}^s$, where $\widetilde{p} := c - \alpha \left(\frac{\alpha}{2\beta} + v - \frac{\tau}{2} \right) < c$.

The first result confirms the coincidence result obtained under pure profit-oriented behaviour and linear production costs, as shown in the previous model variants. The second and third results show that in regulated markets where motivated providers compete dynamically on quality, steady-state quality is higher in the feedback solution than in the open-loop one, if and only if the price is above a certain threshold level. Notice, however, that the threshold value of p is such that the price markup on marginal cost is negative, i.e. $\widetilde{p} < c$, and a lump-sum transfer is necessary to cover total costs.

The intuition for these results is related to how the presence of motivated providers affects the strategic nature of competition. Suppose that Provider i increases its quality. This reduces the number of consumers patronising Provider j and therefore also reduces j's marginal benefit of quality investments for altruistic reasons (i.e. $\partial B_j / \partial q_j$ is reduced). Consequently, Provider j responds by reducing its quality. Hence, qualities are *strategic substitutes* at each point in time. If the price is sufficiently high, $p > \widetilde{p}$, this strategic substitutability makes dynamic competition tougher in the feedback solution, where players can set their quality choices according to the evolution of demand and taking into account the strategic interaction at each instant of time. By increasing its quality today, Provider i can provoke a quality reduction from its competitor tomorrow. Due to this strategic nature of the dynamic competition and the lack of any form of commitment over time, steady-state quality turns out to be higher in the feedback solution. However, this conclusion holds only if each provider has a sufficiently strong incentive to increase demand.

If the price is sufficiently low, such that the price-cost margin is well below zero ($p < \widetilde{p} < c$), the incentive to attract more consumers for altruistic reasons is counteracted by incentives to dampen demand for profit reasons. In such a case, the incentives to compete for consumers are weak, and a more collusive outcome with lower steady-state quality levels is achieved under the feedback rule behaviour.[17]

---

[17]Note that provider motivation ($\alpha > 0$) still ensures that quality levels are positive even if providers face negative price-cost margins. With purely profit-oriented providers, interior solutions would not exist if $p < c$.

4.4   The Socially Optimal Solution

We would like to mention here that the model lends itself to finding the solution to the social optimum problem; that is, the optimal plans, if the objective function were the maximum social welfare, and a Social Planner would be able to control the choice variables of all players. Such a dynamic problem does not involve, of course, strategic interaction among providers. Limiting the attention to the last variant of the model, with sluggish demand and motivated providers, the Social Planner's problem would be given by

$$\underset{q_i,q_j}{\text{Maximise}} \int_0^{+\infty} W(t)\, e^{-\rho t} dt,$$

subject to (5) and initial conditions. This is a simple optimal control problem, without strategic interaction. $W(t)$ is the instantaneous social welfare, whose definition is not trivial. For instance, one can wonder whether, in the presence of motivated providers, the altruistic part of provider preferences should be included in the definition of social welfare (implying that consumer utility is to some extent 'double-counted') or not. Let us therefore make W flexible enough to incorporate both alternatives, by introducing a binary parameter $\eta = \{0, 1\}$, which can account for both double counting ($\eta = 1$) and no double counting ($\eta = 0$). Letting the opportunity cost of public funds be denoted by λ, social welfare can be expressed as

$$W = (1 + \eta\alpha) \left[\int_0^D (v + q_i - \tau x)\, dx + \int_D^1 (v + q_j - \tau(1 - x))\, dx \right]$$
$$- (1 + \lambda) \left[c + \frac{\beta}{2} (q_i^2 + q_j^2) + 2F \right].$$

Notice that the treatment price does not appear in this problem, since it only redistributes between patients and providers. The problem can be easily solved analytically, and the steady state, along with the dynamic path off steady state, can be found (see Siciliani et al. [46], for the details). In the steady state, demand and quality are, respectively, $D^* = 1/2$ and $q_i^* = q_j^* = \frac{1+\alpha\eta}{2\theta(1+\lambda)}$ (with stars denoting the social optimum) and are stable in the saddle sense. During the transition, quality and demand for a given provider move in the same direction on the socially optimal path. For $D > 1/2$, the quality for Provider i is higher than Provider j. Intuitively, if initially Provider i has more than half of the market, his marginal benefit from quality is higher (than for Provider j), as quality affects a larger number of consumers. As demand for Provider i reduces over time, the optimal quality for Provider i (j) reduces (increases). This process continues until the steady state where quality and demand differences vanish.

The socially optimal quality can be implemented by a regulator setting prices which lead individual providers (under the open-loop or feedback rule) to replicate the social optimum. Confining the attention to the steady state, one can compute the optimal price that ensures steady-state quality being at the first-best level. For instance, if providers follow the open-loop rule, the optimal price (i.e. the price p^{OL} such that $q^{OL}\left(p^{OL}\right) = q^*$) is

$$
p^{OL} = c - \alpha v + \frac{(1 + \alpha\eta)\,(2\theta\tau\,(\gamma + \rho) - \alpha\gamma) - \theta\alpha\tau\,(1 + \lambda)\,(2\rho + \gamma)}{2\gamma\theta\,(1 + \lambda)}. \tag{37}
$$

A different optimal price is appropriate if providers adopt feedback behaviour. The optimal regulated price is higher (lower) under the open-loop than the feedback behaviour if the motivation degree α is smaller (larger) than a threshold level. Interestingly, for sufficiently high motivation, the optimal regulated price can be lower than the marginal cost: for high level of motivation, the regulator still has to incentivise the provider, but the power of the incentive scheme is low. In turn, the pricing below the marginal cost requires that the provider is compensated for the losses with a positive lump-sum transfer (T) to complement the revenues from the low unit prices (p) and to ensure unit costs are fully covered. With no double counting $(\eta = 0)$, it is easy to show that the space for an optimal price below unit costs is larger if the providers use feedback rules, as compared to the open-loop solution.

In principle, one could also consider the whole dynamic path to the steady state: comparing the time path of the socially optimal quality with the time path of quality in the individual provider's problem, one could compute the path of the price, fixed by the regulator, that leads individual solutions to replicate the socially optimal dynamic path. In this case, the price regulation regime requires the regulator to be able to choose provider-specific and time-varying prices. Differently, one can explore the characteristics of more realistic regulatory regimes where, e.g., the regulated price is the same for both providers and is only allowed to vary with time.

5 Extensions

The framework laid out in the previous sections may also be extended and used to analyse related issues. Here we will briefly mention two such extensions.

Cellini and Lamantia [13] focus on the quality variable and take into account that in several circumstances minimum quality standards are in operation. This is relevant for the health sector where minimum quality standards are present (e.g. physician malpractice rules or requirements set out by regulators for offering hospitals a certification). This means that a constraint on q (such as $q_i(t) \geq q_{min} > 0$ for any i and t) has to be taken into account. This may affect the market equilibrium outcome, its stability property and the dynamic adjustment to equilibrium (the analysis is under discrete time). When there are multiple equilibria, they can coexist with more complicated attractors of the system that are created through so-called border collision bifurcations (Nusse and Yorke 1992). In some circumstances, minimum quality standards are responsible for the outcome of maximal differentiation in quality across providers. In this application, a very simple set of assumptions and a standard analytical context can produce a rich variety of outcomes.

Cellini et al. [14] introduce price competition, along with quality competition, allowing prices to be endogenously determined by providers. This can also be relevant for the health sector in some countries, for example, in the USA, where private market health-care providers compete on price and quality, or in the UK under the 'internal market' in the 1990s when hospitals competed on price to obtain a contract with a health authority. The interesting point to analyse is the interaction between price and quality competition. What are the effects of price competition on quality provision? Is it appropriate to allow providers to compete on price? The model provides three main results. First, steady-state quality under the feedback information structure is increasing in the degree of competition, as measured by a reduction in transportation costs. This is in contrast to the open-loop solution, where steady-state quality does not depend on the degree of competition (as would be the case in an equivalent static game). Second, steady-state quality is lower in the feedback than in the open-loop solution. The reason is that, in the former case, each firm has an incentive to reduce current quality investments in order to dampen future price competition. This incentive is absent in the open-loop solution, where the players do not interact strategically over time. Third, quality provision is socially optimal in the open-loop solution, while the feedback solution is characterised by under provision of quality in steady state. Thus, the dynamic interaction of feedback behaviour represents a source of inefficiency in quality provision.

6   Conclusions

In this paper we have offered a review of differential game models of quality competition in health-care markets under fixed (regulated) prices. We have proposed a unified framework with possible variants concerning the way in which the dynamics is introduced. This framework allows us to grasp relevant institutional and behavioural features of health-care markets, keeping the model simple with a linear-quadratic structure, which enables analytical solutions under the assumption of open-loop and feedback closed-loop information.

Demand sluggishness or the stickiness of quality is of particular relevance in the health-care sector, and hence a dynamic framework analysis is appropriate for the analysis of such markets. However, and more importantly, the differential game approach allows us to highlight how quality choices interact when providers make their decisions over time and to draw results which are relevant for policy. This is important for several countries whose governments have introduced policies aimed at increasing competition among health-care providers to improve quality of care. Static models generally formalise the increase in competition through an increase in the number of providers, and they generally drive to the conclusion that more competition fosters higher quality (though the opposite may arise when provider's altruism is relatively high). The differential game approach permits to investigate alternative measures of competition which are due to the environment in which providers operate and to the frequency of their interactions (open vs closed loop), and it allows to investigate intertemporal strategic interactions between providers. In a dynamic analysis, the intertemporal complementarity or substitutability across providers can reverse some of the conclusions obtained within a static framework.

The mechanisms at work are as follows: when a provider increases the current quality of its service, a larger demand is attracted which implies a reduction in demand for the competitors. With fixed regulated prices, competitors face higher or smaller marginal benefit from quantity and quality, depending on several factors, including cost structure, and possibly altruistic motivation. Thus, an increase of quality today may entail higher or lower quality tomorrow from the competitors.

We highlight the result that when the marginal treatment cost is increasing (i.e. in the presence of smooth capacity constraints), more competitive environments (as captured by the closed-loop solution) lead to lower quality in equilibrium. Moreover, if policymakers try to stimulate quality competition by publishing quality indicators to make demand less sluggish, this will have the intended effect only if the providers receive a sufficiently high unit price. Otherwise, if the price is too low (and combined with a lump-sum transfer, to ensure the provider does not make losses, which is a common feature of health-care payment system in many countries) and providers are altruistic, such policy measures may be counterproductive and lead to lower levels of quality in equilibrium.

The analysis of the transition to the steady state also shows that it can be usual to observe that efforts to increase quality investments, and the quality level itself, can move in opposite directions over time; the same may occur for quality and demand for a given provider, so that efforts and results move in opposite directions. These co-movements occur during the transition to the steady state and imply that caution needs to be exercised in evaluating policies through empirical observation.

Overall, results obtained from differential game models can help explaining the rational basis of specific forms of regulation and may help policymakers and regulation authorities to improve the efficiency of health-care systems and their evaluation.

References

1. Beitia, A., 2003. Hospital quality choice and market structure in a regulated duopoly. *Journal of Health Economics*, 22, 1011–36.
2. Bloom, N., C. Propper, S. Seiler, J. Van Reenen. 2015. The impact of competition on management quality: evidence from public hospitals. *The Review of Economic Studies*, 82, 457–489.
3. Brekke, K.R., R. Cellini, L. Siciliani, O.R. Straume, 2010, Competition and quality in health care markets: a differential-game approach. *Journal of Health Economics*, 29, 508–523.

4. Brekke, K.R., R. Cellini, L. Siciliani, O.R. Straume, 2012. Competition in regulated markets with sluggish beliefs about quality. *Journal of Economics & Management Strategy*, 21, 131–178.
5. Brekke, K.R., R. Nuscheler, O.R. Straume, 2006. Quality and location choices under price regulation. *Journal of Economics & Management Strategy*, 15, 207–27.
6. Brekke, K.R., R. Nuscheler, O.R. Straume, 2007. Gatekeeping in health care. *Journal of Health Economics*, 26, 149–70.
7. Brekke, K.R., L. Siciliani, O.R. Straume, 2010. Price and quality in spatial competition. *Regional Science and Urban Economics*, 40, 471–80.
8. Brekke, K.R., L. Siciliani, O.R. Straume, 2011. Hospital competition and quality with regulated prices. *Scandinavian Journal of Economics*, 113, 444–69.
9. Brekke, K.R., L. Siciliani, O.R. Straume, 2012. Quality competition with profit constraints, *Journal of Economic Behavior & Organization*, 84, 642–59.
10. Brekke, K.R., L. Siciliani, O.R. Straume, 2014. Can competition reduce quality? CEPR Discussion Paper No. 9810
11. Buratto, A., R. Cesaretto, R. Zamarchi, 2015. HIV vs. the Immune System: A Differential Game. *Mathematics*, 3, 1139–70.
12. Calem, P.S., Rizzo, J.A., 1995. Competition and specialization in the hospital industry: an application of Hotelling's location model. *Southern Economic Journal*, 61, 1182–98.
13. Cellini, R., F. Lamantia, 2015. Quality competition in markets with regulated prices and minimum quality standards. *Journal of Evolutionary Economics*, 25: 345–70.
14. Cellini, R, L. Siciliani, O.R. Straume, 2015. A dynamic model of quality competition with endogenous prices. NIPE working paper 2015/08.
15. Cooper, Z., S. Gibbons, S. Jones, S., A. McGuire, 2011. Does hospital competition save lives? Evidence from the NHS patient choice reforms. *Economic Journal*, 121(554), 228–260.
16. Cutler, D., R. Huckman, M.B. Landrum, 2004. The role of information in medical markets: an analysis of the piblicly reported outcomes in cardiac surgery. *American Economic Review*, 94, 342–46.
17. Dockner, E.J, S. Jørgensen, N. Van Long, G. Sorger, 2000. *Differential Games in Economics and Management Science*, Cambridge, Cambridge University Press.
18. Dranove, D., A. Sfekas, 2008. Start spreading the news: a structural estimate of the effects of New York hospital report cards. *Journal of Health Economics*, 27(5), 1201–07.
19. Eggleston, K., 2005. Multitasking and mixed systems for provider payment. *Journal of Health Economics*, 24, 211–23.
20. Ellis, R.P., T. McGuire, 1986. Provider behavior under prospective reimbursement: Cost sharing and supply. *Journal of Health Economics*, 5, 129–151.
21. Francois, P., 2000. Public service motivation' as an argument for government provision. *Journal of Public Economics*, 78, 275–299.
22. Gaynor, M., 2006. What do we know about competition and quality in health care markets? *Foundations and Trends in Microeconomics*, 2, 6.
23. Gaynor, M., R. Moreno-Serra, C. Propper, C. 2013. Death by market power: Reform, competition and patient outcomes in the British National Health Service. *American Economic Journal: Economic Policy*, 5, 134–166.
24. Glazer, A., 2004. Motivating devoted workers. *International Journal of Industrial Organization*, 22, 427–440.
25. Gowrisankaran, G., Town, R., 2003. Competition, payers, and hospital quality. *Health Services Research*, 38, 1403–1422.
26. Graf, J., 2016. The effects of rebate contracts on the health care system. *The European Journal of Health Economics*, 15, 477–87.
27. Gravelle, H., 1999. Capitation contracts: access and quality. *Journal of Health Economics*, 18, 315–40.
28. Gravelle, H., Masiero, G., 2000. Quality incentives in a regulated market with imperfect competition and switching costs: capitation in general practice. *Journal of Health Economics*, 19, 1067–88.

29. Jun, B., X. Vives, 2004. Strategic incentives in dynamic oligopoly. *Journal of Economic Theory*, 116, 249–81.
30. Kaarbøe, O., Siciliani, L., 2011. Multitasking, quality and Pay for Performance. *Health Economics*, 2, 225–238.
31. Karlsson, M., 2007. Quality incentives for GPs in a regulated market. *Journal of Health Economics*, 26, 699–720.
32. Kessler, D., McClellan, M., 2000. Is hospital competition socially wasteful? *Quarterly Journal of Economics*, 115, 577–615.
33. Kuhn, M., S. Wrzaczek, A. Prskawetz, G. Feichtinger,2015. Optimal choice of health and retirement in a life-cycle model, *Journal of Economic Theory*, 186–212.
34. Kuhn, M., C. Ochsen, 2009. Demographic and geographic determinants of regional physician supply. Thuenen Series of Applied Economic Theory Working paper 105, University of Rostock.
35. Léonard, D., Van Long, N., 1992. *Optimal control theory and static optimization in economics*. Cambridge: Cambridge University Press.
36. Lyon, T.P., 1999. Quality competition, insurance, and consumer choice in health care markets. *Journal of Economics & Management Strategy*, 8, 545–80.
37. Ma, C.A., Burgess, J.F., 1993. Quality competition, welfare, and regulation. *Journal of Economics*, 58, 153–173.
38. McGuire, T.G., 2000. Physician agency. In Culyer, A J., J.P. Newhouse (Eds.), *Handbook of Health Economics*. Amsterdam: Elsevier, 461–536.
39. Montefiori, M., 2005. Spatial competition for quality in the market for hospital care, *European Journal of Health Economics*, 6, 131–135.
40. Mukamel, D.B., D.L. Weimer, J. Zwanziger, et al., 2005. Quality report cards, selection of cardiac surgeons, and racial disparities: a study of the publication of the New York state cardiac surgery reports. *Inquiry*, 41, 435–46.
41. Nuscheler, R., 2003. Physician reimbursement, time-consistency and the quality of care. *Journal of Institutional and Theoretical Economics*, 159, 302–22.
42. OECD, 2016. *Health expenditure and financing*. www.oecd.stat. Accessed on July, 10, 2016.
43. Salop, S.C., 1979. Monopolistic competition with outside goods. *Bell Journal of Economics* 10, 141–156.
44. Shen, Y.S., 2003. The effect of financial pressure on the quality of care in hospitals. *Journal of Health Economics*, 22, 243–69.
45. Shortell, S.M., E.F. Hughes, 1988. The effects of competition, regulation, and ownership on mortality rates among hospital inpatients. *New England Journal of Medicine*, 318, 1100–07.
46. Siciliani, L., O.R. Straume, R. Cellini, 2013. Quality competition with motivated providers and sluggish demand. *Journal of Economic Dynamics and Control*, 37, 2041–71.
47. Tay, A., 2003. Assessing competition in hospital care markets: the importance of accounting for quality differentiation. *RAND Journal of Economics*, 34, 786–814.

Open-Loop Nash Equilibria for Dynamic Games Involving Volterra Integral Equations

Dean A. Carlson

Abstract In this paper, we consider a class of finite horizon dynamic games in which the state equation is a Volterra integral equation with infinite delay. This may be viewed as a game with coupled state and control constraints and is viewed as a game with coupled constraints in the spirit of Rosen. Our existence result is obtained by a fixed point argument using normalized equilibria. The results rely on convexity and seminormality conditions popularized by L. Cesari in the 1960s and 1970s as well as many others. As an example to which our results are applicable, we consider a competitive economic model originally appearing in the works of C. F. Roos in 1925.

1 Introduction

Differential games are normally thought of as being first studied in the late 1940s and early 1950s with the work of Rufus Isaacs [1]. These games were primarily two-player, zero-sum games. The study of nonzero sum games with $p \geq 2$ players, while studied earlier, was further promoted at approximately the same time when the Nobel laureate John Nash presented the first existence theorem for an equilibrium (now referred to as a Nash equilibrium) for nonzero sum p player games. Since that time, there have been great advancements in the theory of p-player nonzero sum differential games as well as the existence results for Nash equilibria of p-player games. The notion of a Nash equilibria is well known to have it roots in the work of A. Cournot in the 1830s, but the notion of a dynamic game had to wait until much later. Perhaps the earliest discussion of a dynamic game was in the 1920s when C. F. Roos considered competitive economic models in continuous time. In particular he considered the following model in 1925.

D.A. Carlson (✉)
Mathematical Reviews, American Mathematical Society, 416 Fourth Street,
Ann Arbor, MI 48103, USA
e-mail: dac@ams.org

© Springer International Publishing AG 2017
J. Apaloo, B. Viscolani (eds.), *Advances in Dynamic and Mean Field Games*,
Annals of the International Society of Dynamic Games 15,
https://doi.org/10.1007/978-3-319-70619-1_8

Example 1 *Assume that there are two producers, each manufacturing the same product at rates $u_1(t)$ and $u_2(t)$ subject to the same cost $q(u) = Au^2 + Bu + C$, in which $A, B, and C$ are positive constants with the linear demand law given by*

$$u_1(t) + u_2(t) = cp(t) + h\dot{p}(t) + d, \tag{1}$$

where $c < 0$, $d > 0$, $h < 0$, and $p(t)$ denotes the unit price of the product at time $t > 0$. Given a time interval $[a, b]$, the profit of each firm is given by the integral expression

$$\Pi_i(u_i(\cdot)) = \int_a^b p(t)u_i(t) - q(u_i(t))\, dt, \quad i = 1, 2.$$

With this model Roos considered two problems:

1. *Given $p(a) = p_a$, choose $u_i(\cdot)$ so that $\Pi_1(u_1(\cdot))$ is a maximum when $u_2(\cdot)$ is regarded as not subject to variation, and at the same time choose $u_2(\cdot)$ so that $\Pi_1(u_2(\cdot))$ is a maximum when $u_1(\cdot)$ is regarded as not subject to variation.*
2. *Given $p(a) = p_a$ and $p(b) = p_b$, choose $p(\cdot)$ so that $\Pi(u_1(\cdot))$ is maximized when $u_1(\cdot)$ varies only with respect to $p(\cdot)$ so that $\Pi_2(u_2(\cdot))$ is maximized when $u_2(\cdot)$ varies only with respect to $p(\cdot)$.*

The first of these problems is that of the producers choosing open-loop strategies $u_i^*(\cdot)$, $i = 1, 2$ so the dynamic game is in Nash equilibrium. Clearly this is a differential game with state variable $p(\cdot)$ and strategies $u_i^*(\cdot)$. The second problem is a dynamic game in which the price $p(\cdot)$ is the state variable and the strategies are $u_i(\cdot)$, but now the price is required to have a specific value at the end of the game. Again a Nash equilibrium is sought.

In today's parlance, we recognize these two games as linear quadratic open-loop games. In Roos's analysis, each of the unknowns $u_1(\cdot)$, $u_2(\cdot)$, and $p(\cdot)$ is treated as state variables, and the solution is obtained by applying calculus of variation techniques. From this he derives the corresponding "Euler-Lagrange equations" and solves them explicitly, thus arriving at the solutions to these games. For the details we direct the reader to Roos' paper [2].

In the above, the "new feature" is that the demand for the total amount of goods at time t is not only dependent on the price $p(t)$ but also on its rate of change $\dot{p}(t)$. In the same paper, Roos proposes the other possibility that the demand at time t for the total amount of goods depends on the price $p(t)$ and the past history of the price up to time t. That is, he proposes the linear demand law

$$u_1(t) + u_2(t) = ap(t) + b + \int_{-\infty}^t \Phi(t - s)p(s)\, ds, \tag{2}$$

in which $\lim_{t \to -\infty} p(t)$ is finite and $\Phi : [0, +\infty) \to \mathbb{R}$ is an appropriate kernel and considers the same two questions. After this paper, Roos [3] continues his study by introducing what he refers to as a "generalized problem of Lagrange." This problem is a dynamic p-player game in which the payoffs of each player are given by an integral functional of the form

$$I_i = \int_a^b F_i(t, u_1(t), \ldots, u_p(t), \dot{u}_1(t), \ldots, \dot{u}_p(t)) \, dt, \quad i = 1, 2, \ldots, p,$$

and the functions $u_i(\cdot)$ are required to satisfy the side constraint

$$G_k(t, u_1(t), \ldots, u_p(t), \dot{u}_1(t), \ldots, \dot{u}_p(t)) =$$

$$\int_a^b P_k(s, t, u_1(t), \ldots, u_p(t), \dot{u}_1(t), \ldots, \dot{u}_p(t)) \, ds, \quad k = 1, 2, \ldots, m.$$

For this problem, primarily considering the case $P \equiv 0$, he develops the classical theory for open-loop games including analogues of the Euler-Lagrange equations, the Weierstrass and Legendre necessary conditions, transversality conditions, as well as the Weierstrass sufficiency condition for a strong local equilibria and sufficient conditions for weak local equilibria. At the end of the paper, he indicates that the case when $P \not\equiv 0$ can be studied with the same technique of proof as for the case when $P \equiv 0$.

The papers of Roos motivated the developments given below in that there are no known existence theorems for open-loop Nash equilibria when the dynamics are governed by a controlled Volterra integral equation. Moreover, it is easy to see that our formulation includes the classical ordinary differential equation case as a special case. The plan of our paper is as follows. In Section 2, we formulate the p-player game we wish to consider, introducing the relevant notions of admissible trajectory/strategy pairs, describe the auxiliary parametrized optimal control problem, and show that the existence of a fixed point of a related set-valued mapping provides us with a Nash equilibrium. We conclude the section by stating the Bohnenblust-Karlin extension of Kakutani's fixed point theorem which will be used to establish our existence result. In Section 3 we study the auxiliary optimal control problem and discuss the growth conditions and lower closure results required to establish the existence of an optimal solution to the auxiliary optimal control problem. In Section 4 we give our main result which establishes conditions sufficient for the existence of an open-loop Nash equilibrium. In Section 5 we return to Roos's examples and show that under appropriate hypotheses, these models can be cast in the form considered here so that our existence result is applicable. In Section 6 we give some perspectives and some future directions for further study. Conclusions are given in Section 7. Finally, the proof of Theorem 6 is given in Appendix A and some results on measurable set-valued mappings and measurable selections are given in Appendix B.

2   The Dynamic Game Model

We consider a p-player continuous time dynamic game in which the dynamics is a control system described by the following Volterra integral equation

$$Y(t) = Y(a) + P(t) + \int_a^t h(t,s)Y(s) + g(t,s)f(s, Y(s), V(s))\, ds, \tag{3}$$

for $t \in (a, b)$, where $a < b$ are fixed given constants, $Y : [a, b] \to \mathbb{R}^N$ is the state variable and $V(t) = (v_1(t), v_2(t), \ldots, v_p(t))$ is a vector in $\mathbb{R}^M$, with $M = \sum_{k=1}^p m_k$, in which $v_k(t) \in \mathbb{R}^{m_k}$ represents the strategy of the kth player at time t. We also impose the constraints

$$(a, Y(a), b, Y(b)) \in \mathscr{B} \subset \mathbb{R}^{2N+2}, \tag{4}$$

$$(t, Y(t)) \in \mathscr{A} \subset \mathbb{R} \times \mathbb{R}^N \quad \text{for} \quad t \in [a, b], \tag{5}$$

$$V(t) \in \mathscr{U}(t, Y(t)) \subset \mathbb{R}^M \quad \text{a.e.} \quad t \in [a, b]. \tag{6}$$

We assume that the projection of $\mathscr{A}$ onto its first component is the bounded interval $[a, b]$ and that its projection on $\mathbb{R}^N$ is compact. This means that we can write $\mathscr{A} = [a, b] \times A$, in which $A \subset \mathbb{R}^N$ is compact. Define $\mathscr{M} = \{(t, Y, V) \in [a, b] \times \mathbb{R}^N \times \mathbb{R}^M : (t, Y) \in \mathscr{A}, \ V \in \mathscr{U}(t, Y)\}$ and assume that $\mathscr{M}$ is closed and the functions $f : \mathscr{M} \to \mathbb{R}^d$, $h : [a, b] \times [a, b] \to \mathbb{R}^{N \times N}$ and $g : [a, b] \times [a, b] \to \mathbb{R}^{N \times d}$ are all continuous. for all $i = 1, 2, \ldots, N$ and $j = 1, 2, \ldots L$.

To complete our game formulation, we define the objective of the kth player by the integral functional

$$J_k(Y, v_k) := \int_a^b f_k^0(t, Y(t), v_k(t))\, dt, \tag{7}$$

where $f_k^0 : \mathscr{M}_k \to \mathbb{R}$ is a continuous function on

$$\mathscr{M}_k = \{(t, Y, v) \in [a, b] \times \mathbb{R}^N \times \mathbb{R}^{m_k} : \exists\, u_j \in \mathbb{R}^{m_j} \text{ for } j \neq k$$

$$\text{such that } (t, Y, [U^{-k}, v]) \in \mathscr{M}\}.$$

Here the notation $[U^{-k}, v]$ denotes the vector in $\mathbb{R}^M$ whose components are u_j for $j \neq k$ and whose kth component is v. That is,

$$[U^{-k}, v] = (u_1, u_2, \ldots, u_{k-1}, v, u_{k+1}, \ldots u_p).$$

Remark 1 *Observe that when $P(t) \equiv 0$, $h(t, s) \equiv 0$, $g(t, s) = I_n$ the $n \times n$ identity matrix, and $d = N$ equation (3) becomes*

$$Y(t) = Y(a) + \int_a^t f(s, Y(s), V(s)) \, ds,$$

which is equivalent to the ordinary differential equation

$$\dot{Y}(t) = f(t, Y(t), V(t)), \quad Y(a) = a.$$

Thus, our formulation covers this important case as well.

Remark 2 *The above formulation includes the case when the components of the state variable, say $Y = (y_1, y_2, \ldots, y_p)$, represent the dynamics of each player. That is, $y_k : [a, b] \to \mathbb{R}^{n_k}$ is the dynamic for player k and $N = \sum_{k=1}^p n_k$.*
For cases other than this scenario we assume we can partition, perhaps artificially, the state vector $Y \in \mathbb{R}$ into a vector $(y_1, y_2, \ldots, y_p)$ with $y_k \in \mathbb{R}^{n_k}$ and $N = \sum_{k=1}^p n_k$. With this notation we have the following definitions.

Definition 1 *We say the pair of functions $\{Y, V\} : [a, b] \to \mathbb{R}^N \times \mathbb{R}^M$ is admissible for the game if $Y : [a, b] \to \mathbb{R}^N$ is continuous, $V = (v_1, v_2, \ldots, v_p) : [a, b] \to \mathbb{R}^M$ is Lebesgue measurable, the integral equation (3) and the constraints (4)–(6) are satisfied, and the functions $t \mapsto f_k^0(t, Y(t), v_k(t))$ are Lebesgue integrable on $[a, b]$ for $k = 1, 2, \ldots, p$. We denote the set of admissible pairs for the game by Ω which we will assume is nonempty.*
For an admissible pair $\{Y, V\}$ for the game we refer to Y as an admissible trajectory, V as an admissible strategy for the game and to v_k as an admissible strategy for the kth player, $k = 1, 2, \ldots, p$.

Definition 2 *For a fixed admissible pair $\{X, U\} \in \Omega$ we say that $v : [a, b] \to \mathbb{R}^{m_k}$ is an admissible strategy for player k relative to the pair $\{X, U\}$ if there exists a function $y_k : [a, b] \to \mathbb{R}^n$ such that the pair $\{[X^{-k}, y_k], [U^{-k}, v]\}$, defined on the interval $[a, b]$, is in Ω. We denote the set of admissible strategies for player k relative to $\{X, U\}$ by $\Omega_k(X, U)$ (i.e., $v \in \Omega_k(X, U)$).*
This allows us to give the following definition of an open-loop Nash equilibrium.

Definition 3 *An admissible pair $\{X^*, U^*\} \in \Omega$ is called a Nash equilibrium for the game if for each $k = 1, 2, \ldots, p$ and every $v \in \Omega_k(X^*, U^*)$ one has*

$$J_k(X^*, u_k^*) = \int_a^b f_k^0(t, X^*(t), u_k^*(t)) \, dt$$

$$\leq \int_a^b f_k^0(t, [X^*(t)^{-k}, y_k(t)], v(t)) \, dt = J_k([X^{*-k}, y_k], v_k),$$

where $\{[X^{*-k}, y_k], [U^{*-k}, v]\} \in \Omega$.

To proceed further we need to define a parametrized optimal control problem as follows. We begin with a set of p positive weights $r_1, r_2, \ldots, r_p$ and define the function $G_r : [a, b] \times \mathbb{R}^N \times \mathbb{R}^N \times \mathbb{R}^M \to \mathbb{R}$ by the formula

$$G_r(t, X, Y, V) = \sum_{k=1}^{p} r_k f_k^0(t, [X^{-k}, y_k], v_k), \tag{8}$$

where $Y = (y_1, y_2, \ldots, y_p) \in \mathbb{R}^N$ and $V = (v_1, v_2, \ldots, v_p) \in \mathbb{R}^M$. For a fixed admissible trajectory-strategy pair $\{X, U\} \in \Omega$, consider the optimal control problem consisting of minimizing the integral functional

$$\mathscr{J}_r(Y, V; X) = \int_a^b G_r(t, [X(t)^{-k}, y_k(t)], V(t)) \, dt, \tag{9}$$

over all pairs $\{Y, V\} \in \Omega$ defined on $[a, b]$ such that the map

$$t \mapsto G_r(t, X(t), Y(t), V(t)), \quad t \in [a, b]$$

is Lebesgue integrable. We will refer to these pairs as admissible for the parametrized optimal control problem and denote the set of all such pairs by $\Omega(X, U)$.

Remark 3 *We observe that since $\{X, U\} \in \Omega$ it follows that the map*

$$t \mapsto G_r(t, X(t), X(t), U(t)), \quad t \in [a, b]$$

is Lebesgue integrable on $[a, b]$ which implies $\{X, U\} \in \Omega(X, U)$, so that $\Omega(X, U) \neq \emptyset$.

Related to this optimal control problem, we define the set-valued mapping $\Gamma(\cdot)$ on the set of admissible trajectories, X, for the game (i.e., there exists U such that $\{X, U\} \in \Omega$).

$$\Gamma(X) = \Big\{ Y = (y_1, \ldots, y_p) : \text{ there exists a strategy } V = (v_1, v_2, \ldots, v_p)$$

$$\text{such that } \{Y, V\} \in \Omega(X, U)$$

$$\text{and } \mathscr{J}_r(Y, V; X) = \inf_{\{Z, W\} \in \Omega(X, U)} \mathscr{J}_r(Z, W; X) \Big\}. \tag{10}$$

With this set-valued mapping, we have the following theorem:

Theorem 1 *Suppose that $X^* = (x_1^*, \ldots, x_p^*) : [a, b] \to \mathbb{R}^N$ is a fixed point of $\Gamma(\cdot)$. Then there exists a strategy $U^* = (u_1^*, \ldots, u_p^*) : [a, b] \to \mathbb{R}^M$ such that $\{X^*, U^*\} \in \Omega$ and, moreover, the admissible pair is a Nash equilibrium for the dynamic game.*

Proof We first observe that since $X^* = (x_1^*, \ldots, x_p^*) \in \Gamma(X^*)$ there exists a strategy vector $U^* : [a, b] \to \mathbb{R}^M$ so that $\{X^*, U^*\} \in \Omega(X^*, U^*)$. Suppose that the pair $\{X^*, U^*\}$ is not a Nash equilibrium. This means that for at least one of the players, say player k, there exists an admissible strategy $v \in \Omega_k(X^*, U^*)$, and a corresponding $y : [s_0, s_1] \to \mathbb{R}^{n_k}$, such that

$$J_k([X^{*-k}, y], v) = \int_a^b f_j^0(t, [X^*(t)^{-k}, y(t)], v(t)) \, dt$$

$$< \int_a^b f_k^0(t, X^*(t), u_k^*(t)) \, dt = J_k(X^*, u_k^*).$$

This means that

$$\mathscr{J}_r([X^{*-k}, y], [U^{*-k}, v]; X^*) = \sum_{j \neq k} r_j J_j(X^*, u_j^*) + r_k J_k([X^{*-k}, y], v)$$

$$< \sum_{j=1}^p r_j J_j(X^*, u_j^*)$$

$$= \mathscr{J}_r(X^*, U^*; X^*).$$

This implies that $\{[X^{*-k}, y], [U^{*-k}, v]\}$ is admissible for the parametrized optimal control problem and additionally violates the assumption that X^* is a fixed point of $\Gamma(\cdot)$. Hence, $\{X^*, U^*\}$ is a Nash equilibrium. $\square$

Remark 4 *Clearly we plan to establish existence of a Nash equilibrium by showing the set-valued mapping $\Gamma(\cdot)$ defined in (10) has a fixed point. To do this we will apply the Bohnenblust–Karlin extension of the Kakutani fixed point theorem.*

Theorem 2 *Let S be a closed convex set in a Banach space X and let $\Psi : S \rightrightarrows S$ be a set-valued mapping satisfying*

(i) $\Psi(x)$ is nonempty and convex for each $x \in S$;
(ii) the graph of Ψ, $\mathrm{gr}\Psi := \{(y, x) \in S \times S : y \in \Psi(x)\}$ is closed; and
(iii) $\bigcap_{x \in S} \Psi(x)$ is contained in a sequentially compact set T.

   Then, the map Ψ has a fixed point.

Proof See [4]. $\square$

Remark 5 *The notation $\rightrightarrows$ indicates a set-valued mapping. That is, for example, in the above $\Psi : S \rightrightarrows S$ means that for each $s \in S$, its image under the map Ψ, i.e., $\Psi(s)$, is a subset of S.*

3   The Auxiliary Optimal Control Problem

To reiterate, for a fixed trajectory/strategy pair $\{X, U\}$, we consider the parametrized optimal control problem OCP(X) as

$$\min \left\{ \mathscr{J}_r(Y, V; X) = \int_a^b G_r(t, X(t), Y(t), V(t)) \, dt \right.$$

$$\left. = \int_a^b \sum_{j=1}^p r_j f_j^0(t, [X(t)^{-k}, y_k(t)], v_k(t)) \, dt \right\},$$

subject to

$$Y(t) = Y(a) + P(t) + \int_a^t h(t, s) Y(s) \, dt$$

$$+ \int_a^t g(t, s) f(s, Y(s), V(s)) \, dt \quad t \in (a, b),$$

$$(t, Y(t)) \in \mathscr{A}, \quad \text{for } t \in [a, b], \quad (a, Y(a), b, Y(b)) \in \mathscr{B},$$

$$V(t) \in \mathscr{U}(t, Y(t)), \quad \text{a.e. } t \in [a, b].$$

Here the minimization is over all pairs $\{Y, V\} \in \Omega$ defined on $[a, b]$ and such that the map $t \mapsto G_r(t, X(t), Y(t), V(t))$ is Lebesgue integrable.

Optimal control problems (unparameterized versions) of this form have been studied previously in Angell [5] so that known conditions are available to assert the existence of an optimal control pair. To state such a result one needs conditions which ensure that a minimizing sequence $\{Y_k, V_k\}_{k \in \mathbb{N}}$ is such that the sequence of admissible trajectories $Y_k : [a, b] \to \mathbb{R}^N$, $k \in \mathbb{N}$, is a relatively compact set in an appropriate topology and that the integral functional $\mathscr{J}_r(\cdot, \cdot; V)$ is sequentially lower semicontinuous with respect to this topology. The appropriate topology in this case is the space of continuous functions $Y : [a, b] \to \mathbb{R}^N$, denoted as $C([a, b]; \mathbb{R}^N)$, endowed with the metric

$$\rho(Y_1, Y_2) = \sup_{t \in [a, b]} |Y_1(t) - Y_2(t)|, \tag{11}$$

for $Y_i : [a, b] \to \mathbb{R}^N$ in $C([a, b]; \mathbb{R}^N)$ for $i = 1, 2$. In addition we require the following definition.

Definition 4 *Let* $H : \mathscr{M} \to \mathbb{R}$ *be continuous and nonnegative. A continuous function* $L : M \to \mathbb{R}^N$ *is said to satisfy the growth condition* (γ_H) *if for each* $\varepsilon > 0$ *there exists a locally integrable function* $\psi_\varepsilon : [a, b] \to [0, +\infty)$ *such that*

$$|L(t, x, u)| \leq \psi_\varepsilon(t) + \varepsilon H(t, x, u) \quad \text{for all } (t, x, u) \in M. \tag{12}$$

To present a compactness result, we need to restrict our admissible pairs for OCP(X) to those that satisfy an isoperimetric constraint of the form

$$\int_a^b H(t, Y(t), V(t)) \, dt \leq K, \tag{13}$$

where $K \geq 0$ is a given constant.

The function H is called a comparison functional, and the growth condition (γ_H) is a somewhat standard hypothesis for the treatment of optimal control problems for ordinary differential equations (see Cesari [6]). A typical candidate for H for problems with an integral functional, as is the case here, would be $H(t, Y, V) = G_r(t, X(t), Y, V)$. In this case, since $\{X, U\} \in \Omega$ is an admissible trajectory/control pair for (OCP(X)), one could choose any $K \geq 0$ satisfying

$$K \geq \int_a^b H(t, X(t), U(t)) \, dt = \int_a^b G_r(t, X(t), X(t), U(t)) \, dt.$$

With this notation, we have the following results from Angell [5], appropriately adapted for the case considered here.

Theorem 3 *Let R denote the set of all measurable functions $z : [a, b] \to \mathbb{R}^d$ defined by the formula $z(t) = f(t, Y(t), V(t))$ for some $\{Y, V\} \in \Omega(X, U)$ that satisfies the isoperimetric constraint (13) and assume that f satisfies the growth condition (γ_H). Then $R \subset L^1([a, b]; \mathbb{R}^d)$ is an equiabsolutely integrable set of functions which is relatively weakly compact in $L^1([a, b]; \mathbb{R}^d)$.*

Theorem 4 *Let C denote the set of all functions $w : [a, b] \to \mathbb{R}^N$ defined by the formula*

$$w(t) = Y(a) + P(t) + \int_a^t h(t, s) Y(s) + g(t, s) f(s, Y(s), V(s)) \, ds,$$

for some $\{Y, V\} \in \Omega(X, U)$ that satisfies the isoperimetric constraint (13). Then, if $\mathscr{A}$ is compact and f satisfies the growth condition (γ_H), the set C is relatively compact in $C([a, b]; \mathbb{R}^N)$.

Remark 6 *If $\mathscr{A}$ is not compact, then Theorem 4 remains true if $\mathscr{A}$ is closed and contained in a slab $[a, b] \times \mathscr{Y}$ (possibly unbounded) provided $h(t, s) \equiv 0$ and g is bounded. Further, Theorem 3 holds if instead of the growth condition (γ_H) one assumes the set R is bounded. Thus if $\mathscr{A}$ is compact the conclusion of Theorem 4 holds without assuming the growth condition (γ_H).*

Assuming some Lipschitz conditions on g and h, one also can prove the following theorem:

Theorem 5 *Let g and h be Lipschitzian in their first argument, assume $\mathscr{A}$ is compact and that there exists a constant N such that for all $z \in R$ one has $|z(t)| \leq N$ a.e. on $[a, b]$. Then the set C is a set of equi-Lipschitzian functions.*

To discuss the appropriate lower semicontinuity properties, we introduce the set-valued mapping[1] $\tilde{Q}_r : [a, b] \times A \times A \rightrightarrows \mathbb{R}^2 \times \mathbb{R}^n$ by the formula

$$\tilde{Q}_r(t, X, Y) \doteq \{(\mathscr{L}, Z^0, Z) : \mathscr{L} \geq H(t, Y, V), \ Z^0 \geq G_r(t, X, Y, V),$$
$$Z = f(t, Y, V), \ V \in \mathscr{U}(t, Y)\}. \quad (14)$$

With respect to this function, we make the following assumptions:

A1. For each $(t, X, Y) \in [a, b] \times A \times A$, the set $\tilde{Q}_r(t, X, Y)$ is closed and convex.

A2. The set-valued map $\tilde{Q}_r$ satisfies the following upper semicontinuity condition with respect to (X, Y) given as,

$$\tilde{Q}_r(t, X, Y) = \bigcap_{\delta > 0} \mathrm{cl} \left(\bigcup \{\tilde{Q}_r(t, X', Y') : |X - X'| + |Y - Y'| < \delta\} \right),$$

for all $(X, Y) \in A \times A$ and almost all $t \in [a, b]$.

Remark 7 *The assumption A2 is commonly referred to as the upper semicontinuity property (K). As written above, the union is taken over all pairs (X', Y') satisfying $|X - X'| + |Y - Y'| < \delta$, the notation cl denotes the closure of the union, and the intersection is taken over all $\delta > 0$.*

Remark 8 *The conditions placed on the set-valued mapping $\tilde{Q}_r$ are standard in the existence theory of optimal control. These conditions are required to ensure that appropriate closure and lower closure theorems are applicable. For completeness we state one such theorem that we will use several times in the discussion that follows.*

Theorem 6 (*Lower Closure Theorem*) *For $t \in [a, b]$ let $A(t) \subset \mathbb{R}^n$, let $\mathbf{A} = \{(t, x) : x \in A(t)\}$ and let $R : \mathbf{A} \rightrightarrows \mathbb{R}^{2+m}$ be a given set-valued mapping. For almost all $\bar{t} \in [a, b]$ assume that $A(\bar{t}) \subset \mathbb{R}^n$ is closed and that the sets $R(\bar{t}, x)$ are closed and convex and satisfy the upper semicontinuity condition known as property (K) with respect to $x \in A(\bar{t})$. That is, one has*

$$R(\bar{t}, x) = \bigcap_{\delta > 0} \mathrm{cl} \left(\bigcup \{R(\bar{t}, x') : |x - x'| < \delta\} \right)$$

for all $x \in A(\bar{t})$. Now, for $k \in \mathbb{N}$ and $i = 1, 2$ let ξ, x, η_k^i, ξ_k, x_k, λ^i, and λ_k^i be measurable functions on $[a, b]$ satisfying

(a) *$\xi, \xi_k \in L^1([a, b]; \mathbb{R}^m)$ and $\xi_k \to \xi$ weakly in $L^1([a, b]; \mathbb{R}^m)$ as $k \to \infty$,*
(b) *$\eta_k^i \in L^1([a, b]; \mathbb{R})$,*
(c) *$x, x_k : [a, b] \to \mathbb{R}^n$ and $x_k \to x$ in measure as $k \to \infty$ on $[a, b]$,*
(d) *$\lambda^i, \lambda_k^i \in L^1([a, b]; \mathbb{R})$ and $\lambda_k^i \to \lambda^i$ weakly in $L^1([a, b]; \mathbb{R})$ as $k \to \infty$,*
(e) *$\eta_k^i(t) \geq \lambda_k^i(t)$ for a.e. $t \in [a, b]$ and all $k \in \mathbb{N}$.*

---

[1]The notation $\doteq$ indicates the left-hand side of the expression is defined to be the right-hand side.

Further suppose there exist a constant $E \in \mathbb{R}$ such that the orientor field equations

$$x_k(t) \in A(t), \quad (\eta_k^1(t), \eta_k^2(t), \xi_k(t)) \in R(t, x_k(t)), \quad a.e.\ t \in [a, b],\ k \in \mathbb{N},$$

and the conditions

$$\int_a^b \eta_k^1(t)\, dt \le E, \quad for\ all\ k \in \mathbb{N}$$

$$-\infty < \tau^2 = \liminf_{k \to \infty} \int_a^b \eta_k^2(t)\, dt < +\infty,$$

hold. Then there exists functions $\eta^i \in L^1([a, b]; \mathbb{R})$ such that

$$x(t) \in A(t), \quad (\eta^1(t), \eta^2(t), \xi(t)) \in R(t, x(t)), \quad a.e.\ t \in [a, b],$$

$$\int_a^b \eta^1(t)\, dt \le E \quad and \quad \int_a^b \eta^2(t)\, dt \le \tau^2.$$

Proof See Appendix A. □

Remark 9 *We will apply Theorem 6 to the set-valued map $\tilde{Q}_r$ in two different settings. The first is for the parametrized optimal control problem OCP(X), in which for a fixed pair $\{X, U\}$ the lower closure theorem is applied to the sets $R(t, Y) = \tilde{Q}_r(t, X(t), Y)$, and the second will be applied to the full set-valued mapping $\tilde{Q}_r(t, X, Y)$.*

With the above preliminaries out of the way, we now state an existence theorem for OCP(X) and provide a brief sketch of its proof.

Theorem 7 *Fix $\{X, U\} \in \Omega$ and assume that the control system (3)–(6) satisfies the following:*

(a) *The set $\mathscr{A} = [a, b] \times A \subset \mathbb{R} \times \mathbb{R}^N$ is compact.*
(b) *The set $\mathscr{M} = \{(t, Y, V) \in [a, b] \times \mathbb{R}^N \times \mathbb{R}^M : (t, Y) \in \mathscr{A},\ V \in \mathscr{U}(t, Y)\}$ is closed.*
(c) *The functions $f : \mathscr{M} \to \mathbb{R}^d$, $h : [a, b] \times [a, b] \to \mathbb{R}^{N \times N}$, $g : [a, b] \times [a, b] \to \mathbb{R}^{N \times d}$ and $f_k^0 : \mathscr{M}_k \to \mathbb{R}\ (k = 1, 2, \dots, p)$, in which $\mathscr{M}_k = \{(t, Y, v) \in [a, b] \times \mathbb{R}^N \times \mathbb{R}^{m_k} : \exists u_j \in \mathbb{R}^{m_j}\ for\ j \ne k\ such\ that\ (t, Y, [U^{-k}, v]) \in \mathscr{M}\}$ are all continuous.*

Further assume there exists a function $H : \mathscr{M} \to \mathbb{R}$ continuous and nonnegative so that $f : \mathscr{M} \to \mathbb{R}^N$ satisfies the growth condition (γ_H) and that there exists a Lebesgue integrable function $\psi : [a, b] \to [0, +\infty)$ such that

$$G_r(t, X(t), Y, V) = \sum_{j=1}^p f_j^0(t, [X(t)^{-j}(t), y_j], v_j) \ge -\psi(t)$$

for almost all $t \in [a, b]$, $(t, Y, V) \in \mathscr{M}$. Then if the set-valued mapping $\tilde{Q}_r$ defined by (14) satisfies assumptions A1 and A2, there exists a pair $\{Y^*, V^*\} \in \Omega(X, U)$

that is a minimizer for OCP(X) in the subset of $\Omega(X, U)$ *of pairs satisfying the isoperimetric constraint (13) for any* $K \in \mathbb{R}$ *for which the fixed pair* $\{X, U\} \in \Omega$ *satisfies*

$$\int_a^b H(t, X(t), U(t))\, dt \leq K \tag{15}$$

Proof We begin by fixing $K \in \mathbb{R}$ so that (15) holds. Also we observe that since $\{X, U\} \in \Omega(X, U)$, the set of admissible controls for OCP(X) is nonempty. Further since $G_r(t, X(t), Y, V) \geq -\psi(t)$ for almost every $t \in [a, b]$, it follows that the optimal control problem has a finite infimum. That is, one has

$$v = \inf_{\{Y,V\}} \mathscr{I}_r(Y, V; X) > -\int_a^b \psi(t)\, dt > -\infty,$$

where the infimum is taken over all admissible pairs $\{Y, V\} \in \Omega(X, Y)$, satisfying the isoperimetric inequality

$$\int_a^b H(t, Y(t), V(t))\, dt \leq K.$$

This means that there exists a sequence of pairs $\{Y_k, V_k\} : [a, b] \rightarrow \mathbb{R}^N \times \mathbb{R}^M$ admissible for OCP(X) and satisfying the above isoperimetric inequality constraint such that

$$\lim_{k \to \infty} \mathscr{I}_r(Y_k, V_k; X) = v.$$

By Theorem 3 the sequence of functions $Z_k(\cdot) = f(\cdot, Y_k(\cdot), V_k(\cdot))$ is a relatively weakly compact subset of $L^1([a, b]; \mathbb{R}^d)$ so that we can assume without loss of generality (by extracting a subsequence if necessary) that there exists an integrable function $Z^* : [a, b] \rightarrow \mathbb{R}^d$ such that $\{Z_k\}$ converges weakly to Z^* in $L^1([a, b]; \mathbb{R}^d)$ as $k \to \infty$. Moreover, by Theorem 4 it follows that the sequence $\{Y_k\}$ is relatively compact sequence in $C([a, b]; \mathbb{R}^N)$, so that we can further assume (by extracting another subsequence if necessary) that there exists a function $Y^* : [a, b] \rightarrow \mathbb{R}^N$ such that $Y_k \rightarrow Y^*$ uniformly on $[a, b]$ as $k \to \infty$. Now define $\mathscr{Z}_k(t) = H(t, Y_k(t), V_k(t))$ and $Z_k^0(t) = G_r(t, X(t), Y_k(t), V_k(t))$, and observe that one has

$$(\mathscr{Z}_k(t), Z_k^0(t), Z_k(t)) \in \tilde{Q}_r(t, X(t), Y_k(t)), \quad \text{a.e. } t \in [a, b].$$

By appealing to the lower closure theorem (Theorem 6), with $A(t) = \{Y : (t, Y) \in \mathscr{A}\}$, $R(t, \cdot) = \tilde{Q}_r(t, X(t), \cdot)$, $\eta_k^1 = \mathscr{Z}_k$, $\eta_k^2 = \mathbb{Z}_k^0$, $\xi_k = Z_k$, $\xi = Z^*$, $x_k = Y_k$, $x = Y^*$, $i = v$, and $\lambda^1, \lambda_k^1 = -\psi_1$ (use (γ_H) with $\epsilon = 1$), $\lambda^2, \lambda_k^2 = -\psi$, we can conclude that there exists an integrable function $\mathscr{Z}^* : [a, b] \rightarrow \mathbb{R}$ and $Z^{0*} : [a, b] \rightarrow \mathbb{R}$ such that

$$Y^*(t) \in A(t), \quad (\mathscr{Z}^*(t), Z^{0*}(t), Z^*(t)) \in \tilde{Q}_r(t, X(t), Y^*(t)), \quad \text{a.e } t \in [a, b],$$

$$-\int_a^b \psi_1(t)\, dt \leq \int_a^b \mathscr{Z}^*(t)\, dt \leq K \doteq E$$

and

$$\int_a^b Z^{0*}(t)\, dt \leq \nu = \liminf_{k \to \infty} \int_a^b G_r(t, X(t), Y_k(t), V_k(t))\, dt.$$

It remains to show that there exists a measurable control $V^* : [a, b] \to \mathbb{R}^M$ so that the pair $\{Y^*, V^*\} \in \Omega(X, U)$ and such that the isoperimetric constraint (13) holds. To this end define the set-valued mapping $\Delta : [a, b] \rightrightarrows \mathbb{R}^M$ by the formula

$$\Delta(t) \doteq \{V \in \mathscr{U}(t, Y^*(t)) : \mathscr{Z}^*(t) \geq H(t, Y^*(t), V),$$

$$Z^{0*}(t) \geq G_r(t, X(t)Y^*(t), V), \ Z = f(t, Y^*(t), V)\}.$$

Under our continuity assumptions, it follows from Theorem B.2 in Appendix B that Δ is a Lebesgue measurable set-valued mapping and consequently has a Lebesgue measurable selection. That is, there exists a measurable function V^* : $[a, b] \to \mathbb{R}^M$ such that $V^*(t) \in \mathscr{U}(t, Y(t))$, $Z^*(t) = f(t, Y^*(t), V^*(t))$ and $Z^{0*}(t) \geq G_r(t, X(t), Y^*(t), V^*(t))$ for almost every $t \in [a, b]$. Moreover, since $Z_k \to Z^*$ weakly in $L^1([a, b]; \mathbb{R}^d)$ and $Y_k \to Y^*$ uniformly on $[a, b]$ as $k \to \infty$, it follows that

$$Y^*(t) = Y^*(a) + P(t) + \int_a^t h(t, s)Y^*(s)\, ds + g(t, s)f(s, Y^*(s), V^*(s))\, ds$$

for $t \in (a, b)$,

$$\int_a^b H(t, Y^*(t), V^*(t))\, dt \leq K$$

and

$$\int_a^b G_r(t, X(t), Y^*(t), V^*(t))\, dt \leq i = \liminf_{k \to \infty} \int_a^b G_r(t, X(t), Y_k(t), V_k(t))\, dt$$

which immediately implies that $\{Y^*, V^*\} \in \Omega(X, U)$ and that

$$\int_a^b G_r(t, X(t), Y^*(t), V^*(t))\, dt \leq \nu = \inf_{\{Y, V\}} \mathscr{J}_r(Y, V; X),$$

where the infimum is taken over all pairs $\{Y, V\} \in \Omega(X, U)$ satisfying

$$\int_a^b H(t, Y(t), V(t)) \, dt \leq K.$$

$\square$

Remark 10 *In this section we have presented an existence theorem for the optimal control problem OCP(X). This ensures that the set-valued map $\Gamma(X)$ is nonempty for each admissible trajectory X.*

4 Existence of a Nash Equilibrium

To apply Theorem 2 to the set-valued mapping $\Gamma(\cdot)$ defined by (10), we must first define the appropriate Banach space. Clearly the Banach space X should be the space of functions defined on closed subintervals of $[a, b]$ endowed with the ρ metric given by (11). The set $S \subset X$ should be the set of admissible trajectories for the game with the property that the isoperimetric constraint

$$\int_a^b H(t, X(t), U(t)) \, dt \leq K, \tag{16}$$

for some fixed $K > 0$. The existence of such a K is ensured by the fact that the set Ω of admissible trajectory-strategy pairs $\{X, U\}$ is assumed to be nonempty. Assumptions must be imposed so that S is convex and $\mathrm{gr}\Gamma(\cdot) = \{(X, Y) \in S \times S : Y \in \Gamma(X)\}$ is closed. To do this, we make the following assumption, in addition to **A1** and **A2**.

A3. The graph of the set-valued mapping $Q_r(t, \cdot, \cdot)$ is closed and convex.

With these preliminaries, we have the following lemmas:

Lemma 1 *Under the assumptions of Theorem 7, the set $\Gamma(X)$ is nonempty for each $X \in S$.*

Proof The proof is an immediate consequence of Theorem 7. $\square$

Lemma 2 *Under the assumptions **A1**, **A2**, and **A3**, if the function f satisfies the growth condition (γ_H), then the set S is closed and convex.*

Proof Let $\lambda \in [0, 1]$ and let $X_i \in S$ for $i = 0, 1$. This means that $X_i : [a, b] \to \mathbb{R}^N$ and there exists measurable strategies $U_i : [a, b] \to \mathbb{R}^M$ such that $\{X_i, U_i\} \in \Omega$ and

$$\int_a^b H(t, X_i(t), U_i(t)) \, dt \leq K.$$

For $i = 0, 1$ and $t \in [a, b]$, define the functions $\mathscr{Z}_i(t) = H(t, X_i(t), U_i(t))$, $Z_i^0(t) = G_{\mathrm{r}}(t, X_i(t), X_i(t), U_i(t))$, and $Z_i(t) = f(t, X_i(t), U_i(t))$. Then we have for almost every $t \in [a, b]$ that

$$(X_i(t), X_i(t), \mathscr{Z}_i(t), Z_i^0(t), Z_i(t)) \in \mathrm{gr}\tilde{Q}_{\mathrm{r}}(t, \cdot, \cdot),$$

which is a convex set by hypotheses. Therefore, we have

$$(\mathscr{Z}_\lambda(t), Z_\lambda^0(t), Z_\lambda(t)) \in \tilde{Q}_{\mathrm{r}}(t, X_\lambda(t), X_\lambda(t),),$$

for almost every $t \in [a, b]$, where for any pair of functions $F_i : [a, b] \to \mathbb{R}^k$, $i = 0, 1$ and $\lambda \in [0, 1]$ the notation $F_\lambda(\cdot) = \lambda F_1(\cdot) + (1 - \lambda)F_0(\cdot)$. Now define the set-valued mapping $\Delta_\lambda : [a, b] \rightrightarrows \mathbb{R}^M$ by the formula

$$\Delta_\lambda(t) = \{V \in \mathscr{U}(t, X_\lambda) : \mathscr{Z}_\lambda(t) \geq H(t, X_\lambda, V),$$

$$Z_\lambda^0(t) \geq G_{\mathrm{r}}(t, X_\lambda(t), X_\lambda(t), V), \ Z_\lambda(t) = f(t, X_\lambda(t), V)\}.$$

Applying Theorem B.2 in Appendix B, it is clear that Δ_λ is a Lebesgue measurable set-valued mapping and consequently has a measurable selection. This means that there exists a measurable function $U^\lambda : [a, b] \to \mathbb{R}^M$ such that for almost every $t \in [a, b]$ one has that

$$\mathscr{Z}_\lambda(t) \geq H(t, X_\lambda(t), U^\lambda(t)), \quad Z_\lambda^0(t) \geq G_{\mathrm{r}}(t, X_\lambda(t), X_\lambda(t), U^\lambda(t)),$$

$$\text{and} \quad Z_\lambda(t) = f(t, X_\lambda(t), U^\lambda(t)).$$

Clearly we have that

$$0 \leq \int_a^b H(t, X_\lambda(t), U^\lambda(t)) \, dt \leq \int_a^b \lambda \mathscr{Z}_1(t) + (1 - \lambda)\mathscr{Z}_0(t) \, dt \leq K,$$

$$-\int_a^b \psi_1(t) \, dt \leq \int_a^b G_{\mathrm{r}}(t, X_\lambda(t), X_\lambda(t), U^\lambda(t)) \, dt$$

$$\leq \int_a^b \lambda Z_1^0(t) + (1 - \lambda)Z_0^0(t) \, dt < +\infty,$$

and

$$Z_\lambda(t) = \lambda Z_1(t) + (1 - \lambda)Z_0(t) = f(t, X_\lambda(t), U^\lambda(t)) \quad \text{a.e. } t \in [a, b].$$

It remains to show that the pair $\{X_\lambda, U^\lambda\}$ satisfy the integral equation (3). To see this, we observe that for each $t \in [a, b]$ one has

$$X_\lambda(t) = \lambda X_1(t) + (1 - \lambda)X_0(t)$$

$$= \lambda \left\{ X_1(a) + \int_a^t h(t, s)X_1(s)\, ds + \int_a^t g(t, s)f(s, X_1(s), U_1(s))\, ds \right\}$$

$$+ (1 - \lambda) \left\{ X_0(a) + \int_a^t h(t, s)X_0(s)\, ds \right.$$

$$\left. + \int_a^t g(t, s)f(s, X_0(s), U_0(s))\, ds \right\}$$

$$= X_\lambda(a) + \int_a^t h(t, s)X_\lambda(s)\, ds + \int_a^t g(t, s)Z_\lambda(s)\, ds$$

$$= X_\lambda(a) + \int_a^t h(t, s)X_\lambda(s)\, ds + \int_a^t g(t, s)f(t, X_\lambda(s), U^\lambda(s))\, ds,$$

which shows that $\{X_\lambda, U^\lambda\} \in \Omega$ and the isoperimetric constraint (16) is satisfied. That is, $X_\lambda \in S$ so that S is convex.

To show that S is closed, consider a sequence of functions $\{X_n\}_{n \in \mathbb{N}} \subset S$ be such that $X_n \to X^*$ uniformly on $[a, b]$. Since each $X_n \in S$, there exists a measurable function $U_n : [a, b] \to \mathbb{R}^M$ so that $\{X_n, U_n\} \in \Omega$. Setting $\mathscr{Z}_n = H(t, X_n(t), U_n(t))$, $Z_n^0(t) = G_r(t, X_n(t), X_n(t), U_n(t))$, and $Z_n(t) = f(t, X_n(t), U_n(t))$, we have that

$$(\mathscr{Z}_n(t), Z_n^0(t), Z_n(t)) \in \tilde{Q}_r(t, X_n(t), X_n(t)), \quad \text{a.e. } t \in [a, b].$$

From the growth condition (γ_H), we can assume without loss of generality that there exists an integrable function $Z^* : [a, b] \to \mathbb{R}^d$ so that $Z_n \to Z^*$ weakly in $L^1([a, b]; \mathbb{R}^d)$. Further we have that

$$- \int_a^b \psi_1(t)\, dt \leq \int_a^b \mathscr{Z}_n(t)\, dt \leq K,$$

where $\psi_1 : [a, b] \to \mathbb{R}$ comes from setting $\epsilon = 1$ in the growth condition (γ_H), so that by appealing to the lower closure theorem (Theorem 6) there exists integrable functions $\mathscr{Z}^* : [a, b] \to \mathbb{R}$ and $Z^{0*} : [a, b] \to \mathbb{R}$ so that

$$(\mathscr{Z}^*(t), Z^{0*}(t), Z^*(t)) \in \tilde{Q}_r(t, X^*(t), X^*(t)), \quad \text{a.e. } t \in [a, b],$$

$$- \int_a^b \psi_1(t)\, dt \leq \int_a^b \mathscr{Z}^*(t)\, dt \leq K \text{ and } \int_a^b Z^{0*}(t)\, dt \leq \liminf_{n \to \infty} \int_a^b Z_n^0(t)\, dt.$$

Now define the set-valued mapping $\Delta : [a, b] \rightrightarrows \mathbb{R}^M$ by the formula

$$\Delta(t) = \{V \in \mathcal{U}(t, X^*(t)) : \ \mathcal{L}^*(t) \geq H(t, X^*(t), V),$$

$$Z^{0*}(t) \geq G_r(t, X^*(t), X^*(t), V), \ Z^*(t) = f(t, X^*(t), V)\}$$

appealing to Theorem B.2 in Appendix B; this is Lebesgue measurable set-valued mapping, and so there exists a Legbesgue measurable selection $U^* : [a, b] \to \mathbb{R}^M$ such that

$$U^*(t) \in \mathcal{U}(t, X^*(t)), \ \mathcal{L}^*(t) \geq H(t, X^*(t), U^*(t)),$$

$$Z^{0*}(t) \geq G_r(t, X^*(t), X^*(t), U^*(t)) \text{ and } Z^*(t) = f(t, X^*(t), U^*(t))$$

for almost every $t \in [a, b]$. Moreover since $X_n \to X^*$ uniformly on $[a, b]$ and $Z_n \to Z^*$ weakly in $L^1([a, b]; \mathbb{R}^d)$ we have

$$X^*(t) = \lim_{n \to \infty} X_n(t)$$

$$= \lim_{n \to \infty} X_n(a) + P(t) + \int_a^t h(t, s)X_n(s) + g(t, s)Z_n(s) \, ds$$

$$= X^*(a) + P(t) + \int_a^t h(t, s)X^*(s) + g(t, s)Z^*(s) \, ds$$

$$= X^*(a) + P(t) + \int_a^t h(t, s)X^*(s) + g(t, s)f(s, X^*(s), U^*(s)) \, ds$$

implying $\{X^*, U^*\} \in \Omega$ and

$$\int_a^b H(t, X^*(t), U^*(t)) \, dt \leq K.$$

That is, $X^* \in S$ as desired. $\qquad\qquad\square$

Lemma 3 *Under the assumptions A1, A2, and A3, if the function f satisfies the growth condition (γ_H), then the graph of $\Gamma(\cdot)$ is closed.*

Proof Let $\{X_n\}_{n \in \mathbb{N}}$ and $\{Y_n\}_{n \in \mathbb{N}}$ be two sequences of functions in S satisfying $Y_n \in \Gamma(X_n)$ for every $n \in \mathbb{N}$ and such that there are two functions, $X^* : [a, b] \to \mathbb{R}^N$ and $Y^* : [a, b] \to \mathbb{R}^N$, such that $X_n \to X^*$ and $Y_n \to Y^*$ uniformly on $[a, b]$ as $n \to \infty$. We must show that $Y^* \in \Gamma(X^*)$. Since the set S is closed, it follows that we have $X^*, Y^* \in S$ so that there exists measurable functions $U^*, V^* : [a, b] \to \mathbb{R}^M$ so that $\{X^*, U^*\}, \{Y^*, V^*\} \in \Omega$. Further, since each X_n and Y_n is in S, there exists measurable functions $U_n : [a, b] \to \mathbb{R}^M$ and $V_n : [a, b] \to \mathbb{R}^M$ so that $\{X_n, U_n\}, \{Y_n, V_n\} \in \Omega$ for each $n \in \mathbb{N}$. Letting $\mathcal{L}_n(t) = H(t, Y_n(t), V_n(t))$,

$Z_n^0(t) = G_{\mathrm{r}}(t, X_n(t), Y_n(t), V_n(t))$, and $Z_n(t) = f(t, Y_n(t), V_n(t))$, we have that for each $n \in \mathbb{N}$

$$(\mathscr{L}_n(t), Z_n^0(t), Z_n(t)) \in \tilde{Q}_{\mathrm{r}}(t, X_n(t), Y_n(t)), \quad \text{a.e. } t \in [a, b]$$

and

$$0 \leq \int_a^b \mathscr{L}_n(t)\, ds \leq K.$$

Applying the growth condition (γ_H), the lower closure theorem Theorem 6, and using a measurable selection theorem, it follows that there exists integrable function $\mathscr{L}^* : [a, b] \to \mathbb{R}$, $Z^{0*} : [a, b] \to \mathbb{R}$, and $Z^* : [a, b] \to \mathbb{R}^d$, and a measurable function $V^* : [a, b] \to \mathbb{R}^M$ such that

$$(\mathscr{L}^*(t), Z^{0*}(t), Z^*(t)) \in \tilde{Q}_{\mathrm{r}}(t, X^*(t), Y^*(t)), \quad \text{a.e. } t \in [a, b],$$

$$-\int_a^b \psi_1(t)\, dt \leq \int_a^b H(t, Y^*(t), V^*(t))\, dt \leq \int_a^b \mathscr{L}^*(t)\, dt \leq K,$$

$$\int_a^b Z^{0*}(t)\, dt \leq \liminf_{n \to \infty} \int_a^b Z_n^0(t)\, dt,$$

$V^*(t) \in \mathscr{U}(t, Y^*(t))$ and $Z^*(t) = f(t, Y^*(t), V^*(t))$ for almost every $t \in [a, b]$. Since $Y_n \in \Gamma(X_n)$ for all $n \in \mathbb{N}$ we have for any $\{\tilde{Y}, \tilde{V}\} \in \Omega$ that

$$\int_a^b Z_n^0(t)\, dt = \int_a^b G_{\mathrm{r}}(t, X_n(t), Y_n(t), V_n(t))\, dt \leq \int_a^b G_{\mathrm{r}}(t, X_n(t), \tilde{Y}(t), \tilde{V}(t))\, dt.$$

From the conclusion of the lower closure theorem, we immediately have

$$\int_a^b G_{\mathrm{r}}(t, X^*(t), Y^*(t), V^*(t))dt \leq \liminf_{n \to \infty} \int_a^b Z_n^0(t)\, dt$$

$$\leq \lim_{n \to \infty} \int_a^b G_{\mathrm{r}}(t, X_n(t), \tilde{Y}(t), \tilde{V}(t))\, dt = \int_a^b G_{\mathrm{r}}(t, X^*(t), \tilde{Y}(t), \tilde{V}(t))\, dt,$$

which implies $Y^* \in \Gamma(X^*)$ as desired.   □

Lemma 4 *If f satisfies the growth condition (γ_H), then the set $\cap_{X \in S} \Gamma(X)$ is contained in a sequentially compact subset of $C([a, b]; \mathbb{R}^N)$.*

Proof Due to the growth condition (γ_H), the set S is a sequentially compact subset of $C([a, b]; \mathbb{R}^N)$ by Theorem 4. The desired conclusion now follows immediately since $\cap_{X \in S} \Gamma(X) \subset S$.   □

The Lemmas 1–4 allow us to prove the following result.

Theorem 8 *Under the assumptions A1, A2, A3 if the function f satisfies the growth condition (γ_H), then the set-valued map $\Gamma : S \Rightarrow S$ has a fixed point whenever the set $\Omega \neq \emptyset$, which implies there exists an open-loop Nash equilibrium for the dynamic game with a Volterra integral equation describing the dynamics of the game.*

Proof Combining the results from the Lemmas 1–4 shows that the set-valued mapping $\Gamma(\cdot)$ satisfies the hypotheses of the fixed point theorem, Theorem 2, so that $\Gamma(\cdot)$ has a fixed point, implying (by Theorem 1) the considered game has an open-loop Nash equilibrium. $\qquad\square$

5   Roos's Examples

In this section we will reformulate Roos's two examples in the form of the dynamic games considered here. In both cases, we will view the production rates $u_1(\cdot)$ and $u_2(\cdot)$ as the strategies of the players.

5.1   Example 1

Begin by letting $\alpha_i > 0$, $i = 1, 2$ satisfy $\alpha_1 + \alpha_2 = 1$. For an admissible price $p(\cdot)$ and production rates $u_i(\cdot)$, $i = 1, 2$, introduce the two state variables $y_i(t) = \alpha_i p(t)$. With this notation, introduce the two equations

$$u_i(t) = \alpha_i[cp(t) + h\dot{p}(t) + d] = cy_i(t) + h\dot{y}_i(t) + \alpha_i d.$$

Clearly adding these two equation leads to the linear demand law (1). Conversely, if $\{y_1(\cdot), y_2(\cdot), u_1(\cdot), u_2(\cdot)\}$ satisfy the above system of differential equations, then $p(t) \doteq y_1(t) + y_2(t)$ can be interpreted as the price and the $u_i(\cdot)$ as the admissible production rates. Therefore, the demand equation (1) and the above pair of differential equations are in one-to-one correspondence. Moreover, the accumulated profit of each firm is given as

$$\Pi_i(u_i(\cdot)) = \int_a^b (y_1(t) + y_2(t))u_i(t) - q(u_i(t)) \, dt, \quad i = 1, 2.$$

In addition the demand constraints become the ordinary differential equation control system

$$\dot{y}_i(t) = -\frac{c}{h}y_i(t) - \frac{\alpha_i}{h}d + \frac{1}{h}u_i(t), \quad t \in [a, b], \quad i = 1, 2,$$

with the additional constraints $y_i(t) \geq 0$, $u_i \in [0, \bar{u}_i]$ and the fixed initial condition $y_i(a) = \bar{y}_{ia}$ for $i = 1, 2$. We are assuming that the production rates are positive and bounded. Recall $q(u) = Au^2 + Bu + C$ with $A > 0$. In our setting the objective of player i is to minimize the integral functional $J_i(u_i(\cdot)) = -\Pi_i(u_i(\cdot))$. It is an easy matter to see that, since $u_i(t)$ is bounded and the initial condition for the state, $\bar{y}_{ia}$, is fixed, the states $y_i(t)$ are uniformly bounded on $[a, b]$. Thus, the relevant compactness conditions are automatically satisfied since, as $\{(y_1, y_2), (u_1, u_2)\}$ runs over the set of all admissible trajectory/strategy pairs, the measurable functions

$$z(t) = f(t, y_1(t), y_2(t), u_1(t), u_2(t)) = \Big(-(c/h)y_1(t) - (\alpha_i/h)d + (1/h)u_1(t),$$
$$-(c/h)y_2(t) - (\alpha_i/h)d + (1/h)u_2(t) \Big)^\mathsf{T}$$

are uniformly bounded. This means the conclusion of Theorem 4 holds automatically (see Remark 6) and the growth condition γ_H is not required. Moreover, the function H appearing in the definition of $\tilde{Q}_\mathbf{r}$ may be taken to be identically zero and therefore is not required at all. To see that the hypotheses **A1**, **A2**, **A3** all hold we observe that for $r_1, r_2 > 0$ we have that, since $q(u) = Au^2 + Bu + C$ with $A > 0$,

$$G_\mathbf{r}(t, x_1, x_2, y_1, y_2, u_1, u_2) = r_1[q(u_1) - (y_1 + x_2)u_1] + r_2[q(u_2) - (x_1 + y_2)u_2]$$

is a convex function of (u_1, u_2) and that f as defined above is linear in (u_1, u_2). From this it easily follows that the set

$$\tilde{Q}_\mathbf{r}(t, x_1, x_2, y_1, y_2) = \{(z^0, z_1, z_2) : z^0 \geq G_\mathbf{r}(t, x_1, x_2, y_1, y_2, u_1, u_2),$$
$$z_i = f_i(t, y_1, y_2, u_i), \ u_i \in [0, \bar{u}_i], \ i = 1, 2\}$$

is closed and convex and moreover satisfies property (K) with respect to $(X, Y) = (x_1, x_2, y_1, y_2)$. That is, that **A1** and **A2** both hold. Further, since $G_\mathbf{r}(t, \cdot, \cdot, \cdot, \cdot)$ is linear in (x_1, x_2, y_1, y_2), it is also easy to see that its graph, $\mathrm{gr}\tilde{Q}_\mathbf{r}(t, \cdot, \cdot, \cdot, \cdot)$, is closed and convex so that **A3** also holds. Thus, we can conclude that this problem has an open-loop Nash equilibrium by Theorem 8.

5.2 Example 2

Once again we let $\alpha_i > 0, i = 1, 2$ satisfy $\alpha_1 + \alpha_2 = 1$ and let $p(t), t \in (-\infty, T]$ and $u_i(t), i = 1, 2$ and $t \in [0, T]$ denote an admissible price and production strategies of the model. Introduce the state variables $y_i(t) = u_i(t) - a\alpha_i p(t)$ defined for $t \in [0, T]$, $i = 1, 2$. From the integral equation (2), one sees that the state variables y_i and production rates u_i satisfy the integral equation

$$y_i(t) = \alpha_i b + \int_{-\infty}^0 \Phi(t-s)\alpha_i p(s)\,ds + \int_0^t \frac{1}{a}\Phi(t-s)(u_i(s) - y_i(s))\,ds,$$

so that

$$y_i(0) = \alpha_i b + \int_{-\infty}^0 \Phi(-s)\alpha_i p(s)\,ds$$

which allows us to write the above equation in the form

$$y_i(t) = y_i(0) + P_i(t) + \int_0^t \frac{1}{a}\Phi(t-s)(u_i(s) - y_i(s))\,ds, \qquad (17)$$

where

$$P_i(t) = \int_{-\infty}^0 (\Phi(t-s) - \Phi(-s))\alpha_i p(s)\,ds.$$

This is a Volterra equation of the form given in (3) in which $h(t,s) \equiv 0, f(t,Y,V) = (v_1 - y_1, v_2 - z_2)^{\mathsf{T}}$ and $g(t,s) = (1/a)\Phi(t-s)I_2$, where I_2 is the 2×2 identity matrix, and we assume that the kernel $\Phi(\cdot)$ is a bounded function that is integrable on $[0, \infty)$. In the above model, we are assuming that the price $p(t)$ is given as data for $t < 0$. For consistency, since $p(t)$ appears in the definition of both state variables, it must further be assumed that

$$\frac{1}{\alpha_1}(u_1(t) - y_1(t)) = ap(t) = \frac{1}{\alpha_2}(u_2(t) - y_2(t))$$

for almost all $t \in [0, T]$. This gives rise to the strategy constraint

$$\mathcal{U}(t, (y_1, y_2)) = \{(v_1, v_2) \in [0, \bar{u}_1] \times [0, \bar{u}_2] : \alpha_2(v_1 - y_1) = \alpha_1(v_2 - y_2)\}.$$

Assuming that the state variables are bounded, say $y_i \in [0, \bar{y}_i]$, gives us the trajectory constraint set $\mathscr{A} = [0, T] \times [0, \bar{y}_1] \times [0, \bar{y}_2]$ so that the trajectory/strategy pair $\{(y_1(t), y_2(t)), (u_1(t), u_2(t))\}$ satisfies a control system of the form given by equations (3)–(6). Finally the payoff for each of the firms is given by

$$J_i(y_1, y_2, u_i) = \int_0^T \left(q(u_i(t)) - \frac{1}{a\alpha_i}(u_i(t) - y_i(t))u_i(t)\right) dt,$$

where we recall that $q(u) = Au^2 + Bu + C$. It is now an easy matter to see that for the correct choice of parameters (e.g., $A > (1/a)\max\{1/\alpha_1, 1/\alpha_2\}$) ensures that the hypotheses **A1**, **A2**, and **A3** hold without the growth condition (γ_H) being required for the same reasons as in Example 1. Thus, we see that this problem has an open-loop Nash equilibrium.

In the case when $P_i(\cdot)$, $i = 1, 2$, is an arbitrary function, we can still relate the above model to Roos's second example provided one can solve the Fredholm integral equation

$$P_1(t) + P_2(t) = \int_{-\infty}^0 (\Phi(t-s) - \Phi(-s))\tilde{p}(s)\,ds, \quad t \in [0, T]$$

for a function $\tilde{p}(\cdot)$. Then the price $p(\cdot)$ in Roos's example is given by

$$p(t) = \begin{cases} \tilde{p}(t), & t \in (-\infty, 0), \\ \dfrac{1}{a\alpha_i}(u_i(t) - z_i(t)), & t \in [0, T], \end{cases}$$

where i is either 1 or 2.

6 Perspectives

In the above, we have presented a theorem concerning the existence of open-loop Nash equilibria for a general p-player dynamic game in which the dynamics are described by control system governed by a nonlinear Volterra integral equation. Such a system includes nonlinear p-player differential games as a special case. As indicated in the introduction, this model was considered as early as 1925 by C. F. Roos for a competitive economic growth model in which the price at time $t > 0$ was dependent on the past history of the price. This case is not covered by the differential game case. Although this early example is perhaps too simple, many phenomena are dependent on their past history. Indeed, in the 1970s, there were several advertising models, for a single player, which involved integrodifferential systems (see, e.g., Sethi [7]). It is not hard to envision a competition model involving two retailers advertising similar products. Further, the role of delays in economic growth models, formulated as optimal control problems (i.e., the single-player game) is not unknown in the literature so that competitive models more general than the simple models considered by Roos could perhaps be of interest.

One of the first questions asked when one proves a theorem about open-loop Nash equilibria concerns the existence of closed-loop or feedback equilibria. Generally speaking, in the differential game setting, the existence of feedback equilibria involves solving a coupled system of Hamilton-Jacobi equations. This is a notoriously difficult problem to solve. The problem would be even more complicated in the Volterra integral equation setting considered here. As is often done, for example, in the linear quadratic case, feedback solutions are obtained by assuming the strategies have a certain form (e.g., linear feedbacks of the form $v_i = K_i(t)Y(t)$ in which $K_i(\cdot)$ is a matrix to be determined from the necessary conditions). In our formulation, the strategy constraint $V(t) \in \mathscr{U}(t, Y(t))$ is of feedback form in that it depends on the admissible trajectory Y. Thus, it is possible, if the form of

$\mathcal{U}(\cdot, \cdot)$ was known more explicitly that one could find a function $u(t, Y)$ so that the optimal strategy $V^*(\cdot)$ would have the representation $V^*(t) = u(t, Y^*(t))$. This is a feedback form, but it is not necessarily a Nash feedback strategy. On the other hand, one could assume that the strategies are all assumed to be of the form of a given class of functions, say $V(t) = F(t, Y(t))$, where $F : [a, b] \times \mathbb{R}^N \to \mathbb{R}^M$ belongs to some class of function (e.g., continuous functions). In this case the minimizing sequences for (OCP(X)), say $\{Y_k, V_k\}$ ($k = 1, 2, \ldots$), would be such that $V_k(t) = F_k(t, Y_k(t))$ for each k. To ensure that the optimal control/strategy for this (OCP(X)) would have the same form (i.e., $V^*(t) = F(t, Y^*(t))$) suggests that some sort of topology would have to be placed on the functions $F(\cdot, \cdot)$. This would impose a topology on the space of admissible strategies, which would have to be defined a priori so that a subsequence of the sequence $\{V_k\}$ would converge to V^*. In our proof, no such topology is required since we use a measurable selection theorem to obtain our optimal strategy. Thus, it indicates that modifying our approach to include feedback strategies might place a severe restriction on the class of games considered.

7 Conclusion

In this paper we have successfully established conditions which ensure the existence of an open-loop Nash equilibrium for a dynamic game in which the dynamics are described by a system of controlled Volterra integral equation with both state constraints and coupled strategy constraints. As a special case, our result includes the dynamics described by a system of controlled ordinary differential equations. Finally, we show that two competitive economic models due to C. F. Roos [2] have open-loop Nash equilibria under appropriate assumptions.

References

1. R. Isaacs, *Differential games. A mathematical theory with applications to warfare and pursuit, control and optimization* (John Wiley & Sons, Inc., New York-London-Sydney, 1965)
2. C.F. Roos, Amer. J. Math. **47**(3), 163 (1925). DOI 10.2307/2370550. URL http://dx.doi.org/10.2307/2370550
3. C.F. Roos, Trans. Amer. Math. Soc. **30**(2), 360 (1928). DOI 10.2307/1989126. URL http://dx.doi.org/10.2307/1989126
4. H.F. Bohnenblust, S. Karlin, in *Contributions to the Theory of Games, Vol. 1*, ed. by H.W. Kuhn, A.W. Tucker (Princeton University Press, Princeton, New Jersey, 1950), pp. 155–160
5. T. Angell, Journal of Optimization Theory and Applications **19**(1), 29 (1976)
6. L. Cesari, *Optimization-Theory and Applications: Problems with Ordinary Differential Equations, Applications of Applied Mathematics*, vol. 17 (Springer-Verlag, New York, 1983)
7. S. Sethi, SIAM Review pp. 685–725 (1977)
8. R.T. Rockafellar, R.J.B. Wets, *Variational analysis, Grundlehren der Mathematischen Wissenschaften [Fundamental Principles of Mathematical Sciences]*, vol. 317 (Springer-Verlag, Berlin, 1998). DOI 10.1007/978-3-642-02431-3. URL http://dx.doi.org/10.1007/978-3-642-02431-3

Appendix A: Proof of Theorem 6

A.1 Preliminaries

In this section we give a proof of the lower closure result, Theorem 6. To facilitate our presentation we begin by recalling the following two definition.

Definition A.1 *A set-valued map $R : A \rightrightarrows \mathbb{R}^{2+m}$ satisfies the upper semicontinuity property (Q) with respect to x if for almost every $t \in [a, b]$ and all $x \in \mathbb{R}^n$ one has*

$$R(t, x) = \bigcap_{\delta > 0} \mathrm{cl} \left(\bigcup \{R(t, x') : |x - x'| < \delta\} \right).$$

A related upper semicontinuity condition originally defined by L. Cesari is given in the following definition.

Definition A.2 *A set-valued map $R : A \rightrightarrows \mathbb{R}^{2+m}$ satisfies the upper semicontinuity property (K) with respect to x if for almost every $t \in [a, b]$ and all $x \in \mathbb{R}^n$ one has*

$$R(t, x) = \bigcap_{\delta > 0} \mathrm{cl} \left(\mathrm{co} \left(\bigcup \{R(t, x') : |x - x'| < \delta\} \right) \right),$$

where, for a set W, the notation $\mathrm{co}(W)$ denotes the convex hull of W.

Remark A.1 *Clearly if a set-valued mapping $R(\cdot, \cdot)$ satisfies property (Q) with respect to x, it also satisfies property (K) with respect to x. Indeed if $R(\cdot, \cdot)$ satisfies property (Q) with respect to x one has*

$$R(t, x) \subset \bigcap_{\delta > 0} \mathrm{cl} \left(\bigcup \{R(t, x') : |x - x'| < \delta\} \right)$$

$$\subset \bigcap_{\delta > 0} \mathrm{cl} \left(\mathrm{co} \left(\bigcup \{R(t, x') : |x - x'| < \delta\} \right) \right) = R(t, x)$$

for almost every $t \in [a, b]$ and $x \in \mathbb{R}^{2+m}$, since for any set $\mathrm{co}(W)$ is the smallest convex set containing W, which implies the inclusions become equalities.

To present the proof of Theorem 6, we need to introduce an additional set-valued mapping, $R^* : A \rightrightarrows \mathbb{R}^{3+m}$ by the formula

$$R^*(t, x) \doteq \{(\rho, y^1, y^2, z) : \rho \geq h(|z|), \ y^1 \geq \lambda^1(t) - 1,$$

$$y^2 \geq \lambda^2(t) - 1, \ (y^1, y^2, z) \in R(t, x)\}, \qquad (A.1)$$

where $h : [0, \infty) \to [0, \infty)$ is monotone decreasing, continuous, and convex and such that $h(\zeta)/\zeta \to \infty$ as $\zeta \to \infty$, and $\lambda^1, \lambda^2 : [a, b] \to \mathbb{R}$ are Lebesgue integrable functions. This set-valued mapping is clearly related to the set-valued mapping

$R(t, x)$. However, from Cesari [6, 10.5.ii], it follows that the set-valued mapping R^* satisfies the stronger upper semicontinuity condition property (Q) with respect to x as given in the above definition. As we will see, this fact will enable us to prove Theorem 6.

A.2 The Proof

In what follows, we will find the need to extract a number of sequences. To avoid extensive detailed notations, we will always assume the sequences are relabeled with the same index.

To begin our proof, let $T_0 \subset [a, b]$ be the set of Lebesgue measure zero where $A(t)$ is not closed and, for $i = 1, 2$ and $k \in \mathbb{N}$ put

$$j_k^i = \int_a^b \eta_k^i(t) \, dt,$$

and let $\tau^i = \liminf_{k \to \infty} j_k^i$. Observe that for $i = 1, 2$, we have $\tau^i > -\infty$. Since $x_k \to x$ in measure as $k \to \infty$, $\lambda_k^i \to \lambda^i$ weakly in $L^1([a, b]; \mathbb{R})$ as $k \to \infty$ and $\eta_k^i(t) \geq \lambda_k^i(t)$ for almost every $t \in [a, b]$, we can extract subsequence so that $x_k(t) \to x(t)$ pointwise almost everywhere on $[a, b]$ and

$$\lim_{k \to \infty} j_k^i = \lim_{k \to \infty} \int_a^b \eta_k^i(t) \, dt = \tau^i, \quad i = 1, 2.$$

We further observe that under our hypotheses, we have $\tau^1 \leq E$. Now, for $s \in \mathbb{N}$, let $\delta_s^i = \max\{|j_k^i - \tau^i| : k \geq s + 1\}$, and notice that $\delta_s^i \to 0$ as $s \to \infty$.

Now let $T_0' \subset [a, b]$ be the set of Lebesgue measure zero for which $x_k(t) \not\to x(t)$ as $k \to \infty$ and observe that we have

$$x(t) \in A(t) \quad \text{for all } t \in [a, b] \setminus (T_0 \cup T_0').$$

The sequences $\{\lambda_k^i\}$ ($i = 1, 2$) and $\{\xi_k\}$ converge weakly in L^1 to λ_k^i and ξ, respectively. By applying the Dunford-Pettis theorem (see Cesari [6, 10.3.i]), there exists a function $h : [0, +\infty) \to [0, +\infty)$ that is monotone nondecreasing, convex, continuous, and satisfying $h(\zeta)/\zeta \to \infty$ as $\zeta \to \infty$, such that the sequence of functions $\{\rho_k\}$, $\rho_k(t) \doteq h(|\xi_k(t)|)$ converges weakly in $L^1([a, b]; \mathbb{R})$ to some nonnegative integrable function $\rho : [a, b] \to \mathbb{R}$.

For any $s \in \mathbb{N}$, we have that the sequences $\{\rho_{s+k}\}$, $\{\lambda_{s+k}^i\}$ ($i = 1, 2$), and $\{\xi_{s+k}\}$ converge weakly in L^1 as $k \to \infty$, respectively, to ρ, λ^i ($i = 1, 2$) and ξ. Consequently by appealing to the Banach-Saks-Mazur theorem (see [6, 10.1.i]), there is a set of real numbers $c_{Nk}^{(s)} \geq 0$, $k = 1, 2, \ldots, N$, $N \in \mathbb{N}$ satisfying $\sum k = 1^N c_{Nk}^{(s)} = 1$ such that the sequences of functions $\{\rho_N^{(s)}\}$, $\{\lambda_N^{i(s)}\}$ ($i = 1, 2$) and $\{\xi_N^{(s)}\}$, defined, for $t \in [a, b]$ and $N \in \mathbb{N}$ by the formulas

$$\rho_N^{(s)}(t) \doteq \sum_{k=1}^{N} c_N^{(s)} \rho_{s+k}(t), \quad \lambda_N^{i(s)}(t) = \sum_{k=1}^{n} c_N^{(s)} \lambda_{s+k}^i(t) \ (i = 1, 2),$$

and

$$\xi_N^{(s)}(t) = \sum_{k=1}^{N} c_N^{(s)} \xi_{s+k}(t),$$

converges strongly in L^1, respectively, to ρ, λ^i $(i = 1, 2)$ and ξ. Moreover, this is true for every $s \in \mathbb{N}$.

Now, for every $s \in \mathbb{N}$, there exists a set $T_s \subset [a, b]$ of Lebesgue measure zero and a sequence of positive integers $N_l^{(s)} \to \infty$ such that for every $t \in [a, b] \setminus T_s$, one has that $\rho(t)$, $\lambda^i(t)$ $(i = 1, 2)$ and $\xi(t)$ are finite and the sequences $\{\rho_{N_l^{(s)}}^{(s)}(t)\}$, $\{\lambda_{N_l^{(s)}}^{i(s)}(t)\}$ $(i = 1, 2)$ and $\{\xi_{N_l^{(s)}}^{(s)}\}$ converge to $\rho(t)$, $\lambda^i(t)$ $(i = 1, 2)$ and $\xi(t)$ as $l \to \infty$.

Let $T = T_0 \cup T_0' \cup \{T_s : s \in \mathbb{N}\}$ and observe T has Lebesgue measure zero. Define the functions $\eta_N^{i(s)} : [a, b] \to \mathbb{R}$, for $i = 1, 2$ and $s, N \in \mathbb{N}$, by the formulas

$$\eta_N^{i(s)}(t) = \sum_{k=1}^{N} c_{Nk}^{(s)} \eta_{s+k}^i(t),$$

and observe that for almost all $t \in [a, b]$, we have

$$\rho_k(t) = h(\|\xi_k(t)\|), \quad \eta_k^i(t) \ge \lambda_k^i(t), \quad \text{and} \quad \lim_{k \to \infty} \int_a^b \eta_k^i(t) = \tau^i.$$

Consequently, for all $s \in \mathbb{N}$ and all $N \in \mathbb{N}$ we have

$$\eta_N^{i(s)}(t) \ge \lambda_N^{i(s)}(t) \quad \text{and} \quad \tau^i - \delta_s^i \le \int_a^b \eta_N^{i(s)}(t) \, dt \le \tau^i + \delta_s^i.$$

For $s \in \mathbb{N}$ and $i = 1, 2$ define $\eta^{i(s)} : [a, b] \to \mathbb{R}$ by the formula

$$\eta^{i(s)}(t) = \liminf_{l \to \infty} \eta_{N_l^{(s)}}^{i(s)}(t), \quad t \in T,$$

and zero elsewhere, and observe that $\eta^{i(s)}(t) \ge \lambda^i(t)$ $(i = 1, 2$ and $s \in \mathbb{N})$ for almost all t. By Fatou's lemma, we now have

$$\int_a^b \lambda^i(t) \, dt \le \int_a^b \eta^{i(s)}(t) \, dt \le \liminf_{l \to \infty} \int_a^b \eta_{N_l^{(s)}}^{i(s)}(t) \, dt \le \tau^i + \delta_s^i,$$

which implies that $\eta^{i(s)}$ is integrable and finite almost everywhere.

For $s \in \mathbb{N}$ and $i = 1, 2$ let $T_s^{i\prime} \subset [a, b]$ be the set of Lebesgue measure zero where $\eta^{i(s)}(t)$ is not finite and define the function $\eta^i : [a, b] \rightarrow \mathbb{R}$ by the formula $\eta^i(t) = \liminf_{s \rightarrow \infty} \eta^{i(s)}(t)$ for $t \in T_s^{i\prime}$ and zero elsewhere and observe that we have $\eta^i(t) \geq \lambda^i(t)$ for almost all $t \in [a, b]$ and that

$$\int_a^b \eta^i(t)\, dt \leq \tau^i.$$

For $t \in [a, b] \setminus T_s^i$ we have that $\lambda^i(t)$ is finite so that for all $l \in \mathbb{N}$ sufficiently large, say $l \geq l_0(t, s)$, we have $\eta_{N_l^{(s)}}^{i(s)}(t) \geq \lambda(t) - 1$. This means we can drop finitely many terms (depending on t and s) so that $\eta_{N_l^{(s)}}^{i(s)}(t) \geq \lambda(t) - 1$ for all $l \in \mathbb{N}$. Let $T_0^{i\prime\prime} \subset [a, b]$ be the set of measure zero where $\eta^i(t)$ is not finite. Finally let $T^* \subset [a, b]$ be the set of measure zero defined as the union of the sets T_0, T_0', $T_0^{i\prime\prime}$, T_s, and $T_s^{i\prime}$ ($s \in \mathbb{N}$ and $i = 1, 2$) and fix $t_0 \in [a, b] \setminus T^*$ and set $x_0 = x(t_0)$. Then $(t_0, x_k(t_0)) \rightarrow (t_0, x_0) \in A$ as $k \rightarrow \infty$ and for any $\epsilon > 0$ there exists $s_0 \in \mathbb{N}$ such that if $s \geq s_0$ one has $|x_s(t_0) - x_0| < \epsilon$. Now, for any $s \geq s_0$ and $k \in \mathbb{N}$, we have $(\eta_{s+k}^1(t_0), \eta_{s+k}^2(t_0), \xi_{s+k}(t_0)) \in R(t_0, x_{s+k}(t_0))$ and $|x_{s+k}(t_0) - x_0| < \epsilon$. Moreover, we also have for $l \geq l(t_0, s)$ that

$$\eta^{i(s)}(t_0) \geq \lambda^i(t_0) - 1,$$

which means that

$$\left(\sum_{k=1}^{N_l} c_{N_l^{(s)} k}^{(s)} \rho_{s+k}(t_0), \sum_{k=1}^{N_l} c_{N_l^{(s)} k}^{(s)} \eta_{s+k}^1(t_0), \right.$$

$$\left. \sum_{k=1}^{N_l} c_{N_l^{(s)} k}^{(s)} \eta_{s+k}^2(t_0), \sum_{k=1}^{N_l} c_{N_l^{(s)} k}^{(s)} \xi_{s+k}(t_0) \right) \in \mathrm{co} R^*(t_0, x_0; \epsilon), \qquad (\text{A.2})$$

where

$$R^*(t_0, x_0; \epsilon) = \bigcup \{ R^*(t_0, x) : |x - x_0| < \epsilon \},$$

in which $R^*(t_0, x)$ is given by (A.1) with $t = t_0$. The right side of (A.2) is a sequence of points in $\mathbb{R}^{m+3}$ which has the vector $(\rho(t_0), \eta^{1(s)}(t_0), \eta^{2(s)}(t_0), \xi(t_0))$ as an accumulation point. This means

$$(\rho(t_0), \eta^{1(s)}(t_0), \eta^{2(s)}(t_0), \xi(t_0)) \in \mathrm{cl}\, \mathrm{co} R^*(t_0, x_0; \epsilon)$$

$$= \mathrm{cl}\left(\mathrm{co}\left(\bigcup \{ R^*(t, x) : |x - x_0| < \delta \} \right) \right),$$

and since $\eta^i(t_0) = \liminf_{s\to\infty} \eta^{i(s)}(t_0)$ $(i = 1, 2)$ is finite, it follows that

$$(\rho(t_0), \eta^1(t_0), \eta^2(t_0), \xi(t_0)) \in \operatorname{cl} \operatorname{co} R^*(t_0, x_0; \epsilon)$$
$$= \operatorname{cl}\left(\operatorname{co}\left(\bigcup\{R^*(t, x) : |x - x_0| < \delta\}\right)\right).$$

Since this holds for all $\epsilon > 0$, it follows that

$$(\rho(t_0), \eta^1(t_0), \eta^2(t_0), \xi(t_0)) \in \bigcap_{\epsilon > 0} \operatorname{cl}\left(\operatorname{co}\left(\bigcup\{R^*(t, x) : |x - x_0| < \delta\}\right)\right)$$
$$= R^*(t_0, x_0),$$

since the set-valued map R^* satisfies property(Q).

The desired conclusion now follows since $t_0 \in [a, b] \setminus T^*$ is arbitrary, and the definition of R^* implies that for almost all $t \in [a, b]$ that $x(t) \in A(t)$, $(\eta^1(t), \eta^2(t), \xi(t)) \in R^*(t, x(t))$ and that

$$\int_a^b \eta^1(t)\, dt \le \tau^1 \le E \quad \text{and} \quad \int_a^b \eta^2(t)\, dt \le \tau^2,$$

as desired.   □

Appendix B: Measurable Selections

In this section we present some well-known results concerning measurable set-valued maps. We begin with the following definitions.

Definition B.1 *A set-valued map $\mathscr{M} : [a, b] \rightrightarrows \mathbb{R}^m$ is said to be Lebesgue measurable if and only if for every open subset $G \subset \mathbb{R}^m$, the set $\{t \in [a, b] : \mathscr{M}(t) \cap G \neq \emptyset\}$ is Lebesgue measurable.*

Definition B.2 *Given a set-valued mapping $\mathscr{M} : [a, b] \rightrightarrows \mathbb{R}^m$ a Lebesgue measurable function $m : [a, b] \to \mathbb{R}^m$ is called a Lebesgue measurable selection of $\mathscr{M}$ if and only if $m(t) \in \mathscr{M}(t)$ for almost all $t \in [a, b]$.*

Regarding the existence of Lebesgue measurable selections, we have the following theorem.

Theorem B.1 *A closed-valued, Lebesgue measurable mapping $\mathscr{M} : [a, b] \rightrightarrows \mathbb{R}^m$ always admits a Lebesgue measurable selection. That is, there exists a Lebesgue measurable function $m : [a, b] \to \mathbb{R}^m$ such that $m(t) \in \mathscr{M}(t)$ for almost every $t \in [a, b]$*

Proof See Rockafellar and Wets [8, Corollary 14.6]   □

Regarding set-valued maps related to the set-valued maps $\mathscr{M}$ defined in the proofs of Theorems 7 and 8 we have the following theorem.

Theorem B.2 *Let $\mathscr{V} : [a, b] \rightrightarrows \mathbb{R}^m$ be a Lebesgue measurable set-valued mapping such that $\mathscr{V}(t) \subset \mathbb{R}^m$ is closed for almost every $t \in [a, b]$, let $g_1, g_2, \ldots, g_l : [a, b] \times \mathbb{R}^m \to \mathbb{R}$ and $f_1, f_2, \ldots, f_k : [a, b] \times \mathbb{R}^m \to \mathbb{R}$ be continuous mappings, and let $\alpha_i : [a, b] \to \mathbb{R}, i = 1, 2, \ldots, l,$ and $\beta_j : [a, b] \to \mathbb{R}, j = 1, 2, \ldots, k,$ be Lebesgue measurable functions. Then the set-valued map $\mathscr{M} : [a, b] \rightrightarrows \mathbb{R}^m$ defined by the formula*

$$\mathscr{M}(t) \doteq \{v \in \mathscr{V}(t) : \ g_i(t, u) \geq \alpha_i(t), i = 1, 2, \ldots, l,$$

$$f_j(t, u) = \beta_j(t), \ j = 1, 2, \ldots, k\}$$

is a Lebesgue measurable set-valued mapping. Therefore, there exists a Lebesgue measurable function $u^* : [a, b] \to \mathbb{R}^m$ such that $u^*(t) \in \mathscr{M}(t)$ for almost every $t \in [a, b]$.

Proof See Rockafellar and Wets [8, Theorem 14.36]. $\qquad\qquad\qquad\qquad$ $\square$

Remark B.1 In Rockafellar and Wets [8, Theorem 14.36], the theorem is stated for normal integrands. This notion includes continuous functions as a special case.

A Discrete Model of Conformance Quality and Advertising in Supply Chains

Pietro De Giovanni and Fabio Tramontana

Abstract In this paper, we analyze a game in which a retailer adjusts her advertising efforts according to the type of defects (failures) that consumers experience, which depends on a manufacturer's conformance quality investments. Negligible failures are associated with an increase of the advertising expenditures, while catastrophic failures are followed by a decrease in the advertising efforts. Moreover, we take into consideration the bounded rationality of both retailer and manufacturer that do not know the shape of their objective functions. Taking all the relevant factors into account is quite demanding, and the assumption of perfect knowledge of the market appears unrealistic. In our model, the firms adopt the heuristic of adjusting their advertising and quality in the direction indicated by the respective marginal profits. We show that firms can optimize their profits when the defect is negligible and the speed of reaction to the marginal profit signals is not excessive, while a catastrophic failure leads to unstable solutions due to its operational and managerial complexity.

Keywords Discrete game • Advertising • Conformance quality • FMEA

1 Introduction

The recent game theoretical developments in supply chain management highlight the issue of interface between operational and marketing concerns [7]. Their joint consideration allows a decision-maker to improve the decision-making process and to face several implied trade-offs [5]. In particular, the relationships between conformance quality and advertising have received particular attention in the

P. De Giovanni (✉)
Department of Operations Management, ESSEC Business School, Paris, France
e-mail: pietro.degiovanni@essec.edu

F. Tramontana
Department of Mathemtical Sciences, Mathematical Finance and Econometrics, Catholic University of Milan, Milano, Italy
e-mail: fabio.tramontana@unicatt.it

© Springer International Publishing AG 2017

J. Apaloo, B. Viscolani (eds.), *Advances in Dynamic and Mean Field Games*,
Annals of the International Society of Dynamic Games 15,
https://doi.org/10.1007/978-3-319-70619-1_9

literature (e.g., [2, 4]). According to the total quality management (TQM) principles, conformance quality refers to the ability of a product to meet given design specifications. It represents the capacity of a firm to produce at a 100% conformance rate [12, 13]. Although the literature has shed light on the analysis of conformance quality (see [8–11, 15, 20], De Giovanni and Roselli (2012)), some recent business cases have demonstrated that none of these models have considered how the gravity of a nonconformance product influences consumers' experience and willingness to repurchase, as well as companies' use of marketing devices, such as advertising strategies, to overcome the possible loss of consumers.

Yet, the literature on TQM has emphasized the importance for firms to identify the failures and categorize them according to their level of severity. Specifically, failures can be classified on a scale from "negligible" to "catastrophic" defects according to the negative impact on the firms' business through the failure mode and effect analysis (FMEA) method. Negligible failures have no relevant effect on the business; thus, firms can continue to sell products in the market because the nonconformance issues relate to a specific production batch or a market niche. In contrast, a catastrophic failure links to a worst-case scenario, in which products become fully inoperative and the failure may result in unsafe operations and possible full market loss. In this case, nonconformance quality has a strong marketing implication as the companies' brand value can be damaged. The recent case of IKEA meatballs in 2013 is an excellent example. Czech authorities discovered horsemeat labeled as beef and pork in Swedish meatballs produced by the furniture giant. Under such circumstances, IKEA marketers could not support a typical advertising strategy to boost horse-meatball sales. Similarly, Toyota and Nissan have recalled 6.5 millions of Takata's cars because of a defective airbag system that explodes without any motivation causing car accidents. In fact, no company has undertaken a "classical" advertising strategy when catastrophic failures occur. Rather, companies should adjust their advertising messages to retain as many consumers as possible in their portfolio by explaining to them the corrective actions to be implemented. For example, IKEA spokeswoman Ylva Magnusson informed consumers that: "...the horse-meatballs were taken off the shelves in IKEA stores in all countries....IKEA is conducting its own tests of the affected batch..." Also, Ms. Magnusson also sent messages in order to comfort consumers: "We hope that by taking decisive action, we can show our customers that we take their concerns seriously...It's important that our customers feel safe, and if they have concerns they should contact us..." (Mollin 2013) [18].

Because the core message of advertising strategies depends on the type of defects that occur in the market, we wish to model a decisional rule for advertising that also accounts for the amplitude of nonconformity that explains either negligible or catastrophic business situations. To the best of our knowledge, this new decisional rule makes a substantial contribution to the field of conformance quality and advertising interfaces in supply chain management because research in this domain has yet to pay attention to the idea that advertising strategies should also be defined according to the type of defects that consumers experience. Specifically, we have modeled a discrete game of supply chain in which a manufacturer decides the conformance quality rate and the retailer sets the advertising expenditures by adjusting its behavior according to the type of nonconformity. Investing more in a classical advertising campaign will have a positive effect on profits when defects are negligible, while it will have a high deteriorating impact on profits in case of catastrophic defect. Therefore, we design a new decisional rule for decision-makers that will be used according to the impact of conformance quality on the business. Our findings show that when firms set any advertising strategy to recover credibility after a "negligible defect" occurs, our model successfully predicts firms' optimal advertising and quality strategies. In contrast, when firms adjust their advertising efforts after consumers experience a "catastrophic" defect, the solution becomes unstable for increasing speed of adjustment values, thus highlighting the highly complex network relationships that characterize this scenario.

The paper is organized as follows. Section 2 introduces a game of conformance quality and advertising. Section 3 presents the conformance quality and advertising dynamics, through which we show the evolution of firms' strategies according to the type of defects that occurred. Section 4 reports the stability analysis. Section 5 presents the managerial implications of the research, that also concludes and addresses the research limitations and possible future extensions.

2 A Theoretical Game Model of Advertising and Conformance Quality

A channel consists of two players, a manufacturer (player M) and a retailer (player R). From now on we will refer to the manufacturer as he and to the retailer as she. Each player seeks to maximize its profit function through different types of devices. M decides on the operational side of the supply chain that consists of some conformance quality efforts. To model the conformance quality, we allow M to set his optimal conformance quality rate, $q \in [0, 1]$, which can be interpreted as the fraction of defect-free (or conforming) products that M sends to the market. Put in another way, $1 - q$ represents the fraction of defective products that M sells to the market that consists of a percentage of nonconforming goods that the consumers tolerate. R decides the marketing devices of the supply chain, which are represented by the advertising efforts, a, in our model. Both the conformance quality and the advertising efforts influence the demand positively and linearly, according to the following functional form:

$$D(a, q) = \theta a + \gamma q \tag{1}$$

where θ represents the consumers' sensitivity to advertising and gives information about the advertising effectiveness. In fact, it shows the number of consumers that have been subjected to advertising messages that will be willing to purchase the product after internalizing that advertising information. γ represents a market potential that the SC exploits according to the conformance quality efforts. Ideally, all consumers γ will make a purchase when M sets $q = 1$, which implies that all products are defect-free. While this practice seems to be quite convenient and is highly promoted by Crosby with the zero-defects policy, plenty of doubts remain regarding its real operational success. The operational efficiency that links to conformance quality challenges M, who should consider three features when deciding the conformance rate:

- There is a direct relationship between the conformance quality and the appraisal and prevention costs. Appraisal costs consist of cost of inspection, testing, and other tasks to ensure that the product or process is acceptable. Prevention costs groups all costs that M faces to prevent any type of defect. Both appraisal and prevention costs directly increase in the conformance quality [3]; therefore, we assume that they take the following linear form:

$$C_p(q) = c_p q \tag{2}$$

where c_p is the marginal appraisal and prevention cost. Equation (2) triggers a trade-off with the demand; thus, the amplitudes of c_p and γ impose a serious operational reflection on the conformance rate.

- There is an indirect relationships between conformance quality and the cost of failures. The latter can be either internal failures, i.e., the cost for defects incurred within the system, or external failures, i.e., costs for defects that pass through the system and that reach the consumers. We assume that the cost for failures is linear in q and assume the following form:

$$C_d(q) = c_d(1 - q) \tag{3}$$

where c_d is the marginal cost for failures and includes the logistics cost for recalling the defective products in M's production plan.

- According to the zero-defect paradigm, the consumer is the king. Thus, M must reimburse consumers when they experience a nonconforming product. Assume that η is the price that consumers have paid to purchase the good; M also faces the reimbursement costs as it is represented by the following form:

$$C_\eta(q) = \eta(1 - q). \tag{4}$$

Both conformance and advertising generate expenses that are modeled through convex and quadratic cost functions:

$$C_q(q) = \frac{hq^2}{2} \tag{5}$$

$$C_a(a) = \frac{ka^2}{2} \tag{6}$$

where h and k are the conformance quality and advertising effectiveness, respectively. Equation (5) represents the amount of dollars that firms within a supply chain invest in order to reach certain levels of quality and pertains to quality planning, statistical process control, investment in quality-related information systems, quality training and workforce development, product-design verification, systems development and management, test and inspection of purchased materials, acceptance testing, inspection, testing, checking labor, setup for test or inspection, test and inspection equipment, quality audits, and field testing. When M optimally decides the conformance rate q, this decision implies a certain level of investments $C_a(q)$ as well as certain marginal costs described in equations (2), (3), and (4).

Equation (6) relates the efforts that the retailer spends to attract consumers through purchasing some products and includes store advertising, price announcements, and information on the product quality. We assume that R's profits incur the same penalty associated with M's conformance quality investments when $h = k$. In the latter case, marketing and operational strategies have the same efficiency. Therefore, we will refer to h and k as marketing and operational efficiency, respectively, and will measure the impact on the profit functions of each dollar invested in a given strategy.

We model a supply chain that coordinates the flow of money through a sharing contract. In particular, M and R share the revenues through a sharing rule $\pi \in (0, 1)$ so that M's marginal revenues will be $\pi \eta$ and the consumers' refund will be $(1 - q) \eta$ while R's marginal revenues will be $(1 - \pi) \eta$. To save notations and sake the exposition, we fix $\eta = 1$ and ensure that $c_p + c_d < 1$. Considering equations (2), (3), and (4), M's marginal profits will be $\pi - (1 - q) - c_p q - c_d (1 - q)$, while R's marginal profits equal her marginal revenues, as we have assumed that R only manages marketing strategies. Thus, she is not challenged by any type of operational trade-off. Similar coordination mechanisms have been shown to be successful and valid by Jorgensen and Zaccour [14].

Finally, the firms' objective functions assume the following functional forms:

$$\max_q \Pi_M = (\theta a + \gamma q) \left(\pi - (1 - q) - c_p q - c_d (1 - q) \right) - \frac{h q^2}{2} \tag{7}$$

$$\max_a \Pi_R = (\theta a + \gamma q) (1 - \pi) - \frac{k a^2}{2} \tag{8}$$

where $(1 - q)$ is the reimbursement to consumers, c_p is the marginal cost for preventions, and $c_d (1 - q)$ is the marginal cost for defects. We play the game á la Nash, so the firms' marginal profits are given by:

$$\frac{\partial \Pi_M}{\partial q} = q (2\gamma \varpi - h) + \gamma(\pi - 1) + \theta a \varpi - \gamma c_d \tag{9}$$

$$\frac{\partial \Pi_R}{\partial a} = \theta (1 - \pi) - ak \tag{10}$$

where $\varpi = 1 + c_d - c_p > 0$. Interestingly, M's optimal decisions on conformance quality depend on the R's advertising decisions, but the reverse is not true. Therefore, R decides her advertising efforts without considering M's conformance quality investments, while we have to identify once for all the conditions for R to get positive marginal profit; specifically, $\theta (1 - \pi) > ak$, thus, each dollar spent in advertising should be lower than the marginal revenue obtainable from selling one more unit of good to consumers through advertising. Surprisingly, R does not improve her profits if M sells more through conformance quality. The reverse does apply. The larger R's advertising efforts, the higher M's profits. Besides, we have to verify that M's marginal revenues are sufficiently greater than the operational inefficiency cost; then, M will find it convenient to invest more in conformance quality, while we need indeed to ensure that the share of revenue is larger than the cost of failures.

The Nash equilibria for the game are characterized by the optimal a^* and q^* strategies:

$$a^* = \frac{\theta (1 - \pi)}{k}$$

$$q^* = \frac{k\gamma c_d + k\gamma(1 - \pi) - \varpi\theta^2(1 - \pi)}{(2\gamma\varpi - h)\,k}$$

Finally, by inserting these strategies inside the profit functions (7) and (8), we obtain the corresponding maximal profits of manufacturer and retailer, respectively.

3   Discrete Dynamics of Conformance Quality and Advertising

The motivations to investigate our research questions in a dynamic context are driven by the long-term impact that defects might have on a business. Selling nonconform products clearly damages the business in the short term while also having some counterintuitive implications in the long term. This statement derives from the previous experiences of Toyota and IKEA. These firms have faced several issues after selling nonconform products, which related to catastrophic defect as their overall businesses resulted to be affected. In 2015, Toyota recalled 6.5 millions of Takata's vehicles because the airbag systems were not conformed and could also explode without any reason. This defective component boosted the number of accidents and implied multiple deaths, thus resulting in a catastrophic defect because it negatively affected the entire business [19]. Similarly, after the scandal, IKEA spokeswoman Ylva Magnusson sent several messages and underwent numerous interviews to inform consumers about the concerns. Then, she informed consumers about the actions taken to solve these concerns: meatballs from the nonconforming batch were taken off the shelves in IKEA stores in all countries, other shipments

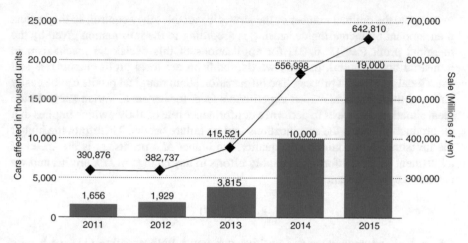

Fig. 1 Takata's recalls and sales (statista.com)

of meatballs were checked and were not affected, there was a reason to extend that guidance globally, and IKEA continued to conduct its tests. IKEA hoped that by taking decisive actions, it could show customers that it took all concerns very seriously and that consumers felt safe and free to contact IKEA for any concerns (Mollin 2013).

These two firms have faced significant issues in the short term, especially to recover the consumers' willingness to purchase from their brands. Nevertheless, both firms have been able to fix properly their strategy to survive in the long term. In fact, after the IKEA's scandals, the stores' food department did not stop working. Rather, they have increased the quality controls and adopted additional stricter practices to ensure people's health and safety (Mollin 2013). Also, after the scandal of airbag explosions, Toyota has intensified the number of recalls in 2016, without stopping its overall business. Although 11 deaths can be linked to nonconform cars [19], a correct policy allows the company to avoid the underperforming business in the long run. As illustrated in Figure 1, Takata's sales and operating incomes have increased over the past years, independent of the operational performance on quality.

Then, we modify our static model to investigate dynamic strategies and payoffs. Accordingly, M decides conformance quality rate by considering a dynamic mechanism that is given by:

$$q(t+1) = q(t) + s_q \frac{\partial \Pi_M}{\partial q(t)} \tag{11}$$

Equation (11) can be interpreted as follows: M evaluates how the conformance quality investments influence its marginal profits. This decisional mechanism relies on the hypothesis of bounded rationality of the firms. Basically, the intuition is the following: the firms are not able to gather and/or use all the information necessary to compute the optimal number of the strategic variables they control, so they adjust their amount by increasing/decreasing it according to the information given by the marginal profit (see [1, 6, 21] for applications of this decisional mechanism in industrial organization). In other words, the firms are local profit maximizers and use a local adjustment process. The information about marginal profits can be easily gathered through some market experiments (see, for instance, Varian [22]). In the ideal situation, M invests to perform a conformance rate of 100%, which implies that all products conform the specifications and no failure occurs. Therefore, the higher the investments in conformance quality, the higher M's profits. s_q is the speed of adjustment of the conformance quality efforts to any change in M's profits, and we assume it to be constant:

$$s_q = \lambda_q > 0$$

where λ_q is an exogenous and positive constant.[1] With equation (11) we have a clear objective in mind: Undertake a proactive approach to conformance quality in which appraisal and prevention cover a major role. Firms should identify their optimal conformance quality policy by investing in appraisal and prevention rather than facing the problem when failures occur. The latter approach has been adopted by previous research on quality management and conformance quality (e.g., [20]), according to which companies invest in conformance only after consumers face some defects. In contrast, according to the zero-defect Japanese principles, failures must be avoided rather than faced. This can be done only through the implementation of an optimal appraisal and prevention strategy.

Similarly, the mechanism that permits to choose the level of advertising at any time period is:

$$a(t+1) = a(t) + s_a(a(t), q(t)) \frac{\partial \Pi_M}{\partial a(t)} \tag{12}$$

but contrary to M, R's speed of adjustment is more complex and it is described as follows:

$$a(t+1) = \begin{cases} a(t) + \lambda_a a(t) [1 - q(t)] \dfrac{\partial \Pi_R}{\partial a(t)} & \text{if } \dfrac{\partial \Pi_R}{\partial a(t)} \geq 0 ==> \text{case i)} \\[4mm] a(t) + \lambda_a a(t) q(t) \dfrac{\partial \Pi_R}{\partial a(t)} & \text{if } \dfrac{\partial \Pi_R}{\partial a(t)} < 0 ==> \text{case ii)} \end{cases} \tag{13}$$

where λ_a is an exogenous and positive constant. The presence of the actual level of advertising $a(t)$ permits to model a higher reactivity of bigger firms (those investing

---

[1]Actually it is technically possible that by using extremely high values of λ_q, the conformance quality level exits the feasible range $[0, 1]$. In the following, we avoid using such values ensuring the feasibility of the conformance quality level.

more in advertising) with respect to smaller ones (see [1]). The intuitions behind the developments of equation (13) derive from the unexplored relationships between conformance quality and advertising strategies. The way companies set and adjust their advertising strategies depends on the type of defect that occurred in the system. In fact, the total quality management literature has developed the failure mode and effect analysis (FMEA), which is a systematic technique for detecting and removing failures while providing a comprehensive report on the resultant effect on the system operations. Failures are classified according to a severity number that goes from 1) no effect to 6) catastrophic effect.

The interpretation of case i) in equation (13) is as follows: When the number of defects is low (high q), defects are sporadic and manageable at a single consumer level. In such situation, the defects are of categories "no/negligible," because they have an isolated negative impact on the entire business. Accordingly, increasing the advertising efforts still has a positive influence on firms' profits because the other consumers do not necessarily need to know of these low amounts of defects, which interests a small piece of the market. Thus, the non-affected consumers, who represent a big proportion, can continue to satisfy their needs even though some other consumers have faced some defects. Typical examples are dents in beverage cans, simple typos in the documentation, defected component in a car, or broken display on a smartphone. When q is high, the decision-maker increases the advertising efforts just enough to recover the low number of nonconforming products inserted in the market. Because most of the products are of high quality, advertising seeks to retain a high-quality impression to consumers, so any experienced nonconformity is negligible.

Instead, when the number of defects is high (low q), defects are systematic and must be managed at a market level. That is, the defects are of categories "catastrophic," because they exert a negative impact on the worldwide business. Consequently, increasing the advertising efforts has a huge negative influence on firms' profits due to the current marginal manufacturer's capability of producing good products. Spending more in advertising compromises the firms' profits because consumers not only are unable to satisfy their needs, but their health or safety can be affected as well, thus negatively influencing the company's worldwide business. In such circumstances, case i) in equation (13) is no longer appropriate for adjusting advertising efforts according to the operational performance. Thus, we have developed case ii) in equation (13) whose interpretation is as follows: Firms must urgently decrease their advertising efforts to benefit of a lumpy decrease of demand when "catastrophic" (systematic, worldwide) defects occur. Firms decrease their advertising efforts when products sold in the worldwide market are of low quality as this may result in a global failure for the business. This claim results from Figures 2 and 3, in which we display the IKEA's and Toyota's advertising efforts over time.

Accordingly, both firms have decreased the advertising efforts in the year of the scandal, thus investing in the advertising according to the operational quality faults. This evidence reinforces our motivations to look at a dynamic model in which firms adjust their advertising efforts according to the operational performance effects on their overall business.

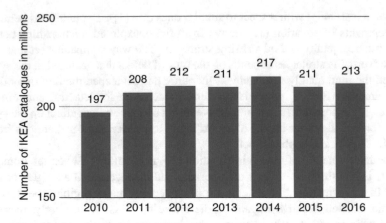

Fig. 2 Ikea's number of catalogues (statista.com)

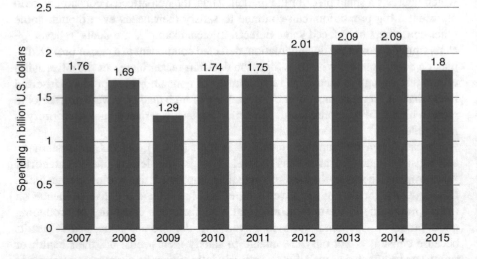

Fig. 3 Toyota's advertising efforts (statista.com)

4 Dynamic Analysis and Stability Conditions

The dynamical system made up by equations (11) and (13) determines the time evolution of these variables:

$$
\begin{cases}
q' = q + \lambda_q \dfrac{\partial \Pi_M}{\partial q} \\[2ex]
a' = \begin{cases}
f_L(q, a) = a + \lambda_a a\,(1 - q)\,\dfrac{\partial \Pi_R}{\partial a} & \text{if } \; a < \dfrac{\theta(1 - \pi)}{k} \\[2ex]
f_H(q, a) = a + \lambda_a a q \dfrac{\partial \Pi_R}{\partial a} & \text{if } \; a \geq \dfrac{\theta(1 - \pi)}{k}
\end{cases}
\end{cases}
\tag{14}
$$

where L(ow) (resp. H(igh)) denotes the regimes where the level of advertising is lower (resp. higher) than its optimal amount.

We first check whether these trajectories will converge toward the steady state by computing the Jacobian matrix of the system at the steady-state values. Each neighborhood of the steady state is characterized by points of the phase plane belonging to both regions of definition of the advertising dynamics. Hence, we need to study two Jacobians referring to low and high values, respectively, which are the following:

$$J_L : \begin{bmatrix} 1 + \lambda_q (2\gamma\varpi - h) & \lambda_q\theta\varpi \\ 0 & 1 - k\lambda_a a^*(1 - q^*) \end{bmatrix} \quad J_H : \begin{bmatrix} 1 + \lambda_q (2\gamma\varpi - h) & \lambda_q\theta\varpi \\ 0 & 1 - k\lambda_a a^* q^* \end{bmatrix}$$
(15)

Note that only the J_{22} elements of the matrices differ.

We will be in one of the following scenarios:

1. The steady state is locally stable in both the regimes;
2. The steady state is unstable in both the regimes;
3. The steady state is locally stable in one regime and unstable in the other.

While we can reasonably ensure that the steady state in the first scenario is convergent to a steady state, the same does not hold true in the second scenario. The third scenario is the most ambiguous and does not permit a clear answer to the question. Since the Jacobian matrices displayed in (15) are similar, the relevance of the third case is negligible.

In a 2D map, a steady state is locally stable provided that the following conditions hold at the same time:

$$\begin{aligned} &(i) \ \ 1 - Tr + Det > 0 \\ &(ii) \ \ 1 + Tr + Det > 0 \\ &(iii) \ \ 1 - Det > 0 \end{aligned}$$
(16)

where Tr and Det denote trace and determinant of the Jacobian matrix calculated at the steady-state values.

Condition (i) gives rise to the same condition in both the regimes[2]:

$$2\gamma\varpi - h < 0$$
(17)

from which is clear the destabilizing effect of an increasing γ or c_d or a decreasing of c_p or h.

---

[2]We only consider cases in which the steady states are feasible, i.e., $a^* > 0$ and $0 \le q^* \le 1$.

Condition (ii) is slightly different in the two regimes:

$$\text{Regime } L \longrightarrow (2\gamma\varpi - h)\,\lambda_q\,[k\lambda_a a^*(1 - q^*) - 2] + 2k\lambda_a a^*(1 - q^*) < 4$$
$$\text{Regime } H \longrightarrow (2\gamma\varpi - h)\,\lambda_q\,(k\lambda_a a^* q^* - 2) + 2k\lambda_a a^* q^* < 4$$

$$(18)$$

If we limit our analysis to cases where condition (i) holds, then we have the simplified conditions:

$$\text{Regime } L \longrightarrow \lambda_a < \widetilde{\lambda}_a^L \cap \lambda_q < \widetilde{\lambda}_q$$
$$\text{Regime } H \longrightarrow \lambda_a < \widetilde{\lambda}_a^H \cap \lambda_q < \widetilde{\lambda}_q$$

$$(19)$$

with:

$$\widetilde{\lambda}_a^L = \frac{2}{ka^*(1 - q^*)} \quad \widetilde{\lambda}_a^H = \frac{2}{ka^* q^*}$$
$$\widetilde{\lambda}_q = -\frac{2}{2\gamma\varpi - h}$$

$$(20)$$

and

$$\widetilde{\lambda}_a^L > \widetilde{\lambda}_a^H \text{ if } q^* > \frac{1}{2}$$
$$\widetilde{\lambda}_a^H > \widetilde{\lambda}_a^L \text{ if } q^* < \frac{1}{2}$$

Condition (iii) gives rise to the following couple of stability rules:

$$\text{Regime } L \longrightarrow \lambda_q < -\frac{k\lambda_a a^*(1 - q^*)}{(2\gamma\varpi - h)\,[k\lambda_a a^*(1 - q^*) - 1]}$$
$$\text{Regime } H \longrightarrow \lambda_q < -\frac{k\lambda_a a^* q^*}{(2\gamma\varpi - h)\,[k\lambda_a a^* q^* - 1]}$$

$$(21)$$

Subsequently, we will only consider scenarios where condition (i) holds; otherwise dynamics become divergent and not economically meaningful.

The optimal way for studying the other conditions is to look at the (λ_a, λ_q) parameters' plane. By fixing the other parameters, we obtain some qualitative result as displayed in Figure 4.

The NS bifurcation curve[3] originated by condition (iii) is never crossed before a violation of condition (ii) leading to a flip bifurcation. In fact, the point $(\widetilde{\lambda}_a^L, \widetilde{\lambda}_q)$ belongs to the NS curve related to the L regime, and the point $(\widetilde{\lambda}_a^H, \widetilde{\lambda}_q)$ is a point of

---

[3]It is the locus of points in a parameter plane when the Jacobian matrix calculated at the steady state has two complex and conjugated eigenvalues whose modulus is equal to 1 (see [16]).

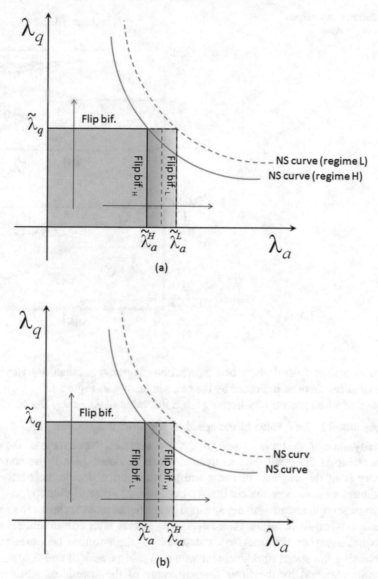

Fig. 4 Bifurcation analysis with $q^* < \frac{1}{2}$ (a) and (b) $q^* > \frac{1}{2}$

the *NS* curve of the *H* regime. Moreover *NS* curves are hyperbolas in the (λ_a, λ_q) parameters' plane with decreasing branches, one of them in a region with negative values of the speeds of adjustment, and thus not interesting.

In other words, when condition (i) holds, the steady state may lose stability only via flip bifurcation.

Fig. 5 Bifurcation analysis
when λ_2 is constant

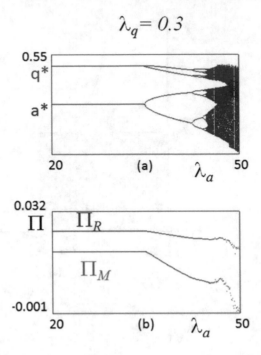

Figures 5(a) and 6(a)[4] show two bifurcation diagrams obtained by varying the speeds of adjustment as indicated by the two red arrows in Figure 1.

Note that when we vary λ_a keeping fixed the value of $\lambda_q < \widetilde{\lambda}_q$, the steady state becomes unstable for a value of the speed of adjustment $\widetilde{\lambda}_a > min\left[\widetilde{\lambda}_a^L, \widetilde{\lambda}_a^H\right]$. Thus, the steady state is unstable in one of the two regimes. Nevertheless, the steady state is stable in the other one, so dynamics can be convergent if the converging trajectory is in this regime. In order words, when firms decide their advertising expenditures to overcome the challenges implied by "no/negligible defects," firms can always set optimal advertising and quality strategies and maximize their profits. In contrast, when firms adjust their advertising efforts after consumers experience defects belonging to the category "catastrophic," the solution becomes unstable for increasing the speed of adjustment values, $\widetilde{\lambda}_a$. Figures 5(b) and 6(b) represent the profits obtained by the firms for any value of the speeds of adjustment in the considered interval. Obviously, the profits are maximal when the speeds of

---

[4]Figures 5 and 6 are both obtained by keeping fixed the parameters $h = 1$, $k = 0.42$, $\theta = 1.15$, $\gamma = 0.2$, $\pi = 0.9$, $c_p = 0.045$, and $c_d = 0.062$. In Figure 5, we see that $\lambda_q = 0.3$ and λ_a varies, while in Figure 6, λ_a is fixed at a value of 2 and λ_q varies. These scenarios are qualitatively similar to what can be obtained with any other parameters' combinations.

Fig. 6 Bifurcation analysis
when λ_a is constant

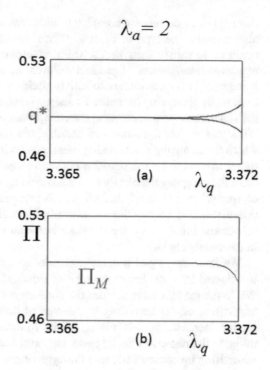

adjustment are low, and the firms learn the optimal strategies. When complicated
dynamics occur, the average profits[5] of both firms decrease.

5 Conclusions

This paper has addressed a novel problem of marketing and operations interface
in supply chain management, in which advertising strategy seeks to mitigate any
possible loss of demand due to underperforming quality. The latter is investigated in
the form of conformance quality, which represents the fraction of defect-free goods
that a manufacturer sells to the market. Because production processes are subject
to systematic variation, some errors and failures during the production may lead
to nonconforming goods, whose performance is not compatible with the expected
product performance. The product failure can have either negligible or catastrophic
consequence on consumers' satisfaction and, in the worst case, on consumers'
lives. As stated by the failure mode and effect analysis, the consequences of

---

[5]We consider average profits because after the flip bifurcation, the firms alternate different values
of advertising and conformance quality, so at each time period, they obtain a different profit. In
Figures 5(b) and 6(b) the average profits are drawn.

these types of defects are completely different, and the advertising strategy should also consider the type of defects. When the defect is negligible, consumers can continue to satisfy their needs while advertising aims at mitigating the possible negative consequences of product failures. In contrast, catastrophic defects make it impossible for consumers to satisfy their needs and even endanger consumers' lives while preserving the entire business. Thus, the advertising strategy cannot have the same message content as in the case of negligible failure. The case of IKEA in 2013 was a typical example of catastrophic consequences of product defects, in which the company's advertising messages no longer aimed at increasing sales but rather at smoothing the negative consequences of selling horse meatballs. In that case, the company drastically cut its advertising campaign seeking to encourage the consumption of its meatballs. All models presented in the literature have disregarded this distinction (negligible vs. catastrophic failures), even though it can have a substantial influence on the interface between marketing and operations strategies in the supply chain.

We have developed a decisional rule according to which advertising strategy is adjusted by considering the type of nonconformity that consumers experience. While the manufacturer decides the conformance quality level, the retailer sets her advertising efforts according to the type of defects. Under negligible failures, the retailer increases the advertising efforts to cover some loss of demand and, thus, mitigate the negative effects; under catastrophic failures, the retailer decreases her advertising investments because this type of nonconformity cannot be overcome by an advertising strategy, independent of the media used and the possible implications for the marketing mix.

We have also considered that firms may not access to all of the information required to compute the optimal amounts of advertising and conformance quality. Thus, they use a myopic adjustment mechanism, following the signal given by the marginal profits.

Our results show that the new decisional rule suggests how to fix advertising strategies according to the type of defects. When the failures are negligible, firms have the possibility to recover business through an advertising strategy. In other words, the inefficiency due to a suboptimal conformance quality strategy can always be overcome by the establishment of a proper advertising strategy. Managerially speaking, our decisional rule shows that the case of negligible product failure can be somehow managed by supply chains as its consequences have a marginal impact on the overall business. When failures are catastrophic, any advertising strategy is ineffective. In such a case, the operational inefficiency cannot be recovered through a simple advertising campaign, and supply chain members have to manage severe facts of demand and goodwill loss. Under these circumstances, it becomes quite difficult to manage the business, while the optimal strategies have been shown to be locally stable only for some values of the parameters. In particular, if the firms are excessively reactive to the signals provided by the marginal profits, they may fail to employ optimal strategies, determining dynamics of advertising and conformance quality that can be periodic or even chaotic. In that case, their profits, on average, decrease.

To the best of our knowledge, this research is the first attempt to look at a decisional rule to adjust advertising strategies according to the type of nonconformity experienced by consumers. Future research can be developed in several directions. Research needs to address the impact of catastrophic events on brand value and goodwill, while the existence of supply chain competition may lead to higher pressure to optimally adjust advertising strategies. Supply chains can coordinate on conformance quality policies to avoid catastrophic failures. The role of contracts can be essential to better coordinate marketing and operations in supply chains. We assume that the consumers experiencing a nonconformance receive a full reimbursement while disregarding the operational implications of such returns. The returned (nonconform) goods can be reprocessed and reused to make new products and thus impact positively on the production cost function. Finally, other dimensions of quality (e.g., durability, quality improvements, etc.) can be considered to further establish the interfaces between marketing and operations.

References

1. Bischi, G.I., Naimzada, A.K., Global Analysis of a Dynamic Duopoly Game with Bounded Rationality, in *Advances in Dynamic Games and Applications*, Filar J.A. et al. (Eds.), Birkhauser, 361–385, 2000.
2. De Giovanni, P., Quality improvement vs. Advertising support: which strategy works better for a manufacturer? European Journal of Operational Research. 208 2 (2011) 119–130.
3. De Giovanni, P., Roselli, M., Overcoming the drawbacks of a Revenue Sharing Contract in a marketing channel through a coop program. Annals of Operations Research, 196 1 (2012) 201–222
4. De Giovanni, P., Should a retailer support a quality improvements strategy? . In: Annals of the International Society of Dynamic Games. , Birkhauser, 125–148 2013.
5. De Giovanni, P., State- and Control-dependent incentives in a Closed-loop supply chain with dynamic returns. Dynamic Games and Applications, 1 4 (2015) 1–35.
6. Dixit, A., Comparative statics for oligopoly. International Economic Review 27 (1986) 107–122;
7. Erickson, G. A differential game model of marketing-operations interface. European Journal of Operational Research. 211 2 (2011) 394–402.
8. El Ouardighi, F., Pasin F., Quality improvement and goodwill accumulation in a dynamic duopoly, European Journal of Operational Research, 175 2 (2006) 1021–32.
9. El Ouardighi, F., Jørgensen, S., Pasin, F., A dynamic game of operations and marketing management in a Supply Chain, International Journal of Game Theory Review, 10 2 (2008) 1–20.
10. El Ouardighi, F., Kim, B., Supply quality management with wholesale price and revenue-sharing contract under horizontal competition. European Journal of Operational Research, 206 2 (2010) 329–340
11. El Ouardighi, F., Kogan, K., Dynamic conformance and design quality in a supply chain: An assessment of contracts' coordinating power, Annals of Operations Research, 211 1 (2013) 137–166
12. Fine, C., Quality improvement and learning in productive systems, Management Science, 32 10 (1986) 1301–1315.
13. Fine, C., A quality control model with learning effects, Operations Research, 36 3 (1988) 437–444.

14. Jørgensen, S. and Zaccour, G., A differential game of retailer promotions, Automatica, 39 7 (2003) 1145–55.
15. Kogan, K. and Tapiero, C., Supply Chain Games: Operations Management and Risk Evaluation, Springer Series in Operation Research, New York Inc. 2007
16. Kuznetsov, Y. Elements of Applied Bifurcation Theory, Springer-Verlag, New York, 1995.
17. Mollin, A. IKEA Chief Takes Aim at Red Tape, The Wall Street Journal, Jan, 22, 2013.
18. Pollak, S. Horsemeat Scandal Spreads to Ikea Swedish Meatballs, Time, Feb. 26, 2013.
19. Tabuchi, H. and Soble, J. (2015). Toyota and Nissan recal 6.5 million more vehicles over Takata airbags, The New York Times, May, 2015.
20. Tapiero, C., Production learning and quality control, IIE Transaction, Vol. 19 (1987)362–370.
21. Tramontana, F., Heterogeneous duopoly with isoelastic demand function. Economic Modelling 27 (2010) 350–357;
22. Varian, H. R. Microeconomic Analysis, 3rd ed. W. W. Norton, 1992.

Sexual Reproduction as Bet-Hedging

Xiang-Yi Li, Jussi Lehtonen, and Hanna Kokko

Abstract In evolutionary biology, bet-hedging refers to a strategy that reduces the variance of reproductive success at the cost of reduced mean reproductive success. In unpredictably fluctuating environments, bet-hedgers benefit from higher geometric mean fitness despite having lower arithmetic mean fitness than their specialist competitors. We examine the extent to which sexual reproduction can be considered a type of bet-hedging, by clarifying past arguments, examining parallels and differences to evolutionary games and presenting a simple model examining geometric and arithmetic mean payoffs of sexual and asexual reproduction. Sex typically has lower arithmetic mean fitness than asex, while the geometric mean fitness can be higher if sexually produced offspring are not identical. However, asexual individuals that are heterozygotes can gain conservative bet-hedging benefits of similar magnitude while avoiding the costs of sex. This highlights that bet-hedging always has to be specified relative to the payoff structure of relevant competitors. It also makes it unlikely that sex, at least when associated with significant male production, evolves solely based on bet-hedging in the context of frequently and repeatedly occupied environmental states. Future work could usefully consider bet-hedging in open-ended evolutionary scenarios with *de novo* mutations.

Keywords Bet-hedging • Environmental fluctuation • Evolutionary games • Geometric mean fitness • Sexual reproduction

X.-Y. Li (✉) • H. Kokko
Department of Evolutionary Biology and Environmental Studies, University of Zurich, Winterthurerstrasse 190, 8057 Zurich, Switzerland
e-mail: xiangyi.li@ieu.uzh.ch; hanna.kokko@ieu.uzh.ch

J. Lehtonen
Evolution and Ecology Research Centre, School of Biological, Earth and Environmental Sciences, University of New South Wales, Sydney, NSW 2052, Australia
e-mail: jussi.lehtonen@iki.fi

© Springer International Publishing AG 2017
J. Apaloo, B. Viscolani (eds.), *Advances in Dynamic and Mean Field Games*,
Annals of the International Society of Dynamic Games 15,
https://doi.org/10.1007/978-3-319-70619-1_10

217

1 Introduction

Evolutionary dynamics in natural populations are under the combined effect of directional selection and randomness that comes from various sources, including environmental fluctuations and demographic stochasticity. Accurate predictions of evolutionary dynamics depend, in principle, on all the moments of the fitness distribution of individuals and their relative weights. In general, populations tend to be driven towards phenotypes that maximise the odd moments (mean fitness being the first moment) while minimising the even moments of their fitness distributions (variance being the second moment) [36]. This implies that the adverse change of one moment can potentially be compensated by the beneficial changes of other moments. Most attention has been placed on the possibility that decreased mean fitness might be sufficiently compensated for by a concomitant decrease of the variance in fitness, such that the strategy with diminished mean fitness outcompetes others over time [35]. Because strategies that gain success by manipulating fitness variance intuitively fit the idea of "hedging one's bets" [40], this has given rise to a precise biological meaning of the phrase "bet-hedging" [39]: it refers to strategies that have diminished arithmetic mean fitness but also reduced variance (and are often studied with the aid of geometric mean fitness).

Bet-hedging bears some similarity to mixed strategies in evolutionary games (the phrase "optimal mixed strategies" [17, 18] has been used near-synonymously with bet-hedging under non-game-theoretical contexts): some forms of bet-hedging imply the production of different kinds of offspring (e.g. different sizes of tubers in the aquatic macrophyte *Scirpus maritimus* [9]). Although both bet-hedging and mixed strategies (in game theory) can lead to a mix of phenotypes in the population, there are two important differences between the concepts: first, the adaptive reasoning is different, and, second, bet-hedging can also occur without phenotypic variation. To explain the first difference: In evolutionary games, the payoff of an individual depends on the action of other individuals in the population. This is not a requirement in bet-hedging, where the payoff is typically thought to be determined by the stochastically varying environment (though, as our examples show, the presence of others can matter too: e.g. sexual reproduction to diversify one's offspring to cope with environmental change would not work if diversity has been lost). A typical context in which bet-hedging is discussed is rainfall that varies over time [37, 40]. Under such conditions it can then be beneficial if an individual can produce both wet-adapted and dry-adapted offspring, so that regardless of the conditions in a given year, some fraction of offspring will survive; a non-bet-hedger's entire genetic lineage might disappear as soon as an environmental condition occurs to which it is not adapted.

The second difference between mixed strategies and bet-hedging is that the latter can work without there being a "mix" of any kind. Instead of diversifying offspring, a so-called *conservative* way of bet-hedging is to produce only one type of offspring that performs relatively well under all different environments, while not being the best under any of them ("a jack of all trades is the master of none"). This can also reduce fitness variance and qualify as bet-hedging if it is achieved at the cost of reduced mean fitness.

One prominent example that seems to have the characteristics of bet-hedging, but is less often mentioned in a bet-hedging context, is sexual reproduction, where offspring are formed using genetic material from two parents (because nature is diverse, there are definitional complications and grey zones regarding what counts as sex [24]). Producing offspring in this way, as opposed to the simpler option of asexual reproduction, incurs costs in many different ways [25, 34]. The best known cost, and the one we focus on here, is the twofold cost of males: if the offspring sex ratio is 1:1 and males and females are equally costly to produce, a mother will use 50% of her resources on offspring that do not themselves contribute material resources to the next generation [30], and this slows the growth of sexual populations compared with asexual ones. Consequently, sexual reproduction – when it involves producing males – is expected to lead to a reduction of mean fitness. But on the other hand, through mixing genetic material from different lineages, sex provides a potent way of producing offspring whose genomes differ from each other. If some always do well no matter what the state of the environment, the variance of reproductive fitness can be reduced compared with an asexual lineage.

Given that effects on genetic diversity are central and much discussed in the sex literature (e.g. [22]), it is surprising that the biological literatures on bet-hedging and on sex are relatively separate. Mixed strategies have been shown to be advantageous in a fluctuating environment [17, 18, 33]. Haccou and Iwasa [17] have shown that the optimal strategy can involve bet-hedging under a fluctuating environment in unstructured populations and showed how to calculate the strategy explicitly for a given payoff function and a given distribution of the environmental parameters. In addition, the optimal bet-hedging strategy is robust against small perturbations of the distribution of environmental conditions and/or the payoff function [18]. Cooperative games between kin can also help maximise the geometric mean fitness of species in fluctuating environments [32]. Furthermore, the strategy that maximises the geometric mean fitness is more likely to evolve in species of nonoverlapping generations compared to species with substantial parental survival. In the latter case, the strategy that maximises the arithmetic mean fitness is more likely to evolve [19]. The review of Grafen [16] discusses different ways of optimising reproductive fitness in a fluctuating environment. None of these studies, however, have explicitly pointed out that sexual reproduction can be a form of bet-hedging.

Williams in his classic book on sex [45] discusses a "lottery model" using the verbal analogy of buying ever more copies of the same number on a lottery ticket (asexual reproduction) vs. buying fewer but a more diverse set of numbers (sexual reproduction). The analogy to a real-life lottery is not perfect, in the sense that asexually produced offspring are often not totally redundant copies of each other, i.e. they do not necessarily have to share the prize if both have a winning number: two asexually produced offspring usually leave more descendants than just one, especially if they disperse to different localities and no longer compete for the same resources ([45], p. 16). The correspondence between Williams' lottery model and bet-hedging, on the other hand, appears perfect. But Williams [45] did not use explicit bet-hedging terminology, probably because it had only very recently been imported to evolutionary terminology [39].

Williams [45] emphasised the need to consider the spatial arrangement of offspring to determine whether, e.g., 10 "winning tickets" can win 10 prizes, which requires dispersal to avoid competition with relatives, or are expected to win less ([45], p. 53). The emphasis in Williams' idea is that the winning numbers vary over time (but not necessarily over space). In a context where dispersal is limited, a similar idea has been formulated emphasising resource diversity rather than its temporal fluctuations. The relevant metaphor is a "tangled bank", a rather poetic phrase that has its origin in Darwin's *On the origin of species*. Darwin contemplated "a tangled bank, clothed with many plants of many kinds, with birds singing on the bushes, with various insects flitting about, and with worms crawling through the damp earth..." [10]. Darwin was not talking specifically about sex, but about life and its evolution in general. Nevertheless, the "tangled bank" has since acquired a specific meaning [4], becoming a metaphor of genetic polymorphisms favoured in environments that might not vary much temporally but that, based on diverse resources present at the same site, offer multiple niches and the resultant higher total carrying capacity for different phenotypes as a whole ("the environment is now more fully utilised ..., the carrying capacity of the diverse population will inevitably exceed that of either single clone" [4], p. 130).

In the "tangled bank" scenario, the carrying capacity of each single clone depends on the distribution of different niches in the environment. The carrying capacity of the entire diversified population in the heterogeneous environment is larger than any of the single clones. Although the "tangled bank" does not require a temporally fluctuating environment, the diversity of different clones is maintained better if the environment changes frequently [4]. In addition, in the "tangled bank", the fitness of a single clone depends not only on the abundance of different niches but also is frequency-dependent when competing for the same niche or invading a new niche [4]. Therefore, the "tangled bank" may capture aspects of the benefits of sexual reproduction, but it does not perfectly correspond to bet-hedging.

2 Bet-Hedging via Heterozygotes and Sexual Reproduction

We examine in the following the conditions under which sexual reproduction might spread as a form of bet-hedging. Our model considers a large well-mixed population where a proportion s of the young produced are male. Note that our assumption of large (infinite) population size allows us to focus on the effects of environmental stochasticity without confounding effects of demographic stochasticity. Asexual individuals are all female. The adaptation to the amount of rainfall in the environment is determined by a diploid genetic locus that has two alleles. The AA genotype

Table 1 The payoff structure under wet and dry years: the arithmetic mean (AMean) and the geometric mean (GMean) of the payoffs of asexual lineages, as well as of a sexual population assumed to be at the Hardy-Weinberg equilibrium

	Wet	Dry	AMean	GMean
asex-*AA*	8	2	5	4
asex-*Aa*	4.5	4.5	4.5	4.5
asex-*aa*	2	8	5	4
Sex-population	4.75(1-*s*)	4.75(1-*s*)	4.75(1-*s*)	4.75(1-*s*)

is well adapted to the wet environment, whereas the *aa* genotype is dry-adapted. The heterozygote *Aa* has intermediate fitness in both environments, but not necessarily exactly the mean of *aa* and *AA*. Example fitness values for each genotype under different environments are shown in matrix (1):

$$\begin{array}{c} \\ AA \\ Aa \\ aa \end{array} \begin{array}{cc} \text{Wet} & \text{Dry} \\ \left(\begin{array}{cc} 8 & 2 \\ 4.5 & 4.5 \\ 2 & 8 \end{array} \right) \end{array} \tag{1}$$

Consider a case where wet and dry environments occur at equal frequencies, and all individuals are asexual females. Table 1 shows the arithmetic mean and geometric mean fitness of the different asexual types. The heterozygote *Aa* has the lowest arithmetic mean fitness, but the highest geometric mean fitness, which predicts higher evolutionary success if we ignore higher moments of the fitness distribution [40]. The asexual heterozygotic form becomes thus a bet-hedging strategy when compared with the two other asexual homozygotic forms. This form of bet-hedging is *conservative*: all *Aa* individuals have the same expected fitness under both environmental conditions.

In contrast to the conservative approach of the asexual heterozygotes, the sexual population as a whole can also be seen to bet-hedge, in this case by producing offspring of different genotypes. It is therefore of interest to ask if sex is a bet-hedger with respect to *AA*, *Aa*, *aa* or perhaps all of them. The comparison is more complicated than the above one, not only because sex produces young that differ from each other (and thus differ in the long-term growth rate impacting the original parent's contribution to the future gene pool) but also because the frequencies of genotypes in the offspring of any given parent depend on the genetic composition of the population as a whole – which in turn depends on how selection has worked on it in the recent past: a run of wet years will have favoured the *A* allele and dry years do the opposite.

We initially assume that the sexual population is always under Hardy-Weinberg equilibrium [21, 43] and that the two alleles are equally abundant. This is a strong assumption that is expected to be violated as soon as selection is applied, but we nevertheless consider it as a useful thought experiment, because the genetic background that an allele faces is then constant across generations (genotypic proportions are always expected to be $x_{AA} = 1/4$, $x_{Aa} = 1/2$, and $x_{aa} = 1/4$). Given that only females contribute directly to offspring production (males only impact the genetic diversity of young she produces), the expected growth rate of the sexual population equals $(8/4 + 4.5/2 + 2/4)(1 - s) = 4.75(1 - s)$, where s is the proportion of males. If the sexual population achieves this growth rate in every year (which requires that it maintains itself at the Hardy-Weinberg equilibrium), and as long as s is not too large, it has performed perfect bet-hedging as the geometric mean now equals the arithmetic mean, which is its maximum value.

But is this geometric mean fitness higher than that of the specialist asexuals (AA and aa)? The answer depends on the cost of sex, which we here model as the proportion s of offspring developing as males. Sex beats AA or aa asexual genotypes if $s < 0.158$, while beating the bet-hedging asexual genotype Aa is harder: it only occurs if $s < 0.0526$.

While the example shows that sexual reproduction can, in principle, be a bet-hedging strategy, it simultaneously shows how difficult it is for sex to evolve based on this benefit alone, especially if competing against asexual types that also bet-hedge (conservatively). The cost of males is captured by s, and the more females produce sons, the higher this cost. Why males exist is a separate evolutionary conundrum from why sex exists: the alternative that is relevant for the "why males?" question is still sex, but without having some individuals specialise in the male strategy that fails to contribute directly to population growth. This question has its own set of game-theoretical answers [5, 23, 26]; the short summary is that (1) males can invade sexual populations despite the reduced growth rate, (2) their existence increases the vulnerability of sexual populations to invasion by asexuals and (3) if a population only consists of (sexual) females and males, sex ratios evolve to $s = 0.5$ under quite general conditions [44].

In Table 1, the arithmetic mean decreases rapidly with an increasing production of males, and any primary sex ratio greater than 15.8% males leads to sexuals being unable to resist invasion by any of the asexual options. Because male presence typically leads to much higher sex ratios, sex is unlikely to persist due to its bet-hedging benefits alone, at least in the simplistic setting of Table 1.

Sexual populations can resist invasions somewhat better (i.e. up to a larger fraction of sons produced) if the dimensionality of bet-hedging increases (i.e. it involves multiple traits). For example, besides the A/a locus that determines an individual's fitness in response to the amount of rainfall, consider another diploid locus that impacts the adaptedness to high or low temperatures. Assume that an individual of the BB genotype is hot-adapted, an individual of the bb type is cold-adapted, and the Bb genotype is intermediate. Also assume the payoff matrices for rainfall, and temperature adaptation has the same structure:

$$
\begin{array}{cc}
\begin{array}{c} \\ AA \\ Aa \\ aa \end{array}
\begin{array}{cc}
\text{Wet} & \text{Dry} \\
\left(\begin{array}{cc} 8 & 2 \\ 4.5 & 4.5 \\ 2 & 8 \end{array} \right)
\end{array}
&
\begin{array}{c} \\ BB \\ Bb \\ bb \end{array}
\begin{array}{cc}
\text{Hot} & \text{Cold} \\
\left(\begin{array}{cc} 8 & 2 \\ 4.5 & 4.5 \\ 2 & 8 \end{array} \right)
\end{array}
\end{array}. \tag{2}
$$

Table 2 Payoff of different genotypes under four different environmental conditions, when there are two traits impacting fitness

Genotype	Freq.	WH	WC	DH	DC
AABB	1/16	64	16	16	4
AABb	1/8	32	32	8	8
AAbb	1/16	16	64	4	16
AaBB	1/8	32	8	32	8
AaBb	1/4	16	16	16	16
Aabb	1/8	8	32	8	32
aaBB	1/16	16	4	64	16
aaBb	1/8	8	8	32	32
aabb	1/16	4	16	16	64

If different traits interact multiplicatively to determine the final fitness, then an AABB individual has payoff of 64 if the environment is both wet and hot (WH), 16 if the environment is wet but cold (WC) or dry but hot (DH) and 4 if the environment is both dry and cold (DC). Table 2 gives the complete list of payoffs of different genotypes under different environments.

For simplicity we may assume that the four environmental conditions occur at equal probabilities (i.e. rainfall does not make the year cooler or vice versa). If we once again assume Hardy-Weinberg equilibrium and equal allele frequencies, the sexual population achieves a growth rate of $22.5625(1 - s)$ in every environmental setting, which also implies a geometric mean of $22.5625(1-s)$. The geometric mean for the asexuals is 16 for homozygote specialists (AABB, AAbb, aaBB, aabb), 18 for those who bet-hedge conservatively with respect to one trait only (AABb, aaBb, AaBB, Aabb) and 20.25 for the asexual genotype that conservatively hedges its bets with respect to both traits (AaBb). The sexual population can beat any asexual genotype if $s < 0.1025$, it can be beaten by the best bet-hedging asexual AaBb but not by others if $0.1025 \leq s < 0.2022$, it can be beaten by all bet-hedging asexuals (AABb, aaBb, AaBB, Aabb and AaBb) but beat the full homozygotes if $0.2022 \leq s < 0.2909$ and remains vulnerable to invasion by any asexual type if s exceeds 0.2909.

We used specific numerical values in the example above, which raises the question how these generalise to other scenarios of allelic dominance, including dominance-recessive, heterosis and inbreeding depression. It has been shown that sexual population can reach all possible phenotypic states if and only if the hereditary system is either dominant-recessive or maternal or the combination of these [15]. We show in Appendix that under the hereditary scheme where one allele is completely dominant over the other allele, the sexual heterozygote ceases to be a bet-hedging strategy since both its arithmetic mean and geometric mean fitness become equal to those of the asexual homozygote. Stronger dominance, on the other hand, improves the geometric mean fitness of the sexual population, making it potentially easier to outcompete asexuals.

3   Numerical Simulations

In the previous section, we used the frequency distribution of different genotypes at Hardy-Weinberg equilibrium for calculating the arithmetic and geometric mean payoff of the sexual population. This is convenient, as it allows us to examine the situation as if the sexual population reached the same growth rate in every environmental setting (it makes sex achieve perfect bet-hedging in the sense that the geometric mean payoff equals the arithmetic mean payoff). However, in reality sex will fail to achieve this perfection, because the genetic environment encountered by a sexual population will be a function of past selection. There will then also be temporal variation in the distributions of genotypes, and sex is likely to fail to achieve perfect bet-hedging. The geometric mean fitness will then drop below the arithmetic mean fitness.

Since the pioneering work of Maynard Smith [28, 29], Hamilton [20] and Bell [4], it has been known that the rate of temporal fluctuations can matter for the evolution of sex. In our setting above, the frequency of switches between wet and dry environments determines how far from equilibrium genotype frequencies will deviate over time. In the following we therefore use numerical simulations to show a more realistic picture of the competition dynamics between sexual and asexual populations.

3.1   Environmental Fluctuations

Here we relax the assumption of Hardy-Weinberg equilibrium: it is only used as a starting state for sexual reproduction, and the following dynamics are computed according to a realised run of fluctuations of the environmental state. Assume that the wet and dry environments follow each other in a manner that can be captured by discrete-time Markov chains (i.e. the transition probability from one state to another does not depend on how long the environment has spent in the current state). The transition probabilities between states can be written in the matrix form:

$$
\begin{array}{c}
\phantom{\text{Wet}} \quad \text{Wet} \qquad \text{Dry} \\
\begin{array}{c} \text{Wet} \\ \text{Dry} \end{array}
\begin{pmatrix} 1 - p_{wd} & p_{dw} \\ p_{wd} & 1 - p_{dw} \end{pmatrix},
\end{array}
\tag{3}
$$

in which p_{wd} denotes the probability that the environment changes from wet to dry in a year and p_{dw} is the probability that the environment changes from dry to wet in a year. The normalized dominant right eigenvector represents the stationary distribution of the environmental states [7] and has the value $(p_{wd}/(p_{wd} + p_{dw}); p_{dw}/(p_{wd} + p_{dw}))$. The subdominant eigenvalue $\rho = 1 - p_{wd} - p_{dw}$ in turn corresponds to the correlation between the environmental states at times t and $t + 1$ [7]. Therefore, consecutive environmental states are negatively autocorrelated if $\rho < 0$, positively autocorrelated if $\rho > 0$ and uncorrelated if $\rho = 0$. In the extreme case where $p_{wd} = p_{dw} = 1$, we have $\rho = -1$ and wet and dry environments alternate, whereas in the other extreme case where $p_{wd} = p_{dw} = 0$, we have $\rho = 1$ and the environment stays in the initial state forever.

3.2 Simulation Results

To focus on the effect of environmental fluctuations, we exclude the effect of demographic stochasticity and drift by assuming that the population size is very large. We use the fixation probability of the invading type as a proxy for the relative advantages of different types. We do this by setting up a population consisting of an initial proportion 0.02 of the invading type, competing against one of the three possible alternative types. We assume that, for sexuals, the growth rate is proportional to $1 - s$ (the frequency of females), and the proportion of AA, Aa and aa young are derived by assuming that both male siring propensity and the female propensity to reproduce are proportional to their genotypes' payoffs (this covers at least two possible biological interpretations: survival probabilities are proportional to payoffs and thereafter mating is random, with each mating producing an equal number of offspring, or the fecundity of females, as well as the siring success of males, is proportional to payoffs). As a caveat, note that the two cases can be mapped to each other directly only in unstructured populations. If the population has overlapping generations, selecting on survival and reproduction have to be treated separately from each other [19, 27].

The invasion is tracked until one of three mutually exclusive events has happened: (a) the invading type has reached frequency 0.9999 or higher (we consider this a successful invasion), (b) the invading type's frequency falls below 0.0001 (we assume that the invasion failed), or (c) neither (a) nor (b) have happened by generation 10^6 (we consider this a coexistence scenario, but in practice event (c) never happened). The Octave codes for all numerical simulations are provided in Supplementary Information. The sexual population starts from the Hardy-Weinberg equilibrium state, with a proportion of 0.25 AA, 0.5 Aa and 0.25 aa types. The payoff of each genotype under different environments follows matrix (1), and fixation probabilities are estimated from 10^4 independent realisations. Because the payoffs of the asexual AA and aa types are symmetric, and the wet and dry environments occur at equal frequencies, they have identical fixation probabilities when invading or being invaded by a sexual population. Therefore, without loss of generality, we use the asexual AA to represent the case of asexual homozygotes in Figure 1.

The figure confirms that sex has a difficult time invading asexual strategies if $s = 0.5$. If we elevate the chances for sexual reproduction to invade others by allowing $s < 0.5$, then cases where sex outcompetes specialist asexuals (AA or aa) still typically do not predict that sex can also outcompete bet-hedging asexuals (comparing the left and right panels: curves are almost invariably higher on the right than on the left when considering an asexual invasion and are always lower

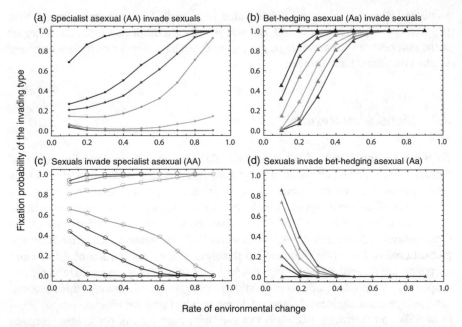

Fig. 1 Fixation probability of the invading types under various rates of environmental change for populations following payoff matrix (1). The x-axis represents the rate of environmental change, assuming $p_{wd} = p_{dw}$. Colours from red to purple to black represent sexual population of different sex ratios (0.01, 0.02, 0.04, 0.08, 0.12, 0.16 and 0.5). The larger the sex ratio, the higher the cost of sex. These figures are based on 10^4 realisations per parameter value and never required stopping the simulation at generation 10^6 (i.e. either fixation is reached or the invader went extinct)

on the right than on the left when considering a sexual invasion). Whether fast or slow environmental fluctuations are best for sex is surprisingly complex. At very small s, sexuals are more likely to invade asexual homozygotes (and also resist their invasion attempts) if the environment changes fast. Other values of s predict the opposite. This complexity contrasts with Hamilton's early work on geometric mean fitness in the context of sex [20], predicting that a fast changing environment is beneficial to the maintenance of sex in general. Note, however, that our model and the one by Hamilton [20] use different payoff structures (for details see Section 4) and that Hamilton did not call his temporal fluctuation model bet-hedging.

The success of invasion is likely to depend on how long allelic diversity persists in the population. If the payoff of the heterozygote is low, and the environment changes relatively slowly, genetic diversity might become extinguished even before the asexual mutant is introduced. When the sexual population exists alone, it is possible that one allele, either a or A, is lost (e.g. Figure 2a–b, mean time to extinction: Figure 2c–d). The better the heterozygote's payoff (Figure 2c) and the faster the environmental fluctuations (Figure 2d), the longer the coexistence time of both alleles. If one allele has already been lost, sex behaves genetically like an asexual homozygote (losing its bet-hedging benefit) but still paying the cost of sex.

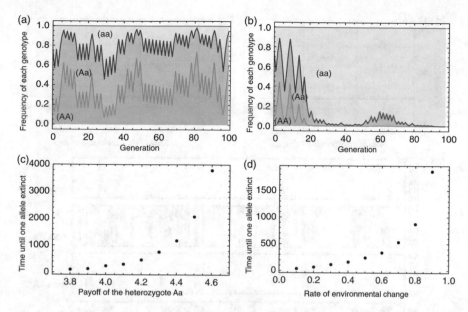

Fig. 2 Examples of genetic diversity in a purely sexual population (no mutation to asexuality), where diversity is maintained (panel a) or lost (panel b) under environmental fluctuations that are tracked for 100 generations. The two trajectories are from simulations with identical parameter settings. In both cases, the rate of environmental change $p_{wd} = p_{dw} = 0.75$, and the payoff of the heterozygote is set to 3.8 under both environmental conditions. The vertical height of regions of various colours represents the frequencies of different genotypes. (c) The mean time to the disappearance of one allele as a function of varying heterozygote payoffs when $p_{wd} = p_{dw} = 0.5$, and (d) the mean time to the disappearance of one allele as a function of the rate of environmental change when $p_{wd} = p_{dw}$ and heterozygote payoff is 4.0 under both environmental conditions. In all simulations, the payoffs of the homozygotes follow payoff matrix (1). In panels (c) and (d), one allele is considered to have gone extinct if the frequencies of both the corresponding homozygote and the heterozygote are smaller than 10^{-4}.

Note that a population that bet-hedges via asexuality (Aa) does not suffer from this risk, as both alleles are kept intact in this lineage in every generation. In this sense, conservative bet-hedging represented by asexuality may perform better than the diversified bet-hedging represented by sexual reproduction.

A key finding is therefore that sex cannot easily outcompete asexual forms based on bet-hedging benefits alone (Figure 3). Sex as bet-hedging requires conditions under which the red symbols are below the dotted line in Figure 3a and above the solid line in Figure 3b. Only four out of the nine cases satisfy the requirements (heterozygote payoff 4.2 or 4.6 in combinations with rate of environmental change 0.5 or 0.75). However, it is possible to construct cases where sex wins in terms of arithmetic mean fitness but loses in terms of geometric mean to the conservative asexual bet-hedger (Figure 3c, where the heterozygote payoff is set to 4.2, and the rate of environmental change is set to 0.5).

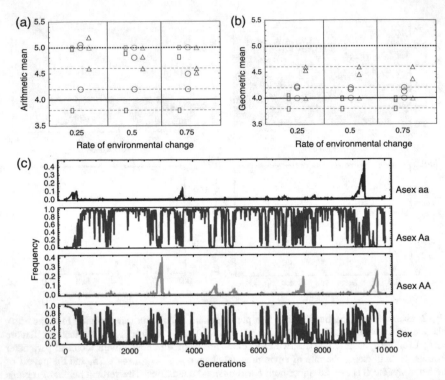

Fig. 3 (a) Arithmetic mean payoffs and (b) geometric mean payoffs of the asexual homozygote (green), asexual heterozygote (blue) and the sexual population (red), computed over 500 generations when the payoffs of the asexual homozygotes follow matrix (1) and the sex ratio of the sexual population is set to $s = 0.01$. Symbols of different shapes represent different payoffs of the heterozygote: square, circle and triangle stand for 3.8, 4.2 and 4.6, respectively. The black dotted line is the expected arithmetic mean payoff of the asexual homozygotes, the black solid line is the expected geometric mean payoff of the asexual homozygotes, and the grey dashed lines are the expected arithmetic and geometric mean payoff of the asexual heterozygote. (c) Frequency dynamics of the sexual population and each asexual genotype under a changing environment over 10000 generations. In each panel, the x-axis is time (the elapsed number of generations), and the y-axis is the frequency of each type. All four panels are from the same instance of simulation. The heterozygote payoff is set to 4.2, and the rate of environmental change is $p_{wd} = p_{dw} = 0.5$. The simulation starts with a pure sexual population with 0.25 AA, 0.5 Aa and 0.25 aa genotypes, but each individual may mutate to being asexual if previously sexual, or sexual if previously asexual, at rate 0.0001 per generation

4   Discussion

There are interesting parallels between sex and bet-hedging theory. Intuitively, the costs of sex reduce the fitness of sexual lineages in every generation that undergoes a sexual life cycle (hence the arithmetic fitness is reduced), but by diversifying the genotypes of offspring, sex can reduce the variance in success: in any given year some offspring will survive, while an asexual specialist proverbially puts "all its eggs in one basket" – leading to very low success if the year features a mismatch between offspring genotype and the state of the environment. However, for this to favour sex over asex, the geometric mean fitness of the former should be elevated above the latter. Although variance reductions have a beneficial effect on geometric mean fitness, arithmetic mean fitness (which is low for sexual types) simultaneously sets an upper limit for it, and hence it is not easy for sex to reach such high bet-hedging benefits that its geometric mean fitness is the best of all competing strategies. In other words, the fact that sexual reproduction shows features of bet-hedging is not the same statement as the claim that bet-hedging provides strong enough benefits for the evolution and maintenance of sex. This is especially true since sex may have to compete against another type of bet-hedger: that of asexual heterozygotes, which avoid paying the cost of sex but may also achieve bet-hedging if their genotype performs reasonably well under all considered environmental conditions. This highlights that (a) it is important to specify that a strategy is performing bet-hedging relative to another strategy, and be explicit about the identity of the relevant competitor, and (b) that it would be premature to consider bet-hedging as a major driving force behind the maintenance of sex, at least under the simplifying assumptions of the current model.

Fast and unpredictable changes of the environment have been found to favour bet-hedging [17] and facilitate the maintenance of sexual reproduction [3, 4, 20, 28, 29, 41, 42], but these authors did not use bet-hedging terminology. Our model shares a similar genetic structure to the model of Hamilton [20], but the payoff structures are different. In our model, the two asexual homozygotes are specialists that adapt to different environmental conditions, and the heterozygote has intermediate payoff under both environmental conditions (this makes it a conservative bet-hedger). In Hamilton's model, the homozygotes receive identical payoffs (that depend on environmental conditions), whereas the payoff of the heterozygote is the reciprocal of this payoff. The heterozygote and homozygotes Hamilton's model [20] thus do not correspond to a bet-hedger and two specialists, and, therefore, although the model shows that sex is beneficial under a fast changing environment, it did not aim to capture the evolutionary dynamics under the bet-hedging context.

Compared to classic bet-hedging scenarios where the bet-hedger always has the same payoff under the same environment [40], sexual reproduction as bet-hedging brings in additional features. In the sexual population, the arithmetic mean payoff in each generation is determined not only by the environment but also the frequency distribution of all genotypes, the sex ratio and possibly other costs or benefits from sexual reproduction. In addition, if mutations between sexual and asexual populations are allowed, more than one type of bet-hedging strategy can (at least temporarily) coexist, and it is insightful to remember that there can be asexual heterozygotes that bet-hedge conservatively, as opposed to the diversified bet-hedging of the sexual population.

Both theoretical and experimental work on the evolution of sex show complications that highlight the simplicity of any two-environment model (indeed, in our model too, increasing the dimensionality of the system helps maintain sex). We have followed a tradition in bet-hedging theory where two (or four) types of environment can be adapted to with one (or two) trait(s). Modern research on genetic variation reveals that there is surprisingly much polygenic variation present in populations [8], and fitness landscapes are often complex. Recent research on sex has revealed the potential importance of processes such as clonal interference [31, 38], which tends to erode the success of asexual lineages over time because they are slow to acquire multiple novel mutations that aid adaptation. Sex improves the rate with which innovations end up in the same organism, while asexual lineages tend to fail in having access to the most "up to date" genetic background, especially if the environment keeps changing. The detrimental interference between competing clones that have acquired one or another beneficial allele (at different loci) eventually makes asexuality an inferior competitor in the adaptive race. While this is a very different situation from what bet-hedging theory traditionally has considered, there is scope to fill this gap: the gist of the argument is that the asexual lineages experience diminishing geometric fitness once timescales become long enough that novel beneficial mutations begin playing a role. Sex and the diversity it creates can help diversify the genetic backgrounds where new mutations can be selected for.

Among the classic literatures, the payoff structure in Treisman [41] is the closest to ours, and it also captures some of the above ideas about the environment changing to something never experienced before. In the model, different alleles interact additively and give the diploid individual a phenotype (in his words, a "genotypical score") that impacts female fertility but not male siring success. Alleles have effects of -0.5 or 0.5, so that homozygotes have phenotypes -1 or 1, and the heterozygote has an intermediate phenotype of 0. Females (both sexual and asexual) can only breed if their phenotype matches, within tolerable range, the environmental conditions (such as temperature). If the environment keeps changing (e.g. increasing temperatures), asexual genotypes cannot keep pace with sexuals that produce diversified offspring through recombination, and extinction can then follow. Like the authors mentioned above, Treisman [41] also did not use the terminology of bet-hedging and hence did not analyse the arithmetic and geometric mean fitness of each genotype.

Given that there is both old and new work on sex that could gain conceptual clarity if researchers routinely reported how the winning strategy (sexual or asexual) performed in terms of arithmetic and geometric mean fitness, we welcome more work in the areas linking sex and bet-hedging. Bet-hedging theory has brought about increased understanding of other evolutionary questions from dispersal evolution [1] and dormancy timing [12–14] to antibiotic resistance [2], microbial population dynamics [11] and phenotypic switching [6]. It would appear timely to add sexual reproduction to this list. Even if sex in simplistic settings (like ours) does not reach the status of a strategy with the highest geometric mean fitness, a bet-hedging perspective can shed light on the precise reasons why it failed. An interesting question would be to use this type of analysis to examine cases where sex, e.g. in situations involving clonal interference and *de novo* mutations, succeeds to maintain itself against asexual competitors.

Acknowledgements We thank the two anonymous reviewers for their comments and suggestions. X.L. and H.K. are grateful to the Swiss National Science Foundation. J.L. was funded by a University of New South Wales Vice-Chancellor's Postdoctoral Research Fellowship. All authors thank the organisers of the 17th International Symposium on Dynamic Games and Applications.

Appendix

Assume that the A allele fully dominates the a allele. The fitness values of each genotype under different environments are shown in matrix (4):

$$
\begin{array}{c}
\quad\quad\text{Wet}\quad\text{Dry} \\
\begin{array}{c} AA \\ Aa \\ aa \end{array}
\left(\begin{array}{cc} 8 & 2 \\ 8 & 2 \\ 2 & 8 \end{array}\right)
\end{array}
\tag{4}
$$

In this case, the payoff of each asexual type and the sexual population is shown in Table 3.

The first observation is that the asexual heterozygote is no longer a bet-hedging strategy, since its payoffs under different environmental conditions become identical to the homozygote AA, and thus its geometric and arithmetic payoffs no longer fit the requirements of bet-hedging. Under Hardy-Weinberg equilibrium, the sexual population would have higher geometric mean payoff and lower arithmetic mean payoff than each asexual type when $0 < s < 0.162$. This range is larger than that under the case of intermediate inheritance, where the sexual population beats any asexual homozygote if $0 < s < 0.158$ and beats the asexual heterozygote if $0 < s < 0.053$.

Similar results hold when populations hedge their bets on multiple traits. Using the case in matrix (2) as an example, if the A allele fully dominates the a allele, and the B allele fully dominates the b allele, the payoff matrices for rainfall and temperature adaptation have the following structure:

Table 3 The payoff structure under wet and dry years when the A allele fully dominates the a allele: the arithmetic mean (AMean) and the geometric mean (GMean) of the payoffs of asexual lineages, as well as of a sexual population assumed to be at the Hardy-Weinberg equilibrium

	Wet	Dry	AMean	GMean
asex-AA	8	2	5	4
asex-Aa	8	2	5	4
asex-aa	2	8	5	4
Sex-population	6.50(1-s)	3.50(1-s)	5(1-s)	4.77(1-s)

Table 4 Payoff of different genotypes under four different environmental conditions under the dominance hereditary system, when two traits determine the fitness together

Genotype	Freq.	WH	WC	DH	DC
$AABB$	1/16	64	16	16	4
$AABb$	1/8	64	16	16	4
$AAbb$	1/16	16	64	4	16
$AaBB$	1/8	64	16	16	4
$AaBb$	1/4	64	16	16	4
$Aabb$	1/8	16	64	4	16
$aaBB$	1/16	16	4	64	16
$aaBb$	1/8	16	4	64	16
$aabb$	1/16	4	16	16	64

$$
\begin{array}{c}
\quad \text{Wet} \quad \text{Dry} \\
\begin{array}{c} AA \\ Aa \\ aa \end{array}
\left(\begin{array}{cc} 8 & 2 \\ 8 & 2 \\ 2 & 8 \end{array} \right)
\end{array}
\qquad
\begin{array}{c}
\quad \text{Hot} \quad \text{Cold} \\
\begin{array}{c} BB \\ Bb \\ bb \end{array}
\left(\begin{array}{cc} 8 & 2 \\ 8 & 2 \\ 2 & 8 \end{array} \right)
\end{array}.
\tag{5}
$$

Again, we assume that different traits interact multiplicatively to determine the final fitness, and the sexual population is under Hardy-Weinberg equilibrium. Table 4 gives the complete list of payoffs of different genotypes under different environments.

In this case the sexual population has a fitness of $42.25(1 - s)$ under the WH environment, $12.25(1 - s)$ under the DC environment and $22.75(1 - s)$ under both WC and DH environments. Therefore, if four different environments occur at equal frequencies, the arithmetic mean payoff of the sexual population is $25(1-s)$, and the geometric mean fitness is $22.75(1 - s)$. The geometric mean for the asexuals is 16 for all asexual types. In this way, the sexual population beats any asexual population if $0 < s < 0.297$. This range is also larger than the condition ($0 < s < 0.102$) for beating any asexual genotype under the intermediate heredity.

References

1. Paul R Armsworth and Joan E Roughgarden. The impact of directed versus random movement on population dynamics and biodiversity patterns. *The American Naturalist*, 165:449–465, 2005.
2. Markus Arnoldini, Ima A Vizcarra, Rafael Peña-Miller, Nicolas Stocker, Médéric Diard, Viola Vogel, Robert E Beardmore, Wolf-Dietrich Hardt, and Martin Ackermann. Bistable expression of virulence genes in salmonella leads to the formation of an antibiotic-tolerant subpopulation. *PLoS Biology*, 12:e1001928, 2014.
3. Roberto Barbuti, Selma Mautner, Giorgio Carnevale, Paolo Milazzo, Aureliano Rama, and Christian Sturmbauer. Population dynamics with a mixed type of sexual and asexual reproduction in a fluctuating environment. *BMC Evolutionary Biology*, 12:49, 2012.
4. Graham Bell. *The Masterpiece of Nature: The Evolution and Genetics of Sexuality*. Croom Helm, London & Canberra, 1982.

5. MG Bulmer and GA Parker. The evolution of anisogamy: a game-theoretic approach. *Proceedings of the Royal Society of London B: Biological Sciences*, 269:2381–2388, 2002.

6. Oana Carja, Robert E Furrow, and Marcus W Feldman. The role of migration in the evolution of phenotypic switching. *Proceedings of the Royal Society of London B: Biological Sciences*, 281:20141677, 2014.

7. Hal Caswell. *Matrix population models*. Sinauer Associates, 2nd edition edition, 2001.

8. Brian Charlesworth. Causes of natural variation in fitness: evidence from studies of drosophila populations. *Proceedings of the National Academy of Sciences, USA*, 112:1662–1669, 2015.

9. Anne Charpentier, Madhur Anand, and Chris T Bauch. Variable offspring size as an adaptation to environmental heterogeneity in a clonal plant species: integrating experimental and modelling approaches. *Journal of ecology*, 100:184–195, 2012.

10. Charles Darwin. *On the origin of species by means of natural selection*. 1859.

11. Imke G de Jong, Patsy Haccou, and Oscar P Kuipers. Bet hedging or not? a guide to proper classification of microbial survival strategies. *Bioessays*, 33:215–223, 2011.

12. Stephen Ellner. Ess germination strategies in randomly varying environments. I. logistic-type models. *Theoretical Population Biology*, 28:50–79, 1985.

13. Margaret EK Evans and John J Dennehy. Germ banking: bet-hedging and variable release from egg and seed dormancy. *The Quarterly Review of Biology*, 80:431–451, 2005.

14. Andrew I Furness, Kevin Lee, and David N Reznick. Adaptation in a variable environment: Phenotypic plasticity and bet-hedging during egg diapause and hatching in an annual killifish. *Evolution*, 69:1461–1475, 2015.

15. József Garay and M Barnabas Garay. Genetical reachability: When does a sexual population realize all phenotypic states? *Journal of Mathematical Biology*, 37:146–154, 1998.

16. Alan Grafen. Formal darwinism, the individual–as–maximizing–agent analogy and bet–hedging. *Proceedings of the Royal Society of London B: Biological Sciences*, 266:799–803, 1999.

17. Patsy Haccou and Yoh Iwasa. Optimal mixed strategies in stochastic environments. *Theoretical Population Biology*, 47:212–243, 1995.

18. Patsy Haccou and Yoh Iwasa. Robustness of optimal mixed strategies. *Journal of Mathematical Biology*, 36:485–496, 1998.

19. Patsy Haccou and John M McNamara. Effects of parental survival on clutch size decisions in fluctuating environments. *Evolutionary Ecology*, 12:459–475, 1998.

20. William D Hamilton, Peter A Henderson, and Nancy A Moran. Fluctuation of environment and coevolved antagonist polymorphism as factors in the maintenance of sex. In R.D. Alexander and D.W. Tinkle, editors, *Natural selection and social behavior*, pages 363–381. Chiron Press, New York, 1981.

21. Godfrey H. Hardy. Mendelian proportions in a mixed population. *Science*, 28:49–50, 1908.

22. Matthew Hartfield and Peter D Keightley. Current hypotheses for the evolution of sex and recombination. *Integrative zoology*, 7:192–209, 2012.

23. Jussi Lehtonen and Hanna Kokko. Two roads to two sexes: unifying gamete competition and gamete limitation in a single model of anisogamy evolution. *Behavioral Ecology and Sociobiology*, 65:445–459, 2011.

24. Jussi Lehtonen and Hanna Kokko. Sex. *Current Biology*, 24:R305–R306, 2014.

25. Jussi Lehtonen, Michael D Jennions, and Hanna Kokko. The many costs of sex. *Trends in Ecology & Evolution*, 27:172–178, 2012.

26. C.M. Lessells, Rhonda R Snook, and David J Hosken. The evolutionary origin and maintenance of sperm: selection for a small, motile gamete mating type. In T.R. Birkhead, D.J. Hosken, and S. Pitnick, editors, *Sperm biology: an evolutionary perspective*, pages 43–67. Academic Press, London, 2009.

27. Xiang-Yi Li, Shun Kurokawa, Stefano Giaimo, and Arne Traulsen. How life history can sway the fixation probability of mutants. *Genetics*, DOI: 10.1534/genetics.116.188409, 2016.

28. John Maynard Smith. What use is sex? *Journal of Theoretical Biology*, 30:319–335, 1971.

29. John Maynard Smith. A short-term advantage for sex and recombination through sib-competition. *Journal of Theoretical Biology*, 63:245–258, 1976.

30. John Maynard Smith. *The Evolution of Sex*. Cambridge University press, 1978.
31. Michael J McDonald, Daniel P Rice, and Michael M Desai. Sex speeds adaptation by altering the dynamics of molecular evolution. *Nature*, 531:233–236, 2016.
32. John M McNamara. Implicit frequency dependence and kin selection in fluctuating environment. *Evolutionary Ecology*, 9:185–203, 1995.
33. John M McNamara, James N Webb, and Edmund J Collins. Dynamic optimization in fluctuating environments. *Proceedings of the Royal Society of London B: Biological Sciences*, 261:279–284, 1995.
34. Stephanie Meirmans, Patrick G Meirmans, and Lawrence R Kirkendall. The costs of sex: facing real-world complexities. *The Quarterly Review of Biology*, 87:19–40, 2012.
35. Tom Philippi and Jon Seger. Hedging one's evolutionary bets, revisited. *Trends in Ecology & Evolution*, 4:41–44, 1989.
36. Sean H Rice. A stochastic version of the price equation reveals the interplay of deterministic and stochastic processes in evolution. *BMC Evolutionary Biology*, 8:1, 2008.
37. J. Seger and J. Brockmann. What is bet-hedging? *Oxford Surveys in Evolutionary Biology*, 4: 181–211, 1987.
38. Nathaniel P Sharp and Sarah P Otto. Evolution of sex: Using experimental genomics to select among competing theories. *BioEssays*, 38:751–757, 2016.
39. Montgomery Slatkin. Hedging one's evolutionary bets. *Nature*, 250:704–705, 1974.
40. Jostein Starrfelt and Hanna Kokko. Bet-hedging – a triple trade-off between means, variances and correlations. *Biological Reviews*, 87:742–755, 2012.
41. Michel Treisman. The evolution of sexual reproduction: a model which assumes individual selection. *Journal of Theoretical Biology*, 60:421–431, 1976.
42. David Waxman and Joel R Peck. Sex and adaptation in a changing environment. *Genetics*, 153: 1041–1053, 1999.
43. Wilhelm Weinberg. Über den nachweis der vererbung beim menschen. *Jahreshefte des Vereins für vaterländische Naturkunde in Württemberg.*, 64:369–382, 1908.
44. Stuart A West. *Sex Allocation*. Princeton University Press, Princeton and Oxford, 2009.
45. George C Williams. *Sex and evolution*. Princeton University Press, Princeton, New Jersey, 1975.

Part III
Stochastic and Pursuit-Evasion Games

On Exact Construction of Solvability Set for Differential Games with Simple Motion and Non-convex Terminal Set

Liudmila Kamneva and Valerii Patsko

Abstract Two-player zero-sum differential games with simple motion in the plane are considered. An explicit formula describing the solvability set is well-known for a convex terminal set. The paper suggests a way of exact construction of solvability set in the case of non-convex polygonal terminal set and polygonal constraints of the players' controls. Some illustrative examples are computed.

Keywords Differential games with simple motion in the plane • Solvability set • Backward procedure

1 Introduction

When considering substantial problems of conflict control, the dynamics of simple motion [1] is used if we have to examine some basic issues of the conflict interaction without involving into the details connected with the real inertial motion of objects.

It is well-known that we can use systems with simple motion for numerical solution (see, e.g. [2]) to linear differential games of the kind

$$\dot{z} = A(t)z + B(t)u + C(t)v, \quad u \in \mathscr{P}, \quad v \in \mathscr{Q}$$

with a fixed terminal time T and given pay-off function φ calculated at the terminal instant.

Let φ be a function of only m special coordinates of z. Then one uses the change of variables $x(t) = Z_m(T, t)z(t)$, where $Z_m(T, t)$ is a matrix consisting of m relevant rows of the fundamental Cauchy matrix of the linear system. As a result, we get the system

$$\dot{x} = Z_m(T, t)B(t)u + Z_m(T, t)C(t)v, \quad u \in \mathscr{P}, \quad v \in \mathscr{Q}$$

L. Kamneva (✉) • V. Patsko
Institute of Mathematics and Mechanics, S. Kovalevskaya str. 16, Ekaterinburg 620990, Russia
e-mail: kamneva@imm.uran.ru; patsko@imm.uran.ru

© Springer International Publishing AG 2017
J. Apaloo, B. Viscolani (eds.), *Advances in Dynamic and Mean Field Games*,
Annals of the International Society of Dynamic Games 15,
https://doi.org/10.1007/978-3-319-70619-1_11

without the state variable x on the right-hand side. Keeping the same terminal time T and the pay-off function φ, we obtain a differential game, which is equivalent (with respect to the value function) to the original linear differential game [3, §40], [4, Section 9.4], [5, p. 89–91].

Let M_c be a sublevel set (Lebesgue's set) of the pay-off function φ in the x-coordinates. We move back in time from the set M_c "freezing" the coefficients of the dynamics on each time interval $[t_i, t_{i+1}]$ of the backward procedure. Thus, we get a recursive procedure for approximate construction of the sublevel sets $W_c(t_i)$ $(W_c(T) = M_c)$ of the value function. Here, we are dealing with the system of simple motion

$$\dot{x} = u + v, \quad u \in P_i, \quad v \in Q_i$$

at each step of the backward procedure, where

$$P_i = Z_m(T, t_i)B(t_i)\mathscr{P}, \quad Q_i = Z_m(T, t_i)C(t_i)\mathscr{Q}.$$

Naturally we are interested in some "good" algorithm of passage from the set $W_c(t_{i+1})$ to the set $W_c(t_i)$. The present state of the theory of differential games says that, to get a good result on the time interval $[t_i, t_{i+1}]$ with dynamics of simple motion, one should divide the interval into a sufficiently large number of parts with step h by the instants

$$t_i^N = t_{i+1}, \quad t_i^{N-1} = t_i^N - h, \quad \ldots, \quad t_i^0 = t_i$$

and then, for each additional interval $[t_i^j, t_i^{j+1}]$, construct approximately the set $W_c(t_i^j)$ on the base of $W_c(t_i^{j+1})$.

The purpose of this article is to show that there is no need to do this for dimension 2 of the state vector x. We describe an algorithm that gives the *exact* solvability set $W_c(t_i)$ on the base of $W_c(t_{i+1})$ taken at the instant t_{i+1} for a fixed dynamics with simple motion on the interval $[t_i, t_{i+1}]$ without any additional partition of the interval.

In Section 2, the statement of the problem is described. Well-known formula [6] for the exact construction of the solvability set in the convex case and the notion of semipermeable surface [1] are recalled in Section 3. In Section 4, the notions of convex and concave sets of half-spaces are introduced, and the weight function for ordered triples of half-spaces is defined. Section 5 discusses construction of a semipermeable surface locally in a neighbourhood of convex or concave vertex of the terminal set. In Section 6, we suggest a way of exact description the solvability set for a non-convex terminal polygon M. The description can be implemented as an algorithm. Section 7 is devoted to numerical examples. The contribution of the paper is briefly discussed in Section 8 in the context of difficulties encountered and future work.

2 Problem Statement

2.1 Differential Game

Let us consider a control dynamical system with simple motion [1]

$$\dot{x} = u + v, \quad u \in P, \quad v \in Q, \quad t \in [0, \vartheta], \quad \vartheta > 0 \tag{1}$$

in the phase plane $x \in \mathbb{R}^2$, where u and v are controls of the first and second players and each of the sets P and Q is either a convex closed polygon or a linear segment.

Assume $M \subset \mathbb{R}^2$ to be a simple polygon, i.e., its boundary is a closed polyline without self-intersections. *Differential game* consists of a problem of M-attainability at the instant ϑ for the first player and a problem of M'-attainability at the instant ϑ for the second player, $M' = \overline{\mathbb{R}^2 \setminus M}$.

Statement of *the problem of M-attainability* at the instant ϑ on the interval $[0, \vartheta]$ is formulated in [7, § 13.1] as follows. The first player tries to guarantee $x(\vartheta) \in M$ from an initial position $(0, x_0)$ provided that the player knows the current position $(t, x(t))$ and generates a feedback control $u(t, x(t)) \in P$.

The problem of M'-attainability at the instant ϑ for the second player is formulated in the similar way.

2.2 Solvability Set

Further, we describe the notion of u-stable bridge to solve the problem of M-attainability for the first player.

The graph $W = \{(t, x) : t \in [0, \vartheta], x \in W(t)\}$ of a set-valued function

$$[0, \vartheta] \ni t \mapsto W(t) \subset \mathbb{R}^2$$

is called *a u-stable bridge* for the problem of M-attainability on the interval $[0, \vartheta]$ if $W(\vartheta) \subset M$, and, for any $v \in Q$, the set W is weakly invariant with respect to the differential inclusion

$$\dot{x} \in P + v. \tag{2}$$

Weak invariance means that, for any $(t_0, x_0) \in W$, there exists a motion

$$x(\cdot) : [t_0, \vartheta] \to \mathbb{R}^2$$

satisfying differential inclusion (2), initial condition $x(t_0) = x_0$, and the following *viability condition*: $x(t) \in W(t)$ for all $t \in [t_0, \vartheta]$. In differential games, the weak invariance is called the u-stability property (in an equivalent formulation).

In the same way, the notion of v-stable bridge to solve the problem of M'-attainability for the second player can be introduced. The original notions of stable bridges in equivalent formulations were presented in [8, p. 52–54], [5, p. 53, 58].

Let W_0 denote *the maximal* (by inclusion) *u-stable bridge* defined on the interval $[0, \vartheta]$ for the problem of M-attainability at the instant ϑ.

We call $W_0(0)$ *the solvability set* for the problem of M-attainability at the instant ϑ on the interval $[0, \vartheta]$.

Our goal is an exact description of the set $W_0(0)$ admitting its construction by a finite number of steps of an algorithm.

Remark 1 It is known [8, § 16] that the set W is the maximal u-stable bridge for the problem of M-attainability at the instant ϑ if and only if the set

$$W' = \{(t, x) : t \in [0, \vartheta], \ x \in \overline{\mathbb{R}^2 \setminus W(t)}\}$$

is the maximal v-stable bridge for the problem of M'-attainability at the instant ϑ. Therefore, if we construct the maximal u-stable bridge, we get the solution of the differential game.

Remark 2 Maximal u-stable bridge is a closed set [8, p. 67], [9, p. 92].

Closedness and u-stability of the set W_0 imply the following lemma.

Lemma 1 *Let $W_0 \subset [0, \vartheta] \times \mathbb{R}^2$ be the maximal u-stable bridge for the problem of M-attainability and $t_* \in [0, \vartheta)$. Then*

$$W_0(t_*) = \{x_* \in \mathbb{R}^2 : \exists (t_n, x_n) \to (t_*, x_*), \ n \to \infty, \ t_n > t_*, \ x_n \in W_0(t_n)\}.$$

3 Some Basic Concepts and Known Results

3.1 Case of a Convex Terminal Set

First suppose that the terminal set M is a convex polygon. In the convex case, the formula

$$W_0(t) = \left(M + \left(-(\vartheta - t)P\right)\right) \stackrel{*}{-} (\vartheta - t)Q =: T_{\vartheta - t}(M), \tag{3}$$

is well-known [6]. Representation (3) describes t-sections $W_0(t)$ of the maximal u-stable bridge W_0 by means of operations of the algebraic (Minkowski) sum and the geometric (Minkowski) difference [10, 11]:

$$A + B = \{a + b : a \in A, \ b \in B\}, \quad A \stackrel{*}{-} B = \{d : d + B \subset A\} = \bigcap_{b \in B}(A - b).$$

We take the right-hand side of the equality in (3) as the definition of an operator T_τ acting on M with $\tau = \vartheta - t$.

Let $\Pi_\eta \subset \mathbb{R}^2$ be a half-plane with a unit outer normal $\eta \in \mathbb{R}^2$ to its boundary. Then, using the definition of the operator T_τ, we calculate

$$T_\tau(\Pi_\eta) = \Pi_\eta - \tau\big(u_0(\eta) + v_0(\eta)\big), \quad \tau > 0, \tag{4}$$

where

$$u_0(\eta) \in \mathrm{Arg}\,\min_{u \in P}\langle u, \eta \rangle, \quad v_0(\eta) \in \mathrm{Arg}\,\max_{v \in Q}\langle v, \eta \rangle.$$

Further, we write ∂A to notate the boundary of A.

It is easy to check that for an arbitrary half-plane $\Pi \subset \mathbb{R}^2$, the set

$$\{(t, x) : t \in [0, \vartheta], x \in \partial T_{\vartheta-t}(\Pi)\}$$

is a planar set in $\mathbb{R}^3 = \{(t, x)\}$. Consequently, any half-plane $\Pi \subset \mathbb{R}^2$ corresponds to the unique *half-space* $\mathcal{T}_\vartheta(\Pi) \subset \mathbb{R}^3$ such that its t-section coincides with $T_{\vartheta-t}(\Pi)$ for any $t \in [0, \vartheta]$.

Given a unit outer normal $\eta \in \mathbb{R}^2$, we denote by $\Pi_\eta[A]$ the supporting half-plane of the set A:

$$\Pi_\eta[A] = \{x \in \mathbb{R}^2 : \langle x, \eta \rangle \le \rho(\eta; A) < +\infty\}.$$

Here, $\rho(\eta; A) = \sup\{\langle a, \eta \rangle : a \in A\}$ is the value of the support function of the set A at the vector η.

Observe that the convex polygon M can be represented as

$$M = \bigcap \big\{\Pi_\eta[M] : \eta \in \mathcal{N}(M) \cup \mathcal{N}(-P)\big\}, \tag{5}$$

where $\mathcal{N}(M)$ and $\mathcal{N}(-P)$ are the sets of unit outer normals to the edges of the polygons M and $-P$, correspondingly. If P is a linear segment, then $\mathcal{N}(-P) = \mathcal{N}(P) = \{v, -v\}$, where v is a unit normal to P.

Note that the half-planes $\Pi_\eta[M]$ corresponding to $\eta \in \mathcal{N}(-P)$ are *inessential* for intersection (5), i.e. the half-planes can be eliminated from the right-hand side of (5) without changing the result of the intersection. But, generally speaking, these half-planes are involved in the construction of the solvability set $W_0(0)$. Therefore, we include them into the representation of the polygon M.

Recall that for the convex polygon M and $t \in [0, \vartheta]$, the equality

$$W_0(t) = T_{\vartheta-t}(M) = \bigcap \big\{T_{\vartheta-t}(\Pi_\eta[M]) : \eta \in \mathcal{N}(M) \cup \mathcal{N}(-P)\big\} \tag{6}$$

holds (see, e.g. [12]), which is an analogue of (3) in terms of half-planes corresponding to the edges of the polygons M, P, and Q.

Equality (6) and the definition of $\mathcal{T}_\vartheta(\cdot)$ imply the following lemma giving a three-dimensional analogue of the "planar" formula (6).

Lemma 2 *For a convex polygon M, we have*

$$W_0 = \Theta \cap \left(\bigcap \{ \mathcal{T}_\vartheta(\Pi_\eta[M]) : \eta \in \mathcal{N}(M) \cup \mathcal{N}(-P) \} \right), \tag{7}$$

where $\Theta := \{ (t, x) : t \in [0, \vartheta] \}$.

Thus, the bridge W_0 can be represented as an intersection of half-spaces in $\mathbb{R}^3 = \{ (t, x) \}$.

3.2 Case of the Convex Complement of the Terminal Set

Now suppose that the set $M' = \overline{\mathbb{R}^2 \setminus M}$ is a convex polygon. In this case, changing the roles of the players, we obtain the convex case discussed above. Namely, similar to (5), we have

$$M' = \bigcap \{ \Pi_\eta[M'] : \eta \in \mathcal{N}(M') \cup \mathcal{N}(-Q) \}.$$

For any half-plane $\Pi \subset \mathbb{R}^2$, we introduce *the half-space* $\mathcal{T}_\vartheta^*(\Pi) \subset \mathbb{R}^3$ such that its t-section coincides with $T_{\vartheta-t}^*(\Pi)$ for any $t \in [0, \vartheta]$. Here, the operator $A \to T_\tau^*(A)$ is defined by

$$T_\tau^*(A) = \left(A + \left(-\tau Q \right) \right) \overset{*}{-} \tau P, \quad \tau > 0.$$

Consequently, the following lemma is true, which is similar to Lemma 2.

Lemma 3 *For the convex polygon* $M' = \overline{\mathbb{R}^2 \setminus M}$, *we have*

$$\Theta \cap (\overline{\mathbb{R}^3 \setminus W_0}) = \Theta \cap \left(\bigcap \{ \mathcal{T}_\vartheta^*(\Pi_\eta[M']) : \eta \in \mathcal{N}(M') \cup \mathcal{N}(-Q) \} \right). \tag{8}$$

Using (4) and the similar representation for the operator T_τ^*, we obtain

$$\overline{\mathbb{R}^3 \setminus \mathcal{T}_\vartheta(\Pi_\eta)} = \mathcal{T}_\vartheta^*(\overline{\mathbb{R}^2 \setminus \Pi_\eta}).$$

This equality allows one to describe the further constructions by means of only the "convex" operators $T_\tau(\cdot)$ and $\mathcal{T}_\vartheta(\cdot)$.

3.3 Simple Polyhedral Tubes

A *polyhedron* in $\mathbb{R}^3$ is a connected set bounded by a finite number of flat polygons (faces). A set $\Gamma \subset \mathbb{R}^3$ is called *simple polyhedral tube* on a time interval $[t_1, t_2]$ $(t_1 < t_2)$ if there exists a polyhedron $D \subset [t_1, t_2] \times \mathbb{R}^2$ such that

$$\Gamma = \{(t, x) \in \partial D : t \in [t_1, t_2], \ x \in \partial D(t)\},$$

where $D(t) = \{x \in \mathbb{R}^2 : (t, x) \in D\}$ is a simple polygon and $\partial D(t)$ is the boundary of $D(t)$, $t \in [t_1, t_2]$. The faces (edges, vertices) of the polyhedron D belonging to Γ are said to be *faces (edges, vertices)* of the polyhedral tube Γ.

Remark 3 Lemma 2 implies that, if the terminal polygon M is convex, the lateral surface

$$\Gamma_0 = \{(t, x) : t \in [0, \vartheta], \ x \in \partial W_0(t)\}$$

of the maximal u-stable bridge W_0 is a simple polyhedral tube on the time interval $[t_1, \vartheta]$, where $t_1 > \min\{t \in [0, \vartheta] : W_0(t) \neq \varnothing\}$.

3.4 Semipermeable Polyhedral Tubes and Surfaces

In the theory of differential games, *a semipermeable surface* [1] is defined by the following property: the first player is capable to prevent an intersection of this surface in one direction by any trajectory of the dynamical system, and the second player is capable to do the same in the opposite direction.

Let us give a formal definition of the semipermeable property for a polygonal tube.

We write $O(x, \varepsilon)$ for an open circle in $\mathbb{R}^2$ with the radius $\varepsilon > 0$ and centre at $x \in \mathbb{R}^2$. Let us also define a cylindrical positive ε-half-neighbourhood $C_\varepsilon^+(z_0)$ of $z_0 = (x_0, t_0)$:

$$C_\varepsilon^+(z_0) = \{(t, x) : t \in [t_0, t_0 + \varepsilon], \ x \in O(x_0, \varepsilon)\}.$$

For $\varepsilon > 0$, we find such a value $\Delta(\varepsilon) > 0$ that for any $x \in \mathbb{R}^2$, all the trajectories of system (1) with the initial point x do not leave the set $O(x, \varepsilon)$ on the time interval $[0, \Delta(\varepsilon)]$.

A simple polyhedral tube Γ is called *semipermeable* on the interval $[t_1, t_2]$ if there exists $\varepsilon > 0$ such that for any $z_0 = (t_0, x_0) \in \Gamma$, $\varepsilon_* = \min\{\Delta(\varepsilon), \ t_2 - t_0\}$, and representation

$$C_\varepsilon^+(z_0) = G^+ \cup S \cup G^-, \quad S \subset \Gamma, \quad G^+ \cap G^- = \varnothing, \quad G^+ \cap \Gamma = \varnothing, \quad G^- \cap \Gamma = \varnothing,$$

the following properties hold:

(1) for any $v \in Q$, there exists a measurable open-loop control $u(t) \in P$ such that the solution $x(t)$ to the equation $\dot{x}(t) = u(t) + v$ with the initial condition $x(t_0) = x_0$ satisfies the inclusion $(t, x(t)) \in G^+ \cup S$ for all $t \in [t_0, t_0 + \varepsilon_*]$;
(2) for any $u \in P$, there exists a measurable open-loop control $v(t) \in Q$ such that the solution $x(t)$ to the equation $\dot{x}(t) = u + v(t)$ with the initial condition $x(t_0) = x_0$ satisfies the inclusion $(t, x(t)) \in S \cup G^-$ for all $t \in [t_0, t_0 + \varepsilon_*]$.

Given the definition, we use notions of *side* $(+)$ adjoining G^+ and *side* $(-)$ adjoining G^-. If the side $(+)$ is internal for the tube, the semipermeable tube is said to be of *type* $\pm$.

Remark 4 If the lateral surface Γ_0 of the maximal u-stable bridge W_0 is a polyhedral tube on the time interval $[t_1, t_2]$, then Γ_0 is a semipermeable tube on the interval.

The definition of a semipermeable tube and Remarks 1, 2, and 4 imply the following properties:

(a) If Γ is a semipermeable tube of type $\pm$ on the time interval $[0, \vartheta]$ and $\Gamma(\vartheta) = \partial M$, then Γ is the lateral surface of the maximal u-stable bridge W_0 on the time interval $[0, \vartheta]$ for the problem of M-attainability.
(b) If Γ_1 is a semipermeable tube of type $\pm$ on the time interval $[t_1, t_2]$, Γ_2 is a semipermeable tube of type $\pm$ on the time interval $[t_2, t_3]$, and $\Gamma_1(t_1) = \Gamma_2(t_1)$, then $\Gamma_1 \cup \Gamma_2$ is a semipermeable tube of type $\pm$ on the time interval $[t_1, t_3]$.

These properties allow us to turn the construction of the maximal u-stable bridge from the terminal set M on the interval $[0, \vartheta]$ into a sequential construction of semipermeable tubes of type $\pm$ on a finite number of adjacent suitable time intervals.

The notion of semipermeability remains the same for an unbounded polyhedral surface in $\mathbb{R}^3$. It is not hard to prove that for any half-plane $\Pi \subset \mathbb{R}^2$, the boundary of the half-space $\mathscr{T}_\vartheta(\Pi)$ is a semipermeable surface with the side $(+)$ adjoining $\mathscr{T}_\vartheta(\Pi)$.

4   Ordered Triples of Half-Spaces in $\mathbb{R}^3$

4.1   Convex and Concave Sets of Half-Spaces: Essential Half-Spaces

A *pair* of different half-planes Π_1 and Π_2 is called *convex* (*concave*) if $\Pi_1 \cap \Pi_2$ is an angle in the plane, and this angle lies on the left-hand side (right-hand side) if one moves first along the straight line $\partial \Pi_1$ up to the intersection with $\partial \Pi_2$ and then along the straight line $\partial \Pi_2$. We say that the half-planes Π_1 and Π_2 are *parallel* if $\Pi_1 \subset \Pi_2$ or $\Pi_2 \subset \Pi_1$.

The properties of convexity, concavity, and parallelism are also defined for any ordered pair of half-spaces in $\mathbb{R}^3 = \{(t, x)\}$ provided the half-spaces are not perpendicular to the time axis t. This is correct since for $t = \text{const}$, we have a pair of different half-planes in $\mathbb{R}^2$, which preserves one of the properties of convexity, concavity, and parallelism for all $t \in \mathbb{R}$.

An ordered list $L_1, L_2, \ldots, L_m$ of half-spaces is said to be *convex* (*concave*) if any pair $(L_i, L_{i+1}), j = \overline{1, m-1}$, is convex (concave).

We call a half-plane $L_i(t), t \in \mathbb{R}$, *inessential* for the intersection

$$L_1(t) \cap L_2(t) \cap \ldots \cap L_m(t)$$

if it can be eliminated without changing the result of the intersection; otherwise, the half-plane $L_i(t)$ is *essential* for the intersection.

4.2 Weight Function for Ordered Triples of Half-Spaces

Let us consider the special case of ordered lists of three half-spaces called *triples of half-spaces* in $\mathbb{R}^3 = \{(t, x)\}$.

We denote by Λ all triples $\lambda = (L_a, L, L_b)$ of half-spaces such that its boundaries (planes in $\mathbb{R}^3$) are not perpendicular to the time axis t and

$$\partial L_a \cap \partial L = l_a, \quad \partial L \cap \partial L_b = l_b,$$

where l_a, l_b are straight lines in $\mathbb{R}^3$, $l_a \neq l_b$. We write $L(t)$ for the section of a half-space L by means of the plane $t = \text{const}$.

Now we define *the weight function* $\mu_t(\lambda)$ of $\lambda = (L_a, L, L_b) \in \Lambda$ at the instant $t \in \mathbb{R}$.

In the case of collinear straight lines l_a and l_b, we set $\mu_t(\lambda) = +\infty$.

Since $l_a \neq l_b$, we consider the remaining case of crossing straight lines l_a and l_b. We write $l_a \cap l_b = \{(t_\lambda^*, x_\lambda^*)\}$.

Let us define auxiliary dihedral angles C_a and C_b in $\mathbb{R}^3$. If the pair L_a, L is convex, we set $C_a = L_a \cap L$; otherwise $C_a = (\overline{\mathbb{R}^3 \setminus L_a}) \cap (\overline{\mathbb{R}^3 \setminus L})$. If the pair L, L_b is convex, we set $C_b = L_b \cap L$; otherwise $C_b = (\overline{\mathbb{R}^3 \setminus L_b}) \cap (\overline{\mathbb{R}^3 \setminus L})$. Write

$$\gamma_\lambda(t) = C_a(t) \cap C_b(t) \cap \partial L(t).$$

We see at once that $\gamma_\lambda(t_\lambda^*) = \{x_\lambda^*\}$ and either $\gamma_\lambda(t)$ is a linear segment for $t < t_\lambda^*$ and an empty set for $t > t_\lambda^*$ or $\gamma_\lambda(t)$ is a linear segment for $t > t_\lambda^*$ and an empty set for $t < t_\lambda^*$. In the first case, we write $\mu_t(\lambda) = +\infty$ for all $t \in \mathbb{R}$. In the second case, we set $\mu_t(\lambda) = t - t_\lambda^*$ for all $t \in \mathbb{R}$. Illustrations of ordered convex triples of half-spaces in $\mathbb{R}^3$ are given in Figure 1.

We denote by Λ_t^0 all the triples $\lambda = (L_a, L, L_b) \in \Lambda$ such that the boundaries of the half-planes $L_a(t), L(t), L_b(t)$ pass through one point.

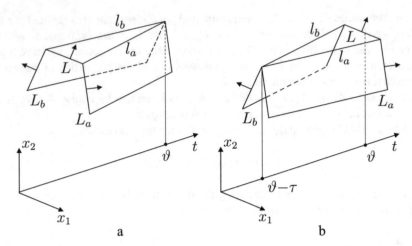

Fig. 1 Illustrations of ordered convex triples $\lambda = (L_a, L, L_b)$ of half-spaces in $\mathbb{R}^3$ for (a) $\mu_\vartheta(\lambda) = +\infty$; (b) $\mu_\vartheta(\lambda) = \tau \in (0, +\infty)$

Lemma 4 *Suppose $t_1 \in \mathbb{R}$ and $\lambda = (L_a, L, L_b) \in \Lambda^0_{t_1}$. Then $\mu_{t_1}(\lambda) \in \{0, +\infty\}$. If the triple λ is convex, then $\mu_t(\lambda) = +\infty$ is equivalent to essentiality of the half-plane $L(t)$ for the intersection $L_a(t) \cap L(t) \cap L_b(t)$ for all $t < t_1$.*

Proof The condition $\lambda \in \Lambda^0_{t_1}$ implies $t_1 = t^*_\lambda$. It is easy to check from the definition of the weight function that $\mu_{t_1}(\lambda) \in \{0, \infty\}$.

In the case of convex triples λ, we have

$$C_a \cap C_b = L_a \cap L \cap L_b, \quad \gamma_\lambda(t) = L_a(t) \cap L_b(t) \cap \partial L(t).$$

Consequently $\mu_t(\lambda) = +\infty$ is equivalent to $\gamma_\lambda(t)$ being a linear segment for $t < t^*_\lambda = t_1$ that means the half-plane $L(t)$ is essential for the intersection $L_a(t) \cap L(t) \cap L_b(t)$ for all $t < t_1$. $\qquad\square$

5 Local Constructions for a Non-convex Terminal Set

It is natural to expect that, in a neighbourhood of a convex (concave) vertex of the terminal set M, the maximal u-stable bridge is locally described by formula (7) (formula (8)) applied to an auxiliary convex set (set with the convex complement) that coincides with the set M in the neighbourhood of the vertex. On this basis, for each convex (concave) pair of consecutive half-spaces L_a, L_b connected with a convex (concave) vertex of the set M, we define convex (concave) list $I_\vartheta = I_\vartheta(L_a, L_b)$ of additional half-spaces corresponding to some normals in $\mathcal{N}(-P)$ (in $\mathcal{N}(-Q)$).

5.1   List $I_\vartheta(L_a, L_b)$ of Additional Half-Spaces

Let L_a, L_b be a convex or concave pair of half-spaces not related to the terminal set M.

First we consider the case of a convex pair L_a, L_b. Let η_a and η_b be the unit outer normals to the half-planes $L_a(\vartheta)$ and $L_b(\vartheta)$ correspondingly. We denote by $\mathscr{A}$ the supporting half-planes to the angle $A = L_a(\vartheta) \cap L_b(\vartheta)$:

$$\mathscr{A} = \{\Pi_\eta[A] : \eta \in \mathcal{N}(-P) \setminus \{\eta_a, \eta_b\}, \quad \rho(\eta; A) < +\infty\}. \tag{9}$$

If $\mathscr{A} = \varnothing$, we write $I_\vartheta(L_a, L_b) = \varnothing$. Otherwise, we introduce a non-empty set I_* of half-spaces:

$$I_* = \{\mathscr{T}_\vartheta(\Pi) : \Pi \in \mathscr{A}\}. \tag{10}$$

Let us examine the intersection

$$L_a(t) \cap L_b(t) \cap \{L(t) : L \in I_*\}, \quad t < \vartheta. \tag{11}$$

We define $I_\vartheta(L_a, L_b)$ to be a set of all the half-spaces from I_* that have essential t-sections (half-planes) for intersection (11). If the set $I_\vartheta(L_a, L_b)$ is not a singleton, we arrange it to be convex.

Finally for the case of a concave pair L_a, L_b, set (9) is formed for the angle $A = \overline{L'_a(\vartheta) \cap L'_b(\vartheta)}$, $L'_a = \overline{\mathbb{R}^3 \setminus L_a}$, $L'_b = \overline{\mathbb{R}^3 \setminus L_b}$, similar to the convex case by replacing $-P$ with $-Q$. The set I_* is given by (10). The additional half-spaces $I_\vartheta(L_a, L_b)$ are defined as all the half-spaces from I_* that have essential t-sections (half-planes) for the intersection

$$L'_a(t) \cap L'_b(t) \cap \{\overline{\mathbb{R}^2 \setminus L(t)} : L \in I_*\}, \quad t < \vartheta.$$

If the list $I_\vartheta(L_a, L_b)$ is not a singleton, we arrange it to be concave.

5.2   Recursive Procedure of Forming $I_\vartheta(L_a, L_b)$

We suggest a procedure of forming the list $I_\vartheta(L_a, L_b)$ for a convex pair L_a, L_b by induction over the number of elements of I_*.

(a) First, assume that the set I_* is a singleton, i.e. $I_* = \{L_1\}$. Let us consider the triple $\lambda_1 = (L_a, L_1, L_b)$. Using Lemma 4, we have $\mu_\vartheta(\lambda_1) \in \{0, +\infty\}$.

If $\mu_\vartheta(\lambda_1) = +\infty$, then the half-plane $L_1(t)$ is essential in (11) by Lemma 4. Thus, we write $I_\vartheta(L_a, L_b) = I_*$. If $\mu_\vartheta(\lambda_1) = 0$, then the half-plane $L_1(t)$ is inessential in (11), and we write $I_\vartheta(L_a, L_b) = \varnothing$.

(b) For clarity, let us consider also the case of two-element list $I_* = \{L_1, L_2\}$. Then the list $I_{ab} = \{L_a, L_1, L_2, L_b\}$ is also convex.

Let us consider the triples $\lambda_1 = (L_a, L_1, L_2)$ and $\lambda_2 = (L_1, L_2, L_b)$. If $\mu_\vartheta(\lambda_1) = +\infty$ and $\mu_\vartheta(\lambda_2) = +\infty$, then the half-planes $L_1(t)$ and $L_2(t)$ are essential in (11). Thus, we write $I_\vartheta(L_a, L_b) = I_*$.

If $\mu_\vartheta(\lambda_1) = 0$, then the half-plane $L_1(t)$ is inessential for the intersection

$$L_a(t) \cap L_1(t) \cap L_2(t), \quad t < \vartheta,$$

and, therefore, for intersection (11). Thus, it remains to examine the half-plane $L_2(t)$ to be essential in (11).

Let us consider the triple $\lambda_3 = (L_a, L_2, L_b)$. Similarly to the case a), we write $I_\vartheta(L_a, L_b) = \{L_2\}$ if $\mu_\vartheta(\lambda_3) = +\infty$ and $I_\vartheta(L_a, L_b) = \varnothing$ if $\mu_\vartheta(\lambda_3) = 0$.

(c) Suppose that there is a procedure of handling I_*^k of k half-spaces. Let us describe the procedure of handling a convex set

$$I_* = \{L_1, L_2, \ldots, L_{k+1}\}$$

of $(k + 1)$ half-spaces. Write

$$I_{ab} = \{L_a, L_1, L_2, \ldots, L_{k+1}, L_b\}.$$

The list I_{ab} is also convex.

If we have $\mu_\vartheta(\lambda) = +\infty$ for any triple λ of sequential half-spaces from I_{ab}, then the t-sections of all half-spaces from I_* are essential in (11). Thus, we write $I_\vartheta(L_a, L_b) = I_*$.

If we find a triple λ_0 of sequential half-spaces from I_{ab} such that $\mu_\vartheta(\lambda_0) = 0$, then we form the set

$$I_*^k = \{L_1, L_2, \ldots, L_{i-1}, L_{i+1}, \ldots, L_{k+1}\}$$

of k elements, where L_i is in the middle of the triple λ_0. The set I_*^k can be handling by the induction hypothesis.

In the case of a concave pair L_a, L_b, the procedure is similar.

5.3 Semipermeable Surfaces Based on a Pair of Half-Spaces

Convex or concave pair L_a, L_b of half-spaces together with the corresponding set $I_\vartheta(L_a, L_b)$ of additional half-spaces is the basis for the construction of a polyhedral semipermeable surface.

Proposition 1 Let L_a, L_b be a convex pair of half-spaces. Then the boundary ∂K of the intersection

$$K = K(L_a, L_b) = L_a \cap L_b \cap \{L : L \in I_\vartheta(L_a, L_b)\} \qquad (12)$$

is a semipermeable surface on the time interval $[0, \vartheta]$, and the set K is at the side $(+)$ of the surface ∂K.

Proof Let us define an auxiliary square with vertices at $(\pm R, \pm R)$, $R > 0$:

$$\Omega_R = \{x = (x_1, x_2) \in \mathbb{R}^2 : |x_1| \le R, \ |x_2| \le R\}.$$

Obviously, we can find $R > 0$ such that the intersection

$$M^R = L_a(\vartheta) \cap L_b(\vartheta) \cap \Omega_R$$

is a convex polygon. Using Lemma 2, we construct the maximal u-stable bridge W^R from the terminal set M^R on the time interval $[0, \vartheta]$. The bridge is an intersection of a final number of half-spaces. Taking into account the definition of the list $I_\vartheta(L_a, L_b)$, we observe that all the half-planes in the intersection

$$L_a(t) \cap L_b(t) \cap \{L(t) : L \in I_\vartheta(L_a, L_b)\}, \quad t < \vartheta, \qquad (13)$$

are essential. The value R can be chosen so large that the half-planes in (13) remain essential for t-sections $W^R(t)$ of the bridge W^R for all $t \in [0, \vartheta]$. Thus, according to Remark 4, we obtain semipermeability of the polyhedral surface ∂K.   □

An illustration of a semipermeable surface ∂K for a convex pair L_a, L_b is given in Figure 2a.

Using Lemma 3, we obtain the following assertion that is similar to Proposition 1.

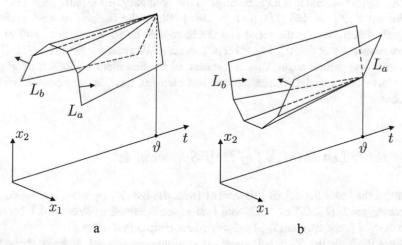

Fig. 2 Illustration of semipermeable surfaces (**a**) $\partial K(L_a, L_b)$ for a convex pair L_a, L_b and (**b**) $\partial K_*(L_a, L_b)$ for a concave pair L_a, L_b

Proposition 2 Let L_a, L_b be a concave pair of half-spaces. Then the boundary ∂K_* of the intersection

$$K_* = K_*(L_a, L_b) = (\overline{\mathbb{R}^3 \setminus L_a}) \cap (\overline{\mathbb{R}^3 \setminus L_b}) \cap \{ \overline{\mathbb{R}^3 \setminus L} : L \in I_\vartheta(L_a, L_b) \} \qquad (14)$$

is a semipermeable surface on the time interval $[0, \vartheta]$, and the set K_* is at the side $(-)$ of the surface ∂K_*.

An illustration of a semipermeable surface ∂K_* for a concave pair L_a, L_b is given in Figure 2b.

6 Construction of a Semipermeable Tube for the Non-convex Case

6.1 List $\widetilde{\mathscr{L}}(M, \vartheta)$ Based on the Terminal Set

Let the terminal set M be a bounded non-convex n-sided polygon, and its edges are numbered from 1 to n, while one moves along the boundary ∂M with the set M lying on the left-hand side. We will form an ordered list $\mathscr{L}(M, \vartheta)$ of half-spaces in $\mathbb{R}^3$ (which are similar to half-spaces in formulas (7) and (8)) combining these formulas in accordance with convex or concave parts of the boundary ∂M. For a basis, we take a list $\widetilde{\mathscr{L}}(M, \vartheta)$ of half-spaces corresponding to the set $\mathscr{N}(M)$ of normals, which is defined as follows.

We assign the set M to a circular list $\widetilde{\mathscr{M}}_1 = \{\widetilde{\Pi}_i\}_{i=1}^n$ of half-planes. Each half-plane $\widetilde{\Pi}_i$ is defined as the half-plane that has the same outer normal as the ith edge of the polygon M, and this edge belongs to the boundary of the half-plane $\widetilde{\Pi}_i$.

We write $\widetilde{\mathscr{L}}_1 = \{\mathscr{T}_\vartheta(\Pi) : \Pi \in \widetilde{\mathscr{M}}_1\}$. Since the list $\widetilde{\mathscr{L}}_1$ of half-spaces is uniquely determined by the set M and the instant ϑ, let us denote the result of this determination by $\widetilde{\mathscr{L}}(M, \vartheta)$, i.e. $\widetilde{\mathscr{L}}(M, \vartheta) = \widetilde{\mathscr{L}}_1$. We consider the list $\widetilde{\mathscr{L}}(M, \vartheta)$ as a circular one, which means that in the case of the first element, its left neighbour is the last element and, in the case of the last element, its right neighbour is the first element.

6.2 Basic List $\mathscr{L}(M, \vartheta)$ of Half-Spaces in $\mathbb{R}^3$

We form the basic list $\mathscr{L}_1$ of half-spaces from the list $\widetilde{\mathscr{L}}_1$ by inserting the convex (concave) set $I_\vartheta(L_a, L_b)$ of additional half-spaces defined in Section 4.1 between each convex (concave) pair $L_a, L_b \in \widetilde{\mathscr{L}}_1$ of sequential half-spaces.

Since the basic list $\mathscr{L}_1$ of half-spaces is uniquely determined by the set M and the instant ϑ, let us denote the result of this determination by $\mathscr{L}(M, \vartheta)$, i.e. $\mathscr{L}(M, \vartheta) = \mathscr{L}_1$.

Each $L \in \mathcal{L}_1$ has the previous L_* and subsequent L^* half-spaces in the list $\mathcal{L}_1$. Since $\lambda = (L_*, L, L^*) \in \Lambda$, the weight function $\mu_t(\lambda)$ is defined for $t \in [0, \vartheta]$. We denote the value $\mu_t(\lambda)$ by $\mu(L; \mathcal{L}_1, t)$.

The definition of $\mathcal{L}_1$ implies $\mu(L; \mathcal{L}_1, \vartheta) > 0$ for any $L \in \mathcal{L}_1$.

6.3   First Time Interval of the Construction

To construct the polyhedral tube based on the list $\mathcal{L}_1$, we define a time interval adjoining the instant ϑ at the left. Write

$$\tau_1 := \min\{ \mu(L; \mathcal{L}_1, \vartheta) : L \in \mathcal{L}_1 \} \in (0, +\infty], \quad \vartheta_1 := \begin{cases} \vartheta - \tau_1, & \tau_1 \in (0, \vartheta), \\ 0, & \tau_1 \in [\vartheta, +\infty]. \end{cases}$$

Note that for any $L \in \mathcal{L}_1$, the plane ∂L intersects the boundaries of the neighbouring half-spaces in the list $\mathcal{L}_1$ by straight lines, which have no common points for $t \in (\vartheta_1, \vartheta)$ and can have a common point at $t = \vartheta_1$ or $t = \vartheta$. It allows us to define a polyhedral surface E_1 for the time interval $(\vartheta_1, \vartheta]$ by successive intersections of the boundaries of the half-spaces from $\mathcal{L}_1$. For $t = \vartheta_1$, the polyhedral surface E_1 is defined by continuity. The faces of the polyhedral surface E_1 on the time interval $[\vartheta_1, \vartheta]$ are triangles and trapezoids.

Let $E_1(t)$ be a section of the surface E_1 by the plane $t = \text{const}$, $t \in [\vartheta_1, \vartheta]$, and M_1 be the bounded set determined by $E_1(\vartheta_1) = \partial M_1$.

An illustration of the sets M and M_1 and the tube E_1 on the first time interval of the construction is given in Figure 3.

Fig. 3 Illustration of a semipermeable tube E_1 on the first segment $[\vartheta_1, \vartheta]$ of the construction

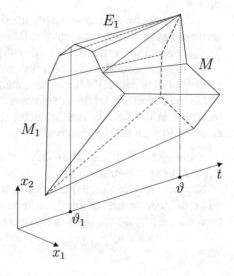

Assumption 1 The surface E_1 has no self-intersection on the time interval $(\vartheta_1, \vartheta]$, i.e. E_1 is a polyhedral tube on the time interval $[\vartheta_1 + \delta, \vartheta]$ for $\delta \in (0, \tau_1)$.

Let us explain the structure of the surface E_1. Suppose the list $\widetilde{\mathscr{L}}_1$ introduced in Section 5.1 has the form

$$\widetilde{\mathscr{L}}_1 = \{\widetilde{L}_1, \widetilde{L}_2, \ldots \widetilde{L}_n\}.$$

The surface E_1 consists of a finite number of pieces of polyhedral surfaces ∂K_i, $i \in \overline{1, n}$, constructed on the basis of convex and concave pairs $\widetilde{L}_i, \widetilde{L}_{i+1}$ of sequential half-spaces from $\widetilde{\mathscr{L}}_1$, $\widetilde{L}_{n+1} = \widetilde{L}_1$. The surface ∂K_i is defined as the boundary of $K(\widetilde{L}_i, \widetilde{L}_{i+1})$ given by (12) (of $K_*(\widetilde{L}_i, \widetilde{L}_{i+1})$ given by (14)) for the convex (concave) pair $L_a, L_b, i \in \overline{1, n}$. Propositions 1 and 2 imply that the polyhedral surfaces ∂K_i, $i \in \overline{1, n}$, are semipermeable. The set $\partial K_i \cap \partial K_{i+1}, i \in \overline{1, n}$, contains exactly one face of E_1 adjoining an edge of the terminal set M, $\partial K_{n+1} = \partial K_1$.

By the definition of ϑ_1, the t-sections of the faces of E_1 do not degenerate to a point for $t \in (\vartheta_1, \vartheta]$. Consequently, for any $\delta \in (0, \tau_1)$, there exist $\varepsilon > 0$ and $j \in \overline{1, n}$ such that $C_\varepsilon^+(z_0) \cap E_1 \subset \partial K_j$ for any $z_0 \in E_1(t), t \in [\vartheta_1 + \delta, \vartheta]$. This yields semipermeability of E_1 on the time interval $[\vartheta_1 + \delta, \vartheta]$. Using Remark 4 and Lemma 1, we obtain

$$\partial W_0(t) = E_1(t), \quad t \in [\vartheta_1, \vartheta]. \tag{15}$$

6.4 Second Time Interval of the Construction

In the case $\vartheta_1 = 0$, from (15), we see $W_0(0) = M_1$. Further, consider the case $\vartheta_1 > 0$.

Assume the set M_1 has an empty interior. Using semipermeability of polyhedral surfaces based on continuation of suitable adjacent faces of E_1 to the time interval $[0, \vartheta_1]$, we can show that $W_0(t) = \varnothing, t \in [0, \vartheta_1)$. The rigorous proof is omitted.

Finally, consider the case of a non-empty interior of the set M_1. If $E_1(\vartheta_1)$ is a closed polyline with self-intersection, then either the set M_1 is not simply connected, or the interior of the set M_1 is not connected, or the set M_1 has "tentacles" of zero thickness in the form of line segments. In this case, we will not consider further constructions on the time interval $[0, \vartheta_1)$ in this article.

Assumption 2 If the set M_1 has a non-empty interior, then the closed polyline $E_1(\vartheta_1)$ has no self-intersection.

Suppose now that $E_1(\vartheta_1)$ is a closed polyline without self-intersection. Then we make the second step of similar construction for the time interval $[0, \vartheta_1]$ and the terminal set M_1. On the basis of the set M_1, we define the list $\mathscr{L}_2 = \mathscr{L}(M_1, \vartheta_1)$ and then find the values

$$\tau_2 = \min\{\mu(L; \mathscr{L}_2, \vartheta_1) : L \in \mathscr{L}_2\} > 0, \quad \vartheta_2 = \begin{cases} \vartheta_1 - \tau_2, & \tau_2 \in (0, \vartheta_1), \\ 0, & \tau_2 \in [\vartheta_1, +\infty], \end{cases}$$

and form the surface E_2 out of the list $\mathscr{L}_2$ on the time interval $[\vartheta_2, \vartheta_1]$ similar to E_1. We write M_2 for the bounded set determined by $E_2(\vartheta_2) = \partial M_2$.

6.5 Construction of $\mathscr{L}_2$ on the Basis of $\mathscr{L}_1$

Let us show that the list $\mathscr{L}_2$ can be found on the basis of $\mathscr{L}_1$ without restoring the set M_1. From the definition of E_1, we see that the list $\widetilde{\mathscr{L}}_2 = \widetilde{\mathscr{L}}(M_1, \vartheta)$ is obtained from the list $\mathscr{L}_1$ by eliminating the "unnecessary" half-spaces that are not involved in forming the edges of the polygon M_1. At first, from the list $\mathscr{L}_1$, we remove all the half-spaces that are inessential with respect to its neighbours at the instant $t = \vartheta_1$, i.e. the half-spaces that have the value τ_1 of the weight function. As a result, we have

$$\mathscr{L}_1^0 = \mathscr{L}_1 \setminus \{L \in \mathscr{L}_1 : \mu(L; \mathscr{L}_1, \vartheta) = \tau_1\}.$$

Note that groups of identical adjacent half-spaces can arise in the list $\mathscr{L}_1^0$. We keep only one half-space in each of these groups and denote the new list by $\mathscr{L}_1^*$. Then the successive intersection of the boundaries of the half-planes $\{L(\vartheta_1) : L \in \mathscr{L}_1^*\}$ forms the polyline $E_1(\vartheta_1)$, i.e. the boundary of the set M_1. Consequently, $\widetilde{\mathscr{L}}_2 = \mathscr{L}_1^*$.

To obtain the list $\mathscr{L}_2$, it remains to attach the sets of additional half-spaces $I_{\vartheta_1}(L_a, L_b)$ between each pair L_a, L_b of adjacent half-spaces from $\mathscr{L}_1^*$.

6.6 Sequence of Adjacent Time Intervals of the Construction

Let us describe the construction for the step $k > 1$. Assuming $\vartheta_{k-1} > 0$ and $E_{k-1}(\vartheta_{k-1})$ is a closed polyline without self-intersection, we write M_{k-1} for the set bounded by the polyline $E_{k-1}(\vartheta_{k-1})$. Then we go to the next step of construction for the time interval $[0, \vartheta_{k-1}]$ and the terminal set M_{k-1}. Using the set M_{k-1}, we form the list

$$\mathscr{L}_k = \mathscr{L}(M_{k-1}, \vartheta_{k-1}),$$

find the values

$$\tau_k = \min\{\mu(L; \mathscr{L}_k, \vartheta_{k-1}) : L \in \mathscr{L}_k\} > 0, \quad \vartheta_k = \begin{cases} \vartheta_{k-1} - \tau_k, & \tau_k \in (0, \vartheta_{k-1}), \\ 0, & \tau_k \in [\vartheta_{k-1}, +\infty], \end{cases}$$

and construct the surface E_k out of half-spaces $\mathscr{L}_k$ on the time interval $[\vartheta_k, \vartheta_{k-1}]$. We write M_k for the bounded set determined by $E_k(\vartheta_k) = \partial M_k$.

The list $\mathscr{L}_k$ can be constructed on the basis of the list $\mathscr{L}_{k-1}$ without restoring the set M_{k-1}.

Thus, we bring the construction of a semipermeable tube $E_1 \cup E_2 \cup \cdots \cup E_k$ of type $\pm$ on the time interval $[\vartheta_k, \vartheta]$ to handling the initial list $\mathscr{L}_1$ that consists of the following two steps repeating for each instant ϑ_s, $s = 1, \ldots, k-1$:

(a) eliminating the "unnecessary" half-spaces with respect to the instant ϑ_s;
(b) attaching the sets of additional half-spaces with respect to the instant ϑ_s.

Now we conclude that, for $k \geq 1$, one of the following options is true:

1. $\vartheta_k = 0$ and the set $W_0(0) = M_k$ is non-empty;
2. $\vartheta_k > 0$ and the interior of the set M_k is empty, which corresponds to degeneration of the maximal u-stable bridge at the instant ϑ_k, which means $W_0(0) = \varnothing$;
3. there is a situation of self-intersection of the closed polyline $E_k(\vartheta_k)$, and further construction is not considered;
4. we can take the next step and construct a semipermeable tube of type $\pm$ on the time interval $[\vartheta_{k+1}, \vartheta]$, $\vartheta_{k+1} < \vartheta_k$.

Since the number of edges of the polygons M, P, and Q is finite, we can prove that there is no instant $\vartheta_* \in [0, \vartheta)$ that would be the limit of the monotonically decreasing sequence of instants $\{\vartheta_k\}$, i.e. the algorithm is finite. The proof is omitted.

7 Numerical Examples

Let us consider some results of numerical construction of the set $W_0(0)$ based on the suggested algorithm. Three examples of the sets M, P, and Q are shown in Figures 4–6. The corresponding solvability sets $W_0(0)$ were computed for different values of ϑ. Each of these sets was obtained independently of the other ones and without restoring any sets for intermediate instants from the time interval $(0, \vartheta)$. The edges, which are not parallel to any edge of the terminal set, correspond to half-spaces added when forming the basic list $\mathscr{L}_1$.

In the first example, the terminal set M is a non-convex nine-polygon, and its boundary is shown in Figure 4 by a solid line; P is a line segment with endpoints $(0, -0.5)$ and $(0, 0.5)$; and Q is a line segment with endpoints $(0, 0)$ and $(1.5, 1.5)$. The set $W_0(0)$ is computed for $\vartheta = 0.5, 1.0, 1.5$. There is only one half-space added at the terminal time instant. Self-intersection of the semipermeable surface occurs for a greater terminal instant ϑ.

In the second example, the terminal set M is a non-convex seven-polygon, and its boundary is shown in Figure 5 by a solid line; P is a line segment with endpoints $(0.5, 0.5)$ and $(-0.5, -0.5)$; and Q is a line segment with endpoints

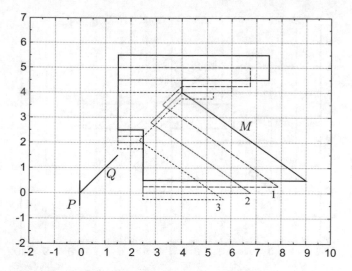

Fig. 4 Example 1 showing the boundary of $W_0(0)$ computed for $\vartheta = 0.5$ (line 1); $\vartheta = 1.0$ (line 2); $\vartheta = 1.5$ (line 3)

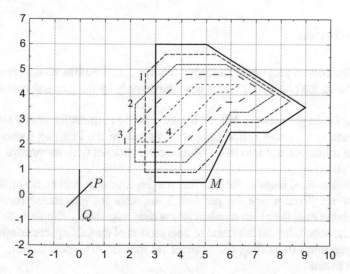

Fig. 5 Example 2 showing the boundary of $W_0(0)$ computed for $\vartheta = 0.8$ (line 1); $\vartheta = 1.6$ (line 2); $\vartheta = 2.4$ (line 3); $\vartheta = 3.2$ (line 4)

$(0, 1)$ and $(0, -1)$. Figure 5 shows the results of computations of the set $W_0(0)$ for $\vartheta = 0.8, 1.6, 2.4, 3.2$. There are two half-spaces added at the terminal time instant.

The last example (Figure 6) illustrates the "evolution" of the symbol "plus" over the set of natural numbers $\vartheta \in \overline{1, 7}$ as the terminal instants. There are six half-spaces added at the terminal time instant. The solvability set becomes empty for $\vartheta > 7$.

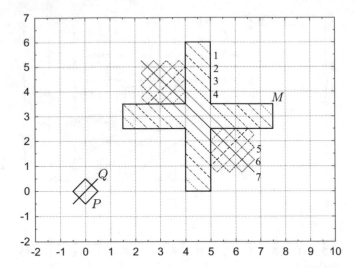

Fig. 6 Example 3 showing the boundary of $W_0(0)$ computed for $\vartheta = 1.0$ (line 1); $\vartheta = 2.0$ (line 2); $\vartheta = 3.0$ (line 3); $\vartheta = 4.0$ (line 4); $\vartheta = 5.0$ (line 5); $\vartheta = 6.0$ (line 6); $\vartheta = 7.0$ (line 7)

8   Conclusion

In a conflict control problem with simple motions, vector of the phase velocity does not depend on the phase state and is determined only by the players' controls at the current instant.

From this, the permanence follows for the velocity of transition of any hyperplane in the phase space under action of control vectors of the first and second players delivering minimum and maximum of the scalar product with the normal vector to the hyperplane.

This, in the turn, leads to the fact that for the case of a convex terminal set, the solvability set (from which the problem is solvable for guaranteed transition for the first player onto the given terminal set) is determined very simple in its essence. Namely, the solvability set is a result of intersection of the half-spaces corresponding to all hyperplanes supporting the terminal set that are extremely transitioned onto the initial instant.

It is difficult to build the mentioned intersection under high dimension of the phase state. But, in principle, it is important that the solvability set can be found without additional partition of the given time interval into small parts and organization of the backward procedure for recalculation of the solvability sets built earlier for the intermediate instants (with the subsequent passage to limit by decreasing the dividing step).

The case of constructing the solvability set is the most simple for problems in the plane when both the terminal set and sets defining the constraints onto the player's controls are convex polygons. In these cases, the solution is reduced to consideration of the finite totality of the straight lines (supporting the terminal set) at the terminal instant, to extreme transition of them onto the initial instant, and to finding intersection of corresponding half-planes. The latter procedure is equal to solving the system of linear inequalities in the plane.

The paper under consideration is devoted to analyse how (for the problems in the plane) the described idea for finding the solvability set by means of extreme transition of the finite number of straight lines could be implemented when the terminal set is not convex.

It has been succeeded to prove that in such a case, it is possible to avoid an additional division of given time interval. But in the algorithm, special instants appear (defined in the backward motion), at which the totality of extremely transitioned lines *has to be rebuilt*.

As a rule, at these rebuilt instants, some lines are removed but there are no added ones. But the cases can appear when new lines take place. Thorough investigation of such cases will be the subject of further research.

If the terminal set is a convex polygon, then it is also possible to say about the rebuilt instants. But under this, the list of the transitioned lines is always decreased. Moreover, each of the transitioned lines can be led to the initial instant, and only then it could be possible to decide whether it is redundant or not. So, it would not be necessary to insert into the algorithm the rebuilt instants in the convex case.

When the terminal instant and terminal polygon are fixed but the initial instant is decreased, the self-intersection of a curve, which is used for the construction of the boundary of the solvability set, can appear if built by the algorithm described in the paper. The algorithm gives correct answer only under the absence of the self-intersection. Note that analysis of possibility of the self-intersection is a separate and difficult problem.

The algorithm described in the paper uses essentially specificity of the plane. For dimensions 3, 4, ... of the phase state and a non-convex polyhedron as the terminal set, notion of the ordered list of transitioned half-spaces and, correspondingly, the notion of the rebuilding principally become more complex.

Acknowledgement The work has been supported by the Russian Foundation for Basic Researches under project no. 18-01-00410.

References

1. Isaacs, R.: Differential Games. Wiley, New York (1965)
2. Kumkov, S.S., Le Menec, S., Patsko V.S.: Zero-sum pursuit-evasion differential games with many objects: Survey of publications. Dynamic Games and Applications, Vol. 7, No. 3, 609–633 (2017)
3. Krasovskii, N.N. Rendezvous Game Problems. Nat. Tech. Inf. Serv., Springfield, VA, 1971.
4. Bryson, A.E., Ho, Y.C. Applied Optimal Control. Optimization, Estimation and Control. Hemisphere publishing corporation. Wiley, 1975.

5. Krasovskii, N.N., Subbotin, A.I.: Game-theoretical control problems. Springer, New York (1988)
6. Pshenichnyy, B.N., Sagaydak, M.I.: Differential games with fixed time. J. Cybernet. **1**, 117–135 (1971)
7. Subbotin, A.I.: Generalized solutions of first-order PDEs: the dynamical optimization perspective. Birkhäuser, Boston (1995)
8. Krasovskii, N.N., Subbotin, A.I.: Positional differential games. Nauka, Moscow (1974) (in Russian)
9. Subbotin, A.I. Minimax inequalities and Hamilton–Jacobi equations. Moscow: Nauka, 1991. (in Russian)
10. Pontryagin, L.S. Linear differential games, II. Soviet Math. Dokl. Vol. 8. 1967. P. 910–912.
11. Hadwiger, H. Vorlesungen uber Inhalt, Oberflache und Isoperimetrie. Berlin: Springer-Verlag, 1957.
12. Kamneva, L.V., Patsko, V.S. Maximal stable bridge in game with simple motions in the plane. Annals of the International Society of Dynamic Games. Vol. 14: Advances in Dynamic and Evolutionary Games Theory, Applications, and Numerical Methods, ed. by Thuijsman F., Wagener F., 139–163, Springer (2016)

Effects of Players' Random Participation to the Stability in LQ Games

Ioannis Kordonis and George P. Papavassilopoulos

Abstract We consider a linear quadratic (LQ) game with randomly arriving players, staying in the game for a random period of time. The Nash equilibrium of the game is characterized by a set of coupled Riccati-type equations for Markovian jump linear systems (MJLS), and the existence of a Nash equilibrium is proved using Brouwer's fixed point theorem. We then consider the game, in the limit as the number of players becomes large, assuming a partially Kantian behavior. We then focus on the effects of the random entrance, random exit, and partial Kantian cooperation to the stability of the overall system. Some numerical results are also presented. It turns out that in the noncooperative case, the overall system tends to become more unstable as the number of players increases and tends to stabilize as the expected time horizon increases. In the partially cooperative case, an explicit relation of the expected time horizon of each player with the minimum amount of cooperation, sufficient to stabilize the closed loop system, is derived.

Keywords LQ games • Random entrance and exit • Stochastic optimal control • Stochastic stability • Cooperative behavior

1 Introduction

In several game situations, the number of players, as well as the time interval during which each one of them is going to stay in the game, is not known a priori. For example, in a competitive market, a firm doesn't know the future number of its opponents, neither for how much time each one of them is going to stay in the game. Another example is the interaction of a bank with its customers. Yet another example

I. Kordonis (✉)
CentraleSupélec, Avenue de la Boulaie, 35576 Cesson-Sévigné, France
e-mail: jkordonis1920@yahoo.com

G.P. Papavassilopoulos
School of Electrical and Computer Engineering, National Technical University of Athens, Iroon Polytechniou 9, Zografou, 15780 Athens, Greece
e-mail: yorgos@netmode.ntua.gr

© Springer International Publishing AG 2017
J. Apaloo, B. Viscolani (eds.), *Advances in Dynamic and Mean Field Games*,
Annals of the International Society of Dynamic Games 15,
https://doi.org/10.1007/978-3-319-70619-1_12

is a pension system, where, in the long run, the future number of working people or their ratio over the retired is uncertain. Motivated by the problems described, this work considers dynamic games with randomly entering and leaving players continuing the work in [22], in which only the case of randomly entering players was considered.

The overall system stability is a very important aspect in systems affected by different self-regarding entities in different instants of time. This is particularly true for the current formulation, since each one of the players is not interested in the value of the state of the system in the far future. Thus, a player does not have enough motivation to act to regulate the system state at an instant of time, in which she is most likely not going to participate in the game. The effects of random entrance and exit to the overall system stability are studied.

1.1   Related Literature

There is a large literature for static games with a random number of players (eg. [26, 27]). In this class of games, the players act before they learn the actual number of their opponents. It is related to the current work, due to the fact that in our case the players do not know the actual number of their future opponents. A complete review of the literature on games with a random number of players is given in the Introduction section of [4].

Another related stream of research is the theory of dynamic games with random duration [14, 23, 31]. The basic idea in these works is to transform the set of random time horizon optimal control problems to a set of appropriately discounted infinite horizon problems and then derive the HJB equations. A similar idea was used in [12] to a differential game with fixed horizon where a player enters at a random instant of time. Kordonis and Papavassilopoulos [20] describes a game where all the players participate in the game at the initial time step, but each one of them may leave the game after a random amount of time.

Yet another related stream of research is the theory of overlapping generation games. This class of games was introduced in [35] (see also [2] and for a differential game formulation [15]). It describes game situations in which a new generation of players enters the game at each time step and stays for a certain period of time. During this time interval, the players of that generation interact with the players of the older and younger generations participating in the game in each time step. In [29] an LQ game involving an infinite horizon (major) player and overlapping generations of players with finite horizons (minor players) was analyzed, sufficient conditions for the existence of an equilibrium were given, and it was shown that if the minor players have long enough horizons, they can stabilize the overall system even in the absence of the major player. Other formulations of the LQ game with a major player and overlapping generations of minor players were given in [16, 30].

A class of dynamic games with randomly entering players was studied in [22] within the LQ framework. In this class of games, there is an infinite horizon player and at each time step, a random number of minor players enter the game. Each one of the minor players stays in the game for a fixed amount of time. The basic tool to characterize Nash equilibrium was the theory of Markovian jump linear systems. The case, where there is a large number of players, was also studied using the mean field approach. Independently, [4] and [6] studied game situations where additional agents are arriving at the system at random time steps and assuming a prespecified behavior of the agents, derived closed formulas for the payoffs. This line of research continued in [5], where the authors considered dynamic games (not necessarily LQ) with randomly arriving or leaving players in discrete and continuous time, with either finite or infinite horizon. They derived the HJB equations and characterized the feedback Nash equilibrium. Then, they focused on the LQ case. For both the discrete and continuous time infinite horizon cases, they provided a characterization of the Nash equilibrium, using Riccati-type equations, and for the continuous time case, they derived sufficient conditions for the existence of an equilibrium.[1] Finally, [10] studied the effects of the number of the players on the stability of the closed loop system, in a class of LQ games.

1.2 Contribution

A dynamic game with random entrance and exit is formulated as follows: We assume (for ease of exposition) a discrete time linear scalar dynamics, affected by the actions of the players participating in the game at each instant of time. In every time step, a random number of players enter the game, and the distribution of this number depends on the number of active players. Furthermore, at each time step, each one of the active players may leave the game with a certain probability, which also depends on the number of the active players.

The basic tool we use to derive optimality and equilibrium conditions for the LQ game with random entrance and exit is the theory of Markovian jump linear systems (MJLS) (eg. [9, 24]). Particularly, the optimal control problem of a player may be restated as an LQ control problem with random horizon for a MJLS or equivalently, as a discounted infinite horizon LQ control problem for a MJLS. We then prove the existence of the equilibrium using Brouwer's fixed point theorem and derive conditions for the stability of the closed loop system.

We then move to limit as the number of players becomes large. In this case, the players may not have a motivation to use actions large enough to stabilize the overall

---

[1] Regarding the relationship of [5] with the current work, [5] has clearly a more general framework. Some of the results of Section 3 of the current work and particularly Proposition 1 can be obtained using a modified version of Theorem 2.4 of [5]. Furthermore, this modification is described in [5]. However, the manipulations leading to Proposition 1 are necessary for the rest of the paper and thus they are included. Let us note that the first version of the current work, including Proposition 1, appeared in the ISDG Symposium paper in Urbino, Italy, July 2016, and we become aware of [5] after the presentation.

system. Therefore, if the open loop system (the system assuming zero control inputs) is unstable, then the closed loop system will remain unstable, leading to a very bad collective performance. Similar phenomena are known in the economics literature as the "tragedy of the commons." We then analyze the same game, assuming partially cooperative players. Specifically, we assume that the participants partially follow Kant's "Categorical Imperative." To this end, we introduce a notion of a partial Kant-Nash equilibrium which is slightly different from the existing notions of the literature [11, 33, 34, 37] (see also [19]). Existence and uniqueness results for the partial Kant-Nash equilibrium are proved.

The effects of random entrance and random exit are studied numerically for the finite number of players case. It turns out that with no cooperation, if the open loop system is unstable, the closed loop system matrix destabilizes when the number of active payers becomes large and stabilizes when the expected horizon of the players becomes large. We also study analytically the stability of the overall system in the large number of players limit. The stability of the closed loop system depends on three factors, namely, the expected time horizon of the players, the amount of cooperation, and the open loop dynamics. It turns out that the minimum amount of cooperation needed to stabilize the overall system is inversely proportional to the expected time horizon of the players. Thus, for long-living players, a very small amount of cooperation is sufficient to improve considerably the collective performance.

1.3   Organization

The rest of this work is organized as follows. In Section 2, we describe the stochastic participation dynamics (random entrance and exit), the system dynamics and the cost. In Section 3, the Nash equilibria are characterized using Riccati-type equations, the existence of a Nash equilibrium is proved, and conditions for the closed loop stability of the system are derived. In Section 4, we study the limiting case as the number of players becomes large, introduce the notion of partial Kant-Nash equilibrium, prove the existence and uniqueness of the equilibrium, and derive analytical conditions for the closed loop stability. Section 5 gives some numerical results on the effects of random entrance and exit to the stability of the overall system. Finally, in Section 6 we summarize the basic contributions of the current work and propose some directions for future research.

2   Description of the Game

Let us first describe the players' entrance/exit stochastic dynamics. There is a countably infinite total number of players. However, only finite of them are active at each instant of time. Denote by $I_k \subset \mathbb{N}$ the set of players participating in the game (active players) at time step k and by $y_k \in \mathbb{N} \cup \{0\}$ their number. At each time step, a random number of new players n_k enter the game. Each player participating in the game at time step k has an equal probability to leave the game (die). This probability may depend on the number of active players y_k and it is denoted by $\pi(y_k)$. A player that has already left the game cannot enter again. We assume that there is a maximum number of active players $N \in \mathbb{N}$, that is, it holds $n_k \leq N - y_k$, with probability 1. Thus, the evolution of the number of active players y_k is described by:

$$y_{k+1} = y_k + n_k - d_k,\tag{1}$$

where $d_k \in \mathbb{N}$ is the number of players leaving the game after time step k and d_k has a binomial distribution $B(y_k, \pi(y_k))$. Assuming that the distribution of the random variables n_k, d_k are allowed to depend only on y_k, the process y_k becomes a Markov chain with state space $\{1, 2, \ldots, N\}$. Let us denote by $C = [c_{ij}]$ the transition probability matrix of the Markov chain, i.e., $c_{ij} = P(y_{k+1} = j | y_k = i)$.

The participation of each Player i can be described by the stochastic process ξ_k^i, where $\xi_k^i = 1$ if Player i participates in the game at time step k and zero otherwise.

There is a single scalar linear discrete time dynamic equation. The evolution of the state vector $x_k \in \mathbb{R}$ is given by:

$$x_{k+1} = a x_k + \sum_{i \in I_k} \frac{b}{y_k} u_k^i,\tag{2}$$

where $u_k^i \in \mathbb{R}$ is the action of Player i at time step k and $a, b \in \mathbb{R}$ are the system parameters. The summation over an empty set is interpreted as 0.

The cost function of a Player i with entrance time t_0^i and exit time t_f^i (both of which are random discrete non-negative variables) is given by:

$$J^i = E \left[\sum_{k=t_0^i}^{t_f^i} \left(q x_k^2 + r(u_k^i)^2 \right) \right].\tag{3}$$

We assume that both q and r are positive scalars.

The interactions of the players are symmetric. Furthermore, for a player i, participating in the game at time step k, the expected cost after that step (known also as the expected "cost to go") is independent of her age. It depends only on the number of the active players y_k and the state variable x_k. Thus, in this work we focus on symmetric linear feedback (no memory) strategies, i.e., strategies of the form:

$$u_k^i = -L(y_k) x_k.\tag{4}$$

Hence, in order to find a Nash equilibrium of this form, we need to determine the N scalar values $L(1), \ldots, L(N)$. We will call L the feedback gain of player i.

Remark 1. Asymmetric equilibria often exist in symmetric games. Furthermore, in some works nonlinear strategies and strategies depending on older values of the state vector constituting a Nash equilibrium have been derived for linear quadratic games [3]. However, in this work we focus only on symmetric linear feedback (no memory) strategies.

3 Nash Equilibrium Characterization and Existence

We first consider the optimal control problem that each player faces, focusing on Player i's point of view. Assume that all the other active players use a strategy in the form given in (4).

In the following lemma, the optimization problem of Player i is expressed as an infinite time LQ control problem for a Markovian jump linear system (MJLS). To do so, we consider an additional random variable w_k, to represent random exit.

Lemma 1. *Under the aforementioned assumptions, a strategy for Player i is optimal if and only if it solves:*

$$\underset{u^i}{minimize} \quad E\left[\sum_{k=0}^{\infty} \xi_k^i \left(qx_k^2 + r(u_k^i)^2\right)\right]$$

$$subject\ to \quad x_{k+1} = \bar{a}(y_k)x_k + \bar{b}(y_k)u_k^i$$

$$\xi_{k+1}^i = \xi_k^i(1 - w_k) \tag{5}$$

$$y_k:\ Markov\ chain\ with\ transition\ matrix\ C$$

$$u_k^i = \gamma(x_k, y_k)$$

where:

$$\bar{a}(y_k) = a - b\frac{y_k - 1}{y_k}L(y_k), \tag{6}$$

$$\bar{b}(y_k) = b/y_k, \tag{7}$$

and w_k is a random variable taking the value 1 with probability $\pi(y_k)$ and the value 0 otherwise.

Proof. The state equation, the participation dynamics and the cost functions do not depend explicitly on time. If $t_0^i \neq 0$ using the time transformation $k' = k - t_0^i$, equations (1), (2), and (3) remain unchanged. Thus, without loss of generality, assume that Player i is present at time step 0, i.e. $t_0^i = 0$. Substituting (4) into (2), we get $x_{k+1} = \bar{a}(y_k)x_k + \bar{b}(y_k)u_k^i$. Equation $\xi_{k+1} = \xi_k^i(1 - w_k)$ is equivalent to the fact that Player i, at each time step k, leaves the game with probability $\pi(y_k)$. Finally, for Player i with entrance time t_0^i and exit time t_f^i it holds:

$$E\left[\sum_{k=t_0^i}^{t_f^i}\left(qx_k^2 + r(u_k^i)^2\right)\right] = E\left[\sum_{k=0}^{\infty}\xi_k^i\left(qx_k^2 + r(u_k^i)^2\right)\right], \tag{8}$$

due to the fact that $\xi_k^i = 1$, if Player i is active at time step k and zero otherwise, which completes the proof. $\qquad\square$

We may observe that the vector $\tilde{y}_k = [y_k\ \xi_k^i]^T$ is a Markov chain as well. Thus, (5) is an LQ control problem for a MJLS. Thus, the optimal control law for player i can be expressed in terms of the positive solution of the following set of coupled scalar Riccati-type equations (see, e.g., [21] or [9]):

$$K'(\tilde{y}) = q + \bar{a}^2(\tilde{y})\left[\Lambda'(\tilde{y}) - \frac{\Lambda'^2(\tilde{y})\bar{b}^2(\tilde{y})}{r + \bar{b}^2(\tilde{y})\Lambda'(\tilde{y})}\right], \tag{9}$$

$$\Lambda'(\tilde{y}) = E\left[K'(\tilde{y}_{k+1})|\tilde{y}_k = \tilde{y}\right]. \tag{10}$$

Equations (9) and (10) correspond to the Bellman equation for the optimal control problem (5). The optimal "cost to go" at time step k is expressed in terms of $K'(\cdot)$ as $K'(\tilde{y}_k)x_k^2$ and the optimal control law by:

$$u_k^i = -\frac{\bar{a}(y_k)\bar{b}(y_k)\Lambda'(\tilde{y}_k)}{r + \bar{b}^2(y_k)\Lambda'(\tilde{y}_k)}x_k. \tag{11}$$

The existence of a positive solution of the Riccati-type equations is guaranteed from the stabilizability of the system.

We may write the Riccati-type equations characterizing the optimal policy in terms of the original Markov chain, based on the following observation:

$$K'\left(\begin{bmatrix} y \\ 0 \end{bmatrix}\right) = \Lambda'\left(\begin{bmatrix} y \\ 0 \end{bmatrix}\right) = 0. \tag{12}$$

Equation (12) holds true, due to the fact that the corresponding "cost to go" is zero. Let us use the notation $K(y) = K'\left(\begin{bmatrix} y \\ 1 \end{bmatrix}\right)$ and $\Lambda(y) = \Lambda'\left(\begin{bmatrix} y \\ 1 \end{bmatrix}\right)$. Then, the Riccati-type equations are expressed in terms of the original Markov chain as:

$$K(y) = q + \bar{a}^2(y)\left[\Lambda(y) - \frac{\Lambda^2(y)\bar{b}^2(y)}{r + \bar{b}^2(y)\Lambda(y)}\right], \tag{13}$$

$$\Lambda(y) = (1 - \pi(y))E\left[K(y_{k+1})|y_k = y, \xi_{k+1}^i = \xi_k^i = 1\right], \tag{14}$$

and optimal control law for (5) is given by $u_k^i = -L^\star(y_k)x_k$, where:

$$L^\star(y) = \frac{\bar{a}(y)\bar{b}(y)\Lambda(y)}{r + \bar{b}^2(y)\Lambda(y)}. \tag{15}$$

Furthermore, if $y_k = 0$, then necessarily $\xi_k^i = 0$. Thus, $K(0)$ and $\Lambda(0)$ are both zero.

Remark 2. The term $(1 - \pi(y))$ in (14) has a role similar to a discount factor. Similar results, where an optimal control problem with random horizon can be restated as a discounted infinite horizon problem, are common in the literature (eg. [8]).

The preceding analysis proves the following proposition.

Proposition 1. *A symmetric set of linear feedback strategies in the form $u_k^i = -L(y_k)x_k$ constitutes a perfect Nash equilibrium if and only if:*

$$L^\star(y) = L(y), \tag{16}$$

for all the possible values of $y = 1, \ldots, N$, where $L^\star$ is computed by (15), using the positive solution of (13), (14). $\qquad\square$

We then prove the existence of a Nash equilibrium.

Proposition 2. *There exists a Nash equilibrium constituting of symmetric linear feedback strategies.*

Proof. Combining (15), (16) and (6), (7) we obtain:

$$L = \frac{\bar{a}\bar{b}\Lambda}{r + \left(\bar{b}^2 + b\bar{b}\frac{y-1}{y}\right)\Lambda}, \tag{17}$$

where the dependence on y is omitted for brevity. Then, consider the mapping $f : \mathbb{R}_+^N \to \mathbb{R}_+^N$ which maps the vector $(\Lambda(1), \ldots, \Lambda(N))$ to the vector $(\Lambda^{\text{New}}(1), \ldots, \Lambda^{\text{New}}(N))$ according to:

$$f : \Lambda \overset{(17)}{\longmapsto} L \overset{(6)}{\longmapsto} \bar{a} \overset{(13)}{\longmapsto} K \overset{(14)}{\longmapsto} \Lambda^{\text{New}}. \tag{18}$$

Observe that, for a given fixed point of f, we may use (17) to obtain a Nash equilibrium. Thus, it is sufficient to prove the existence of a fixed point for f. It is not difficult to show that f is bounded. Observe that due to (17), L is bounded for $\Lambda \in \mathbb{R}_+^N$. Equation (6), implies that $\bar{a}$ is bounded. Furthermore, (13) can be written as:

$$K = q + \bar{a}^2 \frac{r\Lambda}{r + \bar{b}^2\Lambda}, \tag{19}$$

and thus K is bounded. Finally, using (14) we conclude that f is bounded. Denote by $[\bar{M} \ldots \bar{M}]^T$ a bound for f. Then, $[0, \bar{M}]^N$ maps into $[0, \bar{M}]^N$. Applying Brouwer's fixed point theorem, we conclude that f has a fixed point and thus there exists a Nash equilibrium. □

The closed loop system evolves according to:

$$x_{k+1} = a_{cl}(y_k)x_k$$

$$y_k: \text{Markov chain with transition matrix } C$$

where $a_{cl}(y_k) = a - bL(y_k)$ is the closed loop system matrix (evolution coefficient). Let us recall that a linear scalar stochastic system is called mean square exponentially stable, if $E[x_0^2] < \infty$ implies that $E[x_k^2] \to 0$ exponentially, as $k \to \infty$.

Proposition 3. *The closed loop system is mean square exponentially stable if and only if the matrix:*

$$M = \begin{bmatrix} c_{11}a_{cl}^2(1) & \ldots & c_{1N}a_{cl}^2(1) \\ \vdots & \ddots & \vdots \\ c_{N1}a_{cl}^2(N) & \ldots & c_{NN}a_{cl}^2(N) \end{bmatrix} \tag{20}$$

has spectral radius $\sigma(M)$ less than 1 (or equivalently has all the eigenvalues inside the unit disc).

Proof. This property is based on standard results in MJLS (eg. [7] or [24]). □

In the remainder of this section, we study the case where the cost of each player is finite, even if all the players have zero actions.

Lemma 2. *Assume that the exit probability is such that $a^2(1 - \pi(y)) \le \zeta < 1$, for all y. Then in the equilibrium it holds:*

$$K(y) \le q/(1 - \zeta) \tag{21}$$

for every y.

Proof. Using (17) we get $|\bar{a}| \le |a|$. Consider a representative Player i and without loss of generality assume that she enters at time step 0. Further, assume all the players except Player i are using their equilibrium policies and that Player i, instead of her optimal policy, uses $u_k^i = 0$. Then her expected cost would satisfy:

$$E\left[\sum_{k=0}^{\infty} \xi_k^i q x_k^2\right] = E\left[\sum_{k=0}^{\infty} \xi_k^i q x_0^2 \prod_{l=0}^{k-1} \bar{a}^2(y_l)\right] \le E\left[\sum_{k=0}^{\infty} \xi_k^i a^{2k}\right] q x_0^2 \le \frac{q x_0^2}{1 - \zeta}.$$

On the other hand, the minimum cost to go is $K(y_0)x_0^2$. Thus, it holds $K(y) \le q/(1 - \zeta)$ for every y. □

Remark 3. Observe that the bound on K (and thus also to the expected cost to go) does not depend on the number of active players y nor on the entrance and exit dynamics. This bound will be used to show that the feedback gain tends to zero, when y becomes large.

Proposition 4. *Under the assumptions of Lemma 2, the feedback gain $L(y)$, in the equilibrium satisfies:*

$$0 \leq L(y) \leq \frac{b\zeta q}{(1-\zeta)ar} \frac{1}{y}. \tag{22}$$

Proof. The first inequality follows from (15). Using (15), (14), and (21) we get:

$$L(y) = \frac{\bar{a}(y)\bar{b}(y)\Lambda(y)}{r + \bar{b}^2(y)\Lambda(y)} \leq \frac{a\bar{b}(y)\Lambda(y)}{r} \leq \frac{ab}{ry} \frac{\zeta}{a^2} \frac{q}{1-\zeta} = \frac{b\zeta q}{ar(1-\zeta)} \frac{1}{y}, \tag{23}$$

which concludes the proof. □

Remark 4. Proposition 3 shows that, under the stated assumptions, if there is a large number of players, then each one of them will have a very small action. Furthermore, as the number of players tends to infinity, the actions of the players tend to zero. In the case of a large but finite number of players if the open loop system is unstable (i.e., the system (2) with zero inputs: $u_k^i = 0$, for all i and k), the closed loop system will remain unstable. This leads to very inefficient outcomes for the players entering the game at late stages. The phenomenon, where the selfish actions of independent agents lead to inefficient collective results, is known in the literature as the "tragedy of the commons" [13].

4   Game with a Large Number of Players and Partial Kantian Cooperation

In this section we consider the limiting case of the game considered in Sections 2 and 3, as the number of players tends to infinity, assuming a continuum of players. Particularly, each player considers that there is a continuum of players (corresponding to a subset of the real numbers), each one of which has vanishing influence to the system dynamics. Furthermore, each individual player assumes that she has an insignificant effect to the others, and her cost is affected only by her own actions and the distribution of the actions of the other players. Probably the first works studying models with a continuum of players were [39] and [1] (see also [28], Ch. X), while a value for such games was introduced in [25]. Existence results were proved in [36]. Dynamic and stochastic games with a continuum of players were considered in [40–42].

In this section, we first define a notion of a partially cooperative equilibrium, then characterize it, and finally analyze the effects of the random exit and cooperation on the stability of the overall system.

4.1 Partial Kantian Cooperation Definition

Consider the class of games described in Sections 2 and 3. As discussed at the end of Section 3, if the number of players is large, then the equilibrium behavior of the players may lead to very inefficient outcomes. In this section, we assume that the players have some amount of cooperation. Instead of using the usual altruism assumption, i.e., that each player minimizes a weighted sum of her own cost and the cost of the rest of the players, we assume that the players partially follow Kant's "Categorical Imperative" [17]. The most usual form of the "Categorical Imperative" is stated as:

> "Act only according to that maxim whereby you can, at the same time, will that it should become a universal law."

Let us note that there exists already a notion of a Kantian equilibrium in the literature. Roemer in [32–34] defined the Kantian equilibrium as a set of strategies in which no player has a motivation to change her action, assuming that the rest of the other players would change also their actions accordingly (eg. multiplicatively or additively). It turns out that for static games, under weak conditions, the set of Kantian Equilibria coincides with the Pareto front. The papers [11, 37, 38] extended [34] in two directions. First, they consider dynamic games and, second, study game situations with mixed Kantian and Nash players. In this case, they introduce the notions of (inclusive and exclusive) Kant-Nash equilibria, or more generally the virtual co-mover equilibria.

Here, we use a slightly different formulation assuming that each player minimizes her own cost assuming that a proportion of the rest of the players will use a strategy (not an action) identical to hers. Let us first introduce some notation. Fix a Player i and denote by $J_i^\alpha(L, L_1, L_2)$ the cost of Player i, assuming that she follows $u_k^i = -L(y_k)x_k$, a portion α of the others use $u_k^j = -L_1(y_k)x_k$ and the rest of the players i.e. a portion $1 - \alpha$ of the others use a strategy $u_k^j = -L_2(y_k)x_k$.

Definition 1. A linear symmetric set of strategies $u_k^i = -L(y_k)x_k$ is an α-Kant-Nash equilibrium within the class of linear feedback strategies if:

$$J_i^\alpha(L, L, L) = \min_{L'} J_i^\alpha(L', L', L). \tag{24}$$

Since this work focuses on linear strategies, we restrict the α-Kant-Nash equilibrium definition to the set of linear feedback strategies. However, Definition 1 could be trivially generalized to consider also nonlinear strategies. A related definition of partial Kant-Nash equilibrium for static games with various types of players is given in [19].

Remark 5. The use of a Kant-Nash formulation has some differences with the altruism formulation. At first, in the altruism formulation, each player adds the costs of the other players participating in the game, even in different time instances. If the closed loop system is unstable, this sum would diverge. Furthermore, in the large number of players case, a weighted sum of the cost of the other players either diverges or assigns a very small (vanishing) weight to the other players. In the latter case, "the tragedy of the commons" remains.

Remark 6. The notion of α-Kant-Nash equilibrium is different from the notions described in [11, 32–34, 37], and [38]. The basic difference is that (24) assumes that some of the other players will imitate the strategy of each player and not the changes in the strategy.

4.2 Partial Kantian Equilibrium of the Game with a Continuum of Players

Let us first describe the participation dynamics in the game with a continuum of players. Assume that the random participation is described by a Markov chain $\bar{y}_k$ taking values in $[0, 1]$ indicating the number of the active players divided by the maximum possible number of players. Assume further that the probability of each player leaving the game at time step k is given by $\pi(\bar{y}_k)$ and that the dynamics of the Markov chain is described by a transition probability kernel $C(\cdot, \cdot)$, that is $P(\bar{y}_{k+1} \in A | y_k = y) = C(y, A)$. Finally assume that, for any Player i the random variable ξ_k^i (indicator for the presence of Player i in the game at time step k) is independent of y_k.

Let us then characterize the α-Kant-Nash equilibrium of the game. Consider a certain Player i and assume that the rest of the players are using $u_k^j = -L(\bar{y}_k)x_k$. Then Player i, who assumes that a portion α of the rest of the players would follow her, will behave as if the dynamics were given by:

$$x_{k+1} = \bar{a}(\bar{y}_k)x_k + \bar{b}(\bar{y}_k)u_k^i,$$

where:

$$\bar{\alpha}(\bar{y}_k) = a - b(1 - \alpha)L(\bar{y}_k), \tag{25}$$

$$\bar{b} = \alpha b. \tag{26}$$

Thus, Player i faces the optimal control problem:

$$\underset{u^i}{\text{minimize}} \quad E\left[\sum_{k=0}^{\infty} \xi_k^i \left(qx_k^2 + r(u_k^i)^2\right)\right]$$

$$\text{subject to} \quad x_{k+1} = \bar{a}(\bar{y}_k)x_k + \bar{b}(\bar{y}_k)u_k^i,$$

$$\xi_{k+1} = \xi_k^i(1 - w_k),$$

$$\bar{y}_{k+1} \sim C(\bar{y}_k, \cdot), \quad y_k : \text{ Markov chain},$$

$$u_k^i = \gamma(x_k, \bar{y}_k). \tag{27}$$

The following proposition characterizes the α-Kant-Nash equilibrium.

Proposition 5. *A set of symmetric linear feedback strategies is an α-Kant-Nash equilibrium if and only if there exist functions $K(\bar{y})$, $\Lambda(\bar{y})$ such that:*

$$K(\bar{y}) = q + \bar{a}^2(\bar{y})\left[\Lambda(\bar{y}) - \frac{\Lambda^2(\bar{y})\bar{b}^2(\bar{y})}{r + \bar{b}^2(\bar{y})\Lambda(\bar{y})}\right], \tag{28}$$

$$\Lambda(\bar{y}) = (1 - \pi(\bar{y}))\int K(\bar{y}')C(\bar{y}, d\bar{y}'), \tag{29}$$

with $K(y) \geq 0$ for any y and:

$$L(y) = \frac{\bar{a}(\bar{y})\bar{b}(\bar{y})\Lambda(\bar{y})}{r + \bar{b}^2(\bar{y})\Lambda(\bar{y})}. \tag{30}$$

Proof. Problem (27) depends on the Markov chain $\tilde{y}_k = \begin{bmatrix} \bar{y}_k \\ \xi_k^i \end{bmatrix}$. The coupled Riccati-type equations for (27) (see [21]) are given by:

$$K'(\tilde{y}) = q + \bar{a}^2(\tilde{y})\left[\Lambda'(\tilde{y}) - \frac{\Lambda'^2(\tilde{y})\bar{b}(\tilde{y})}{r + \bar{b}^2(\tilde{y})\Lambda'(\tilde{y})}\right], \tag{31}$$

$$\Lambda'(\tilde{y}) = E\left[K'(\tilde{y}_{k+1})|\tilde{y}_k = \tilde{y}\right]. \tag{32}$$

As in Section 3, $K'\left(\begin{bmatrix} \bar{y} \\ 0 \end{bmatrix}\right) = \Lambda'\left(\begin{bmatrix} \bar{y} \\ 0 \end{bmatrix}\right) = 0$ and using a transformation of the Markov chain similar to the one used in Section 3 we obtain (28)–(30). $\qquad\square$

We then examine the degenerate Markov chain case. Particularly, we assume that $C(\bar{y}, \cdot) = \delta_{y_0}$, for every $\bar{y} \in [0, 1]$ where δ denotes the Dirac measure. Furthermore, we assume that the exit probability is independent of the number of players.

Example 1. This example describes a participation dynamics with a large number of players, which in the limit converges to a degenerate Markov chain. At each time step each one of the existing players leaves the game with probability $1/2$, and each one of a very large number of players N enters the game with probability $(1 - \bar{y}_k)/2$, where $\bar{y}_k = |I_k|/N$ and I_k is the set of active players. Then as $N \to \infty$ the Markov chain stochastic kernel converges weakly to $C(\cdot, \cdot)$ with $C(\bar{y}, \cdot) = \delta_{1/2}$ (see [18] for the definition of the weak convergence of stochastic kernels).

From now on we assume a degenerate Markov chain. We may thus drop the dependence of K, Λ, L, $\bar{a}$, and π on $\bar{y}$. Combining (28) and (29) we get:

$$\frac{\Lambda}{1 - \pi} = q + \bar{a}^2 \frac{r\Lambda}{r + \bar{b}^2 \Lambda}. \tag{33}$$

Furthermore, L can be expressed in terms of Λ as:

$$L = \frac{\alpha b \Lambda}{r + \alpha b^2 \Lambda} a. \tag{34}$$

Substituting to (33) we have:

$$\frac{\Lambda}{1 - \pi} = q + a^2 r \frac{r + \alpha^2 b^2 \Lambda}{(r + \alpha b^2 \Lambda)^2} \Lambda = f(\Lambda). \tag{35}$$

We may observe that $f(0) = q$ and that $f(\Lambda)$ is bounded. Hence, there exists a solution to (35) and the game has an α-Kant-Nash equilibrium. In the following proposition, it is shown that the equilibrium is also unique.

Proposition 6. *In the degenerate Markov chain case, there exists an α-Kant-Nash equilibrium. Furthermore, the equilibrium is unique within the class of symmetric linear feedback strategies.*

Proof. The first part is already proved. To prove the second part, consider the solution Λ of (33). Straightforward computations show that the positive solution Λ is strictly increasing in $\bar{a}$. Furthermore, from (34), L is strictly increasing in Λ, and from (25), $\bar{a}$ is strictly decreasing in L. Hence, the following mapping:

$$f : \bar{a} \mapsto \Lambda \mapsto L \mapsto \bar{a}^{\text{New}}, \tag{36}$$

is strictly decreasing. However, an equilibrium corresponds to a fixed point of the mapping, and a decreasing mapping could at most once intersect the identity mapping. Thus, there is a unique value for $\bar{a}$ which may correspond to equilibrium, which implies its uniqueness. □

4.3 The Effects of π and α to the Stability

We then focus on the effects of the expected time horizon of the players (depending on π) and the amount of cooperation α on the stability of the overall system. In this section we assume that $a > 1$ (the following results can be easily adapted for

$a < -1$). Let us denote by $\Lambda_{\alpha,\pi}$ the positive solution of (35) parametrized by α and π and by $L_{\alpha,\pi}$ the corresponding feedback gain. The following lemma studies how $\Lambda_{\alpha,\pi}$, $L_{\alpha,\pi}$ and the corresponding closed loop system gain $a_{cl}^{\alpha,\pi} = a - bL_{\alpha,\pi}$ vary as the exit probability π changes.

Lemma 3. *Consider two values π_1 and π_2 of the exit probability with $\pi_2 < \pi_1$. Then it holds:*

(i) $\Lambda_{\alpha,\pi_1} < \Lambda_{\alpha,\pi_2}$.
(ii) $L_{\alpha,\pi_1} < L_{\alpha,\pi_2}$.
(iii) $a_{cl}^{\alpha,\pi_1} < a_{cl}^{\alpha,\pi_2}$.

Proof. (i) We have proved that (35) has a unique solution. Furthermore, it holds $f(0) = q > 0$ and f is continuous. Thus,

$$\frac{\Lambda}{1 - \pi_2} < \frac{\Lambda}{1 - \pi_1} < f(\Lambda) \tag{37}$$

for $\Lambda < \Lambda_{\alpha,\pi_1}$. Using $\Lambda(1 - \pi_2) < f(\Lambda)$ for $\Lambda < \Lambda_{\alpha,\pi_1}$, we get $\Lambda_{\alpha,\pi_2} \geq \Lambda_{\alpha,\pi_1}$ and the equality case can be ruled out easily. To prove (ii) and (iii) it is sufficient to substitute (i) into (34) and then (ii) into $a_{cl}^{\alpha,\pi} = a - bL_{\alpha,\pi}$. □

We then compute the value of π which makes the system marginally stable. Recall that a system is called marginally stable if it is neither asymptotically stable nor unstable. The feedback gain leading to marginal stability is $L_{ms} = \frac{a-1}{b}$. Substituting this value to (34) and solving for Λ, we obtain:

$$\Lambda_{ms} = \frac{r(a-1)}{\alpha b^2}. \tag{38}$$

Substituting to (35) and solving for π, we conclude to the following proposition:

Proposition 7. *Assume that the players follow the α-Kant Nash equilibrium. Denote by:*

$$\pi_{ms} = 1 - \frac{\Lambda_{ms}}{f(\Lambda_{ms})} = \frac{(qb^2 + r(a-1)^2)\alpha}{(qb^2 + r(a-1)^2)\alpha + (a-1)r}. \tag{39}$$

Then, the closed loop system is:

(i) *Unstable if $\pi > \pi_{ms}$.*
(ii) *Marginally stable if $\pi = \pi_{ms}$.*
(i) *Asymptotically stable if $\pi < \pi_{ms}$.*

Remark 7. Proposition 7 states that for a positive amount of cooperation α, there is an expected life span, given by $1/\pi$, above which the closed loop system is stabilized. It is interesting that the expected life span, which leads to marginal stability, has a linear relationship with the inverse amount of cooperation $1/\alpha$. The value of $1/\alpha$ has the following meaning: each player assumes that one in every $1/\alpha$ players will imitate her strategy.

Remark 8. The situation described in the Proposition 7 is similar with the one occurring in games with overlapping generations with a fixed number of players entering at each time step (one at every time step) staying for a specified amount of time described in [29]. Particularly, in both cases if the players live for a long time, they tend to stabilize the overall system.

5 Numerical Examples

In this section, we present some examples of games with random entrance and exit and study how the number of players and the expected time horizon affect the stability of the closed loop system. The first example studies games having a finite number of players, at each time step, and the second example concerns a game with a continuum of players and partial Kantian cooperation.

Example 2. The maximum possible number of players is 100, and at each time step either 1 or 3 new players enter the game. If the number of active players y_k is larger than or equal to 98, then the number of new players is 1 with probability 1.

The game parameters are $q = r = b = 1$ and $a = 2$. In the first two runs, the parameters are chosen such that the expected number of players in the invariant distribution is 75 and in the third run the expected number of active players is 60. The parameters of the entrance and exit dynamics, as well as the spectral radius of the closed loop system matrix $\sigma(M)$, described in (20), are presented in Table 1. The invariant distribution for the number of players is illustrated in Figure 1. The invariant distributions, for the parameter combinations 1 and 2, are almost identical.

In the first two runs the expected number of players is the same. However, the players in the first run live on average longer. The difference between the second and the third run is that the average number of players in the third run is smaller, while the average life spans are equal. These results show that the system tends to be more unstable as the number of players increases and tends to be more stable as the average time horizon increases.

Example 3. In this example we consider a game with a continuum of players and partial Kantian cooperation. The game parameters are again $q = r = b = 1$. The plot of π_{ms} as a function of α for various values of a is depicted in Figure 2 and the linear dependence of $1/\pi_{ms}$ on $1/\alpha$ in Figure 3. In Figure 2 the stable region for each a is below the curve, and in Figure 3 the stable region for each a is above the straight line.

Table 1 The parameters for the participation dynamics for three different runs

Run	$P(n_k) = 1$	$P(n_k = 3)$	π	Exp. # of pl.	Exp. life span	$\sigma(M)$	Stable
#1	0.7498	0.2502	0.02	75	50	0.9943	✓
#2	0.3747	0.6253	0.03	75	33.33	1.0043	✗
#3	0.6	0.4	0.03	60	33.33	0.9980	✓

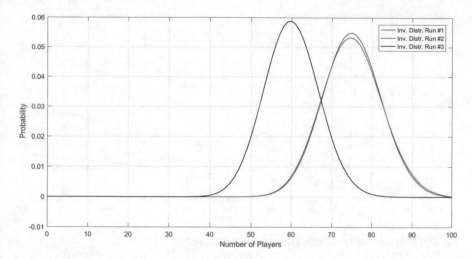

Fig. 1 The invariant distribution of the Markov chain y_k, representing the number of active players, for the runs 1–3. The entrance and exit parameters are described in Table 1

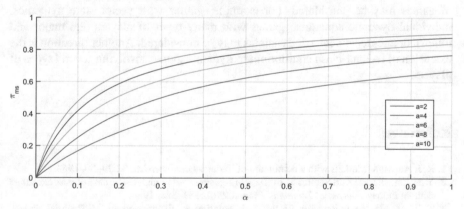

Fig. 2 The relation of π_{ms} with α for various values of a

6 Conclusion

We studied a simple class of dynamic games with randomly entering and randomly leaving players, focusing on stability issues. The equilibria were characterized by coupled Riccati-type equations for MJLS. Through numerical examples, we showed that an increase in the number of players tends to destabilize the closed loop dynamics, whereas an increase in the expected life span of the players tends to stabilize the closed loop dynamics. We also considered games with a very large number of players and studied the equilibria assuming a partial Kantian cooperation. It was proved that there is a life expectancy above which the overall system becomes stable and its value is a linear function of the inverse of the amount of cooperation.

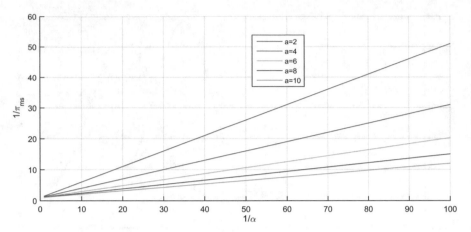

Fig. 3 The relation of $1/\pi_{ms}$ with $1/\alpha$ for various values of a

This work can be extended in several directions. At first, more general classes of games may be considered. For example, games with vector state dynamics, individual dynamic equations, games with many types of players (eg. major and minor), or age-dependent exit dynamics may be considered. Another direction is the extension of α-Kant-Nash equilibrium to dynamic games involving several types of players.

References

1. R. J. Aumann. Markets with a continuum of traders. *Econometrica*, 32:39–50, 1964.
2. Yves Balasko and Karl Shell. The overlapping-generations model, I: The case of pure exchange without money. *Journal of Economic Theory*, 23(3):281–306, 1980.
3. T Basar. On the uniqueness of the nash solution in linear-quadratic differential games. *International Journal of Game Theory*, 5(2–3):65–90, 1976.
4. Pierre Bernhard and Marc Deschamps. On dynamic games with randomly arriving players. *Dynamic Games and Applications*, pages 1–26, 2015.
5. Pierre Bernhard and Marc Deschamps. Dynamic equilibrium in games with randomly arriving players. *[Research Report] RR-8961 Université de la Côte d'Azur-INRIA, 2016, pp.38. <hal-01379644>*, 2016.
6. Pierre Bernhard and Frédéric Hamelin. Sharing a resource with randomly arriving foragers. *Mathematical biosciences*, 273:91–101, 2016.
7. Behram Homi Bharucha. On the stability of randomly varying systems. Technical report, DTIC Document, 1961.
8. El-Kébir Boukas, Alain Haurie, and Philippe Michel. An optimal control problem with a random stopping time. *Journal of optimization theory and applications*, 64(3):471–480, 1990.
9. Oswaldo Luiz Valle Costa, Marcelo Dutra Fragoso, and Ricardo Paulino Marques. *Discrete-time Markov jump linear systems*. Springer Science & Business Media, 2006.
10. Jacob Engwerda. Properties of feedback nash equilibria in scalar lq differential games. *Automatica*, 69:364–374, 2016.

11. A. Ghosh and N. Van Long. Kant's rule of behavior and kant-nash equilibria in games of contributions to public goods. 2015.
12. Ekaterina Viktorovna Gromova and José Daniel López-Barrientos. A differential game model for the extraction of non renewable resources with random initial times. *Contributions to Game Theory and Management*, 8(0):58–63, 2015.
13. Garrett Hardin. The tragedy of the commons. *Science*, 162(3859):1243–1248, 1968.
14. A Haurie. Piecewise deterministic differential games. In *Differential Games and Applications*, pages 114–127. Springer, 1989.
15. Steffen Jørgensen and David WK Yeung. Inter-and intragenerational renewable resource extraction. *Annals of Operations Research*, 88:275–289, 1999.
16. Marc Jungers, George P Papavassilopoulos, and Hisham Abou-Kandil. Feedback nash strategy for games with player-dependent time horizons. *IFAC Proceedings Volumes*, 45(25):45–50, 2012.
17. Immanuel Kant. Groundwork for the metaphysics of morals. 1785.
18. Alan F Karr. Weak convergence of a sequence of markov chains. *Probability Theory and Related Fields*, 33(1):41–48, 1975.
19. Ioannis Kordonis. A model for partial kantian cooperation. *arXiv preprint arXiv:1609.01921*, 2016.
20. Ioannis Kordonis and George P. Papavassilopoulos. Lq nash games with players participating in the game for random time intervals. In *The 15 th International Symposium on Dynamic Games and Applications*, Bysice, Czech Republic, July 18–22, 2012.
21. Ioannis Kordonis and George P Papavassilopoulos. On stability and lq control of mjls with a markov chain with general state space. *IEEE Transactions on Automatic Control*, 59(2): 535–540, 2014.
22. Ioannis Kordonis and George P Papavassilopoulos. Lq nash games with random entrance: An infinite horizon major player and minor players of finite horizons. *IEEE Transactions on Automatic Control*, 60(6):1486–1500, 2015.
23. Jesús Marín-Solano and Ekaterina V Shevkoplyas. Non-constant discounting and differential games with random time horizon. *Automatica*, 47(12):2626–2638, 2011.
24. Michel Mariton. *Jump linear systems in automatic control*. M. Dekker New York, 1990.
25. J. W. Milnor and L. S. Shapley. Values of large games ii: Oceanic games. *Mathematics of Operations Research*, 3(4):290–307, 1978.
26. Roger B Myerson. Population uncertainty and poisson games. *International Journal of Game Theory*, 27(3):375–392, 1998.
27. Roger B Myerson. Large poisson games. *Journal of Economic Theory*, 94(1):7–45, 2000.
28. G. Owen. *Game Theory*. W.B. Saunders Company, 1969.
29. George P. Papavassilopoulos. University-students game. *Dynamic Games and Applications*, 3(3):387–418, 2013.
30. George P Papavassilopoulos, Hisham Abou-Kandil, and Marc Jungers. A nash game with long-term and short-term players. In *52nd IEEE Conference on Decision and Control*, pages 1726–1731. IEEE, 2013.
31. Leon A Petrosjan and EKATERINA V Shevkoplyas. Cooperative solutions for games with random duration. *Game theory and applications*, 9:125–139, 2003.
32. John E Roemer. Kantian equilibrium. *The Scandinavian Journal of Economics*, 112(1):1–24, 2010.
33. John E Roemer. How we (do and could) cooperate. 2015.
34. John E Roemer. Kantian optimization: A microfoundation for cooperation. *Journal of Public Economics*, 127:45–57, 2015.
35. Paul A Samuelson. An exact consumption-loan model of interest with or without the social contrivance of money. *The journal of political economy*, pages 467–482, 1958.
36. David Schmeidler. Equilibrium points of nonatomic games. *Journal of statistical Physics*, 7(4):295–300, 1973.
37. Ngo Van Long. Kant-nash equilibrium in a dynamic game of climate change mitigations. 2015.

38. Ngo Van Long. Kant-nash equilibria in a quantity-setting oligopoly. In *Equilibrium Theory for Cournot Oligopolies and Related Games*, pages 179–201. Springer, 2016.
39. John Glen Wardrop. Road paper. some theoretical aspects of road traffic research. *Proceedings of the institution of civil engineers*, 1(3):325–362, 1952.
40. Agnieszka Wiszniewska-Matyszkiel. Static and dynamic equilibria in games with continuum of players. *Positivity*, 6(4):433–453, 2002.
41. Agnieszka Wiszniewska-Matyszkiel. Static and dynamic equilibria in stochastic games with continuum of players. *Control and Cybernetics*, 32(1):103–126, 2003.
42. Agnieszka Wiszniewska-Matyszkiel. Open and closed loop nash equilibria in games with a continuum of players. *Journal of Optimization Theory and Applications*, 160(1):280–301, 2014.

Interval Computing of the Viability Kernel with Application to Robotic Collision Avoidance

Stéphane Le Ménec

Abstract Viability theory provides a set of concepts and algorithms to study uncertain continuous dynamic systems under viability (or state) constraints. Interval computation is about guaranteed numerical methods for approximating sets. The main tool to be used is based on the idea of enclosing real numbers in intervals and real vectors in boxes. Refined interval techniques as contractor programming and guaranteed integration allow to implement the viability kernel and the capture basin algorithms. Results are provided considering the kinematics of the game of two cars. Viability kernel and capture basin algorithms are used to compute backward reachable sets which are differential game capture zones assuming predefined maximum time horizons. Then, collision avoidance between two noncooperative ground mobile robots is performed based on the backward reachable sets mentioned above.

Keywords Non cooperative differential games • Pursuit-evasion games • Backward reachable sets • Level sets of the value function • Viability theory • Viability kernel • Capture basin • Interval computing • Contractor programming • Collision avoidance • Game of two identical cars • Robotics

1 Introduction

This paper deals with collision avoidance. Set membership techniques make possible to advice when performing robust evasion maneuvers. When talking about noncooperative collision avoidance between two moving vehicles, uncertainties occur at various level. Uncertainties may occur in the kinematics, i.e. in the way we model the vehicle physics (numerical parameters, model simplifications). The second type of uncertainties we have to deal with is the second vehicle behavior, i.e., the second vehicle future trajectory. This project mainly focuses on the second aspect of uncertainty.

S. Le Ménec (✉)
Airbus/MBDA, 1, Avenue Réaumur, 92 358 Le Plessis-Robinson, France
e-mail: stephane.le-menec@mbda-systems.com

© Springer International Publishing AG 2017   279
J. Apaloo, B. Viscolani (eds.), *Advances in Dynamic and Mean Field Games*,
Annals of the International Society of Dynamic Games 15,
https://doi.org/10.1007/978-3-319-70619-1_13

The differential game approach [14] suggests to model the moving obstacle as a second player in a pursuit evasion game framework. According to the pursuit evasion game approach, the evader player is the vehicle performing robust collision avoidance maneuvers, while the second player (the pursuer) tries to catch the first one. Rather than estimating the pursuer trajectory, differential game theory puts assumptions on the pursuer control domain range (bounds). An example is maximum acceleration capability if the pursuer controls its acceleration. Qualitative analysis of such models allows to identify what are the safe and unsafe initial conditions, i.e., safe and unsafe states. Safe means that whatever are the uncertainties, there exists a successful evasion strategy, while unsafe means that no evasion strategy guaranties to avoid collision against all the pursuer behaviors. The unsafe state space area is also called capture zone, while the safe part of the state space is the non-capture zone.

There exist several theories close to differential games and corresponding algorithms to solve set membership problems. In addition to the seminal book of Rufus Isaacs, among various relevant theories, we may refer to the concept of stable bridges [17], to viability theory [4, 5] and to positive set invariance tools [8]. In addition, we acknowledge the fact that other techniques exist to address reachability questions and to perform collision avoidance. However, in the context of this project, we focus on set value analysis and not on the stochastic interpretation of the situation [10], potentially using Markov processes [11].

Algorithms have been implemented in a way to compute sets verifying the properties mathematically defined by the aforementioned theoreticians. Kumkov et al. [18] and Patsko and Turova [22] provide examples about how to compute differential game capture zones using stable bridges. Cardaliaguet et al. [12] are devoted to a survey of results for differential games obtained through viability theory. After recalling the basic theory for differential games (obtained in the 1990s), this article also provides an overview of recent advances on the computing aspects. Other toolboxes as [20] and [24] based on level set techniques allow to compute maximal reachable sets.

The benchmark we use for applying interval-based algorithms is the well-known differential game of two cars defined by Rufus Isaacs and studied in details by Antony Mertz [19]. Other differential game models as the homicidal chauffeur game, also initially described by Rufus Isaacs, has been recently used for collision avoidance purpose [13]. More precisely, it is explained how collision avoidance capability between drones can be seen as a reversed homicidal chauffeur game, hence the name "Suicidal Pedestrian Differential Game."

This paper is organized as follows. After this brief introduction (Section 1), Section 2 recalls basic definitions of viability theory, before to provide set definitions of the viability kernel and capture basin algorithms (Section 3). Then, Section 4 introduces what is interval computing using natural interval arithmetics. Section 5 is again about viability but more interval algorithm oriented. First versions of the viability kernel algorithm (Section 5.1) and of the capture basin algorithm (Section 5.2) are provided using the basic interval techniques introduced in Section 4. Section 6 explains what are contractors, which are interval operators that allow to re-implement the viability kernel algorithm (Section 6.1) and the capture basin algorithm (Section 6.2) in an easy manner. The last section (Section 7) is about applying the contractor-based viability algorithms to the differential game of two identical cars. Backward reachable sets have been computed (Section 7.2), and robotic simulations results are given for illustration purpose (Section 7.3) before to summarize and to conclude (Section 8) what has been done so far in this project.

The main contribution of this paper is to provide contractor-based versions of the viability kernel and of the capture basin algorithms. At the knowledge of the author, it is the first paper published on the topic, i.e., paper mixing viability theory and contractor programming (especially in the context of pursuit evasion games).

2 Viability Theory

2.1 Definitions

Viability theory [4] studies in a systematic manner the viability properties of a system evolving in an environment. Let us consider either a temporal interval $[0, T]$ with finite horizon or $[0, \infty[$ (denoted for simplicity by $[0, T]$ with $T := \infty$) with infinite horizon. We consider an (evolving) environment:

$$K(\cdot) : t \in [0, T], \quad t \rightsquigarrow K(t) \subset X \tag{1}$$

(defined by viability constraints) and, when $T < +\infty$, an (evolving) target:

$$C(\cdot) : t \in [0, T], \quad t \rightsquigarrow C(t) \subset K(t) \tag{2}$$

The usual case is a non-evolving target for which $C(t) = \emptyset$ when $t < T$ and $C(T) = C$. An evolution $x(\cdot) : t \mapsto x(t)$ is said to be viable in $K(\cdot)$ if:

$$\forall t \in [0, T], \quad x(t) \in K(t) \tag{3}$$

Finally, we introduce the right-hand side $F : (x, t) \in X \times [0, T] \rightsquigarrow F(X, t) \in X$ of the differential inclusion:

$$\forall t \in [0, T], \quad x'(t) \in F(x(t), t)) \tag{4}$$

We introduce the capture basin $\text{Capt}_F(K, C)$ of the target $C(\cdot)$ viable in the environment $K(\cdot)$ as the set of initial states $x_0 \in K_0$ such that there exist:

1. a finite horizon $T^\star$,
2. at least one evolution $x(\cdot)$ starting at x_0

   a. defined on $[0, T^\star]$,
   b. governed by the differential inclusion $x'(t) \in F(x(t), t)$,

c. viable in $K(\cdot)$,
d. reaching the target at time $T^\star\colon x(T^\star) \in C(T^\star)$.

Although dynamical games make sense only in finite horizon temporal case, in the infinite (or perennial) case $T = +\infty$ for which the target is empty, we introduce the viability kernel $\text{Viab}_F(K)$ as the set of initial states $x_0 \in K_0$ such that there exist:

a. at least one evolution $x(\cdot)$ starting at x_0

   i governed by the differential inclusion $x'(t) \in F(x(t), t)$,
   ii viable in $K(\cdot)$,

without closed subsets $C \subset K$, regarded as a target contained in an environment K. An evolution $x(\cdot)$ is said to be viable in K on a temporal interval $[0, T]$ if:

$$\forall\, t \in [0, T], \quad x(t) \in K(t) \tag{5}$$

2.2 Two-Player Pursuit Evasion Differential Games

According to viability wording and definitions, a two-player differential game is described by:

- a (state) variable $x(t) \in K \subset X = \mathbb{R}^n$, K being the environment in which the evolution states must belong (viability constraints), n being the problem dimension, i.e., the number of state variables,
- regulated by one or more controls, i.e., $u_p \in \mathscr{U}(x)$, $u_e \in \mathscr{V}(x)$ in the specific case of a two-player game,
- which evolution is ruled by a continuous dynamic law (written in the specific case of a two-player game):

$$x'(t) = f(x(t), u_p(t, x(t)), u_e(t, x(t))) \tag{6}$$

$$u_p(t) \in \mathscr{U}(x(t)) \tag{7}$$

$$u_e(t) \in \mathscr{V}(x(t)) \tag{8}$$

- plus an optimization criterion, i.e., a target C.

3 Viability Algorithms

In a nutshell, once a differential inclusion $x'(t) \in F(x(t))$ has been discretized in time by $x_{m+1} \in \Phi(x_m)$ and "restricted" to grids of the finite dimensional vector space, then the viable capture basin $\text{Capt}_\Phi(K, C)$ of elements of K from which an evolution (x_m) viable in K reaches the target C in finite discrete time. It can be obtained by two algorithms:

1. The *capture basin algorithm*. It is based on the formula:

$$\text{Capt}_\Phi(K, C) = \bigcup_{m \geq 0} C_m \tag{9}$$

where the *increasing* sequence of subsets $C_m \subset \text{Capt}_\Phi(K, C)$ is iteratively defined by:

$$\begin{cases} C_0 = C \\ \forall\, m \geq 1, \ C_{m+1} := K \cap (C_m \cup \Phi^{-1}(C_m)) \end{cases} \tag{10}$$

2. The *viability kernel algorithm*. Whenever $K \setminus C$ is *repeller* (for all $x \in K \setminus C$, all evolutions $x(\cdot)$ leave $K \setminus C$ in finite time), there is another class of general algorithms allowing to compute viable capture basins:

$$\text{Capt}_\Phi(K, C) = \bigcap_{m \geq 0} K_m \tag{11}$$

where the *decreasing* sequence of subsets $K_n \supset \text{Capt}_\Phi(K, C)$ is iteratively defined by:

$$\begin{cases} K_0 = K \\ \forall\, m \geq 1, \ K_{m+1} := C \cup (K_m \cap \Phi^{-1}(K_m)) \end{cases} \tag{12}$$

Naturally, both subsets C_m and K_m are computed at each iteration subsets on a grid of the state space. The convergence of the subsets C_m and the K_m follows from convergence theorems presented in Chapter 19, p. 769, of *Viability Theory. New Directions*, [4] (see, for instance, Theorem 19.3, p. 774).

- *Shooting Methods*

  Basically, viability algorithms are radically different from the *shooting methods*. Shooting methods check state after state x whether or not at least one discrete evolution $(x_m)_k$ governed by $x_{m+1} \in \Phi(x_m)$ is viable in K until it reaches a target. They need much less memory space but demand a considerable amount of time, because the number of initial states of the environment is high and, second, in the case of controlled systems, the set of evolutions starting from a given initial state becomes huge. Furthermore, it is impossible to verify whether an evolution is viable until they reach a target when the duration to reach the target is too high. Since the initial state chosen at random does not necessarily belong to the viable capture basin, the absence of corrections (brought by the regulation map of the viability algorithm) does not allow us to "tame" evolutions which then may leave the environment and very fast for systems which are sensitive to initial states. ∎

- *The Curse of Dimensionality*

    Viability algorithms are prone to the *curse of dimensionality*, as all algorithms defined on grids. This term was coined by *Richard Bellman* [7] in 1957 for describing the computational consequences of the exponential increase in volume of grids and data in vector spaces when their dimension increases. The viability algorithms and their software do not depend on the dimension of the problem, but their implementation on actual computers does.   ■

- *A Posteriori Error Estimates*

    The viability kernel algorithm providing approximations C_n containing the viable capture basin and the viability kernel algorithm approximations K_n contained in it, their "Minkowski differences" $K_m \setminus C_m$ provides actual *a posteriori error estimates*, the only ones that are useful. See, for instance, *A Posteriori Error Estimation in Finite Element Analysis*, [1], and Chapter 10, p. 284, of *Approximation of Elliptic Boundary-Value Problems*, [6].   ■

Handling subsets, *viability algorithms must be stable under set operations*: union, complement, etc. Approximations of subsets by boxes or ellipsoids do not enjoy these stability properties, so that their arithmetical properties are not necessarily appropriate for these topics because they inflict unnecessary losses of information. In the same way, "support vector machines" arising in pattern recognition allow us to smooth subsets defined on grids by smooth boundary. Nevertheless, they are useful for providing a smooth description of a viable capture basin, even if regularizing the iterated subsets V_n at each iteration imposes here again losses of information. On the other hand, there exist efficient C++ libraries dedicated to interval computing that allow to implement algorithms that are running fast. In addition, as explained in Section 4, we claim that by using contractor programming, we accelerate the interval computing-based viability algorithms.

4   Interval Arithmetics

Interval computation [15] is about guaranteed numerical methods for approximating sets and their application to engineering. Guaranteed means here that outer (and inner if needed depending on the application) approximations of the sets of interest are obtained, which can, at least in principle, be made as precise as desired. It thus becomes possible to achieve tasks such as computing (over- and underapproximating) capture basins or capture zones of differential games.

    The main tool to be used is interval analysis, based upon the very simple idea of enclosing real numbers in intervals and real vectors in boxes, i.e., sub-pavings. Interval computation is a special case of computation on sets. The operations on sets fall into two categories. The first one such as union, intersection consists of operations that have a meaning only in a set-theoretic context. The union of two disconnected intervals can be overapproximated by an interval even if not an interval in the set-theoretic sense. The second one consists of the extension of operations that are already defined for numbers (or vectors): addition and multiplication.

Thanks to these properties and fast interval-based algorithms (contractor programming, guaranteed integration of sub-pavings), it may be possible to implement the capture basin algorithm and solve problems such as those described in [17]. Set invariance [9] which is more related to viability kernel computation has also an interval-based implementation.

5   Interval Viability Kernel and Capture Basin Algorithms

5.1   Interval Viability Kernel Algorithm

Consider Φ a discrete flow of trajectories generated by a dynamical system, x being the state vector, u_e being the perturbation (uncertainties) we have in the process evolution we model, and u_p being the controls we optimize for ensuring viability of the system evolution.

$$x \in X, \ u_p \in U, \ u_e \in V \tag{13}$$

We assume that both players have bounded controls that describe the physical limits and uncertainties we want to model and that fit well into the interval formalism. Using interval notations, we write Φ as follow:

$$[x_{m+1}] = [\Phi]\,([x_m], [u_p], [u_e]) \tag{14}$$

In the interval framework, states of dimension n are boxes, and intervals are boxes of dimension one. Both (intervals and boxes) are noticed between brackets ([.]). Brackets around Φ in equation (14) acknowledge the fact that we use interval functions to compute images of function Φ in the world of intervals. An interval function $[\Phi]$ from $\mathbb{IR}^n$ (intervals of real numbers of dimension n, i.e., box of dimension n) to $\mathbb{IR}^p$ (box of dimension p) is an inclusion function of Φ if:

$$\Phi : \mathbb{R}^n \to \mathbb{R}^p \tag{15}$$

$$\forall\, [x] \in \mathbb{IR}^n, \ \Phi\,([x]) \subset [\Phi]\,([x]) \tag{16}$$

Because inclusion functions $[\Phi]$ are non-necessary optimal (optimal inclusion functions are denoted $[\Phi]^*$), i.e., minimal, and because sets are represented by lattices, i.e., collections of boxes that overapproximate (general) sets, the interval image $[\Phi]$ of a set encounters over margins. This phenomenon is commonly called wrapping effect.

$$\Phi\,([x]) \subseteq [\Phi]^*\,([x]) \subseteq [\Phi]\,([x]) \tag{17}$$

Consider a lattice L such as:

$$L = \bigcup_i [.]_i^L \supset X \subset \mathbb{R}^n, \quad [.]_i^L \in \mathbb{IR}^n \tag{18}$$

Then,

$$\forall x \in X, \quad \Phi(x) \in \bigcup_i [\Phi]([.]_i^L) \subset \mathbb{R}^p \tag{19}$$

There exist techniques and refinements to decrease large wrapping effects that occur when using natural interval arithmetics, i.e., basic interval calculus. However, natural interval arithmetics is an easy way to perform fast implementation and fast computation.

Always using intervals to express uncertainties, we rewrite the viability kernel algorithm of [4]:

$$\begin{cases} K_0 = K \\ (\exists u_p \ \forall u_e \ |) \ K_{m+1} = C \cup (K_m \cap \Phi^{-1}(K_m, u_p, [u_e])) \end{cases} \tag{20}$$

Equation (20) explains how viability kernels are computed in an iterative manner. m is the iteration index. At iteration 0, $Viab_\Phi(K, C)$ is K. First iteration of $Viab_\Phi(K, C)$ is K_0. Then, at each iteration, the objective is to compute what states are staying inside K_m by evolution Φ taken into account the adequate quantifiers on controls, i.e., what are the initial conditions invariant to K_m by Φ.

A first (natural) implementation of the viability kernel algorithm using interval arithmetics in the case of a two-player pursuit evasion game is through a forward, test, and bisect process. Interval arithmetics is used to compute overapproximations of $\Phi([x_m], [u_p], u_e)$. Guaranteed interval integration (simulation) techniques are required at this stage. The interval implementation of the viability algorithm is working following an erosion process (see *Morphologie Mathématique*, [23]). It has to be noticed that this algorithm is not computing invariant states but the opposite, i.e., states that are leaving for sure K_m. For this reason, we do not compute $\Phi^{-1}(K_m, u_p, [u_e])$ but $\Phi([x_{m\,out}], [u_p], u_e)$. Moreover, K_m is enclosed by two sub-pavings (lattices, interval representation of K_m).

$$K_m^- \subseteq \{x \in K_m\} \subseteq K_m^- \cup K_m^+ \tag{21}$$

Sub-pavings (K_m^-, K_m^+) are lists of boxes that allow to approximate (general) sets. Over and under sub-pavings make possible to represent in theory sets with whatever precision we need.

At each step, the algorithm removes the initial conditions $[x_{m\,out}]$ for which there exists no control u_p able to keep the state inside K_m for all the u_e uncertainties:

$$\Phi([x_{m\,out}], [u_p], u_e) \iff \exists u_e \ \forall u_p \ | \ [x_{m+1}] = \Phi([x_{m\,out}], .) \not\subset K_m \tag{22}$$

1. if $(\exists u_e \,\forall u_p \,|\, \varPhi([x_m], [u_p], u_e) \not\subset K_m^+)$ then we remove $[x_m]$ from K_m^+

2. if $(\nexists u_e \,\forall u_p \,|\, \varPhi([x_m], [u_p], u_e) \not\subset K_m^+)$ then we do nothing
   (go to next box defining K_m, i.e. to the next box in the list K_m^+)

3. if we can not conclude about $\varPhi([x_m], [u_p], u_e)$ remaining inside K_m^+
   then we split (bisect) $[x_m]$ (and $[u_p]$ as well) into two boxes
   and put the two boxes in the lists K_m^+ of boxes to check.

4. The algorithm stops at iteration m when the size of the "maybe" boxes, i.e. boxes bisected at
   step 3 are below a precision parameter value.

Fig. 1 Viability kernel algorithm based on forward, test, and bisection computing

As we will explain in more details after, the main interest is to compute an (tight) overapproximation of $Viab_\varPhi(K, C)$. For this reason, the $[x_m]$ boxes we check are boxes of the list (sub-paving) K_m^+. Three situations may occur.

When running this type of algorithms, it is of first importance to maintain regular sub-pavings (K_m^- and K_m^+) of K_m when removing $[x_{m\,out}]$ boxes. By regular sub-pavings, we mean lattices with no overlapping boxes. In addition, it is convenient to avoid as much as possible small boxes when it is possible to group them in a larger one. We are looking for a u_e value (quantifier $\exists$) when having $[u_p]$, $[x]$ being both intervals and interval vectors.

Computation time is another difficulty due to the 2^b complexity of the algorithm. When boxes are bisected (b being the number of successive bisections on a box), algorithms as the one described in Figure 1 create two new branches in a tree that has binary branches to explore. However, it has also to be noticed that viability kernel (and capture basin) algorithm presented after (Section 5.2) can be parallelized thanks to sub-pavings (of K_m but also potentially sub-pavings describing the flow $\varPhi$).

5.2 Capture Basin Algorithm

In a similar manner, the capture basin algorithm starting from $C \subset K$ and increasing until an underapproximation of $Capt_\varPhi(K, C)$ is:

$$\begin{cases} C_0 = C \\ (\exists u_p \,\forall u_e \,|\,) \ C_{k+1} = K \cap (C_k \cup \varPhi^{-1}(C_k, u_p, [u_e])) \end{cases} \tag{23}$$

with k being the iteration index and $\varPhi$ the following flow:

$$[x_{k+1}] = \varPhi([x_k], u_p, [u_e]) \iff \exists u_p \,\forall u_e \,|\, x_{k+1} = \varPhi(x_k, u_p, u_e) \tag{24}$$

The meaning of the algorithm is to increase a capture set C_k with states the Pursuer P is able to catch, i.e. to drive into C_k in one step time whatever is the uncertainty evader control u_e. Then, the old target set plus the initial conditions satisfying capture conditions in one time step is considered as the new target set. Of course, the trajectories we consider have to be viable. The process starting with $C_0 = C$, $k \in \mathbb{N}$, i.e., $k \in 0, 1, 2\cdots$.

Therefore, iterative computing allows to perform the following sequence of sets:

$$C = C_0 \subset C_k \subset C_{k+1} \subset \ldots \subset V \subset \ldots \subset K_{m+1} \subset K_m \subset K_0 = K \qquad (25)$$

V is then computed with error:

$$K_m \setminus C_k = K_m \cap C_k^{\complement} \qquad (26)$$

$C_k^{\complement}$ being the complementary set of C_k in X, the "\" character designing the relative complement of C_k.

Sets are represented by two lattices, an under sub-paving and an over sub-paving as explained by equation (21). When computing V, the frontier separating the capture zone from the non-capture zone is only required to compute an underapproximation of C_k, i.e., a lattice C_k^-, and an overapproximation of K_m, i.e., a lattice K_m^+. For this reason, in Section 5.2 in a way to simplify the equations, we write the set iterations C and K with any details about if we consider over or under interval lattices. C is always C^- and K is always K^+.

6 Contractor Programming

Sub-pavings belong to $\mathbb{IR}^n$ (boxes of finite dimension representing bounded continuous values). For compactness reasons, boxes are written $[x]$, x being a state vector with state variables in $\mathbb{R}$. Set membership computation consists to characterize sets $\mathbb{X} \subset \mathbb{R}^n$, $\mathbb{X}$ being a general set (non-necessary a box) described by constraints. The constraints we talk about are geometric conditions on state variables (equalities, inequalities) but can be also more general constraints defined by ordinary differential equations (ODE as those governing differential games). Interval contractors [16] have been used for computing over- and underapproximations of capture zones. The operator $C_{\mathbb{X}} : \mathbb{IR}^n \to \mathbb{IR}^n$ is a contractor for $\mathbb{X} \subset \mathbb{R}^n$ if:

$$\forall [x] \in \mathbb{IR}^n, \begin{cases} C_{\mathbb{X}}([x]) \subset [x] & (contractance) \\ C_{\mathbb{X}}([x]) \cap \mathbb{X} = [x] \cap \mathbb{X} & (completness) \end{cases} \qquad (27)$$

Moreover, contractor arithmetics allow to compose contractors together: $(C_1 \circ C_2)([x]) = C_1(C_2([x]))$. To characterize $\mathbb{X} \subset \mathbb{R}^n$, bisection algorithms (computation and eval algorithms) that bisect all boxes in all directions become rapidly inefficient with the dimension of $\mathbb{X}$. However, contrary to bisection algorithms that have a complexity in 2^b, b being the number of bisections to perform for obtaining results at the precision required, contraction procedures that reduce boxes faster are algorithms with polynomial complexity only [16].

6.1 Viability Kernel with Contractors

The basic idea used for rewriting viability kernel and capture basin algorithm using contractor programming is the following one.

After contraction, we are not sure that all the states inside a contracted box satisfy constraints because a contractor is not necessary optimal. However, we are sure that no state inside the complementary set of the contracted box satisfies the property. First of all, we define a contractor (operator) with quantifiers (on controls). We define $C_{\Phi \exists u_p} ([x_n, x_{n+1}])$ the contractor associated to the flow described hereafter.

$$C_{\Phi \exists u_p} ([x_m, x_{m+1}]) : \exists u_p \in U \ \forall u_e \in V \mid [x_{m+1}] = \Phi ([x_m], u_p, [u_e]) \quad (28)$$

$C_{\Phi \exists u_p}([x_m, x_{m+1}])$ is an ordinary differential equation (ODE) contractor between state vectors x_m and x_{m+1}. An ODE contractor [3] can be easily written assuming an Euler integration scheme. Higher-order Runge-Kutta numerical schemes linking x_m and x_{m+1} could be also considered (please see [2] for more details and for complete description of the resulting margins when applying guaranteed integration techniques). Then, using ODE contractors with quantifiers, the viability kernel K_m is computed, always in an iterative manner as follows:

$$K_0^{\complement} = K^{\complement} = K \setminus X \quad (29)$$

$$K_{m+1}^{\complement} = \left(\bigcup_{[x] \subset K_m / C_k} \bigcap_{[y] \subset K_m} ([x] - Proj_x \left(C_{\Phi \exists u_p} ([x, y]) \right)) \right) \bigcup K_m^{\complement} \quad (30)$$

$$m = 0 .. \left(\left\lfloor \frac{t_f}{dt} \right\rfloor - 1 \right) \quad (31)$$

with $Proj_x ([x, y]) = [x]$, the projection operator along the $[x]$ dimension:

$$Proj_x : \mathbb{IR}^{2n} \to \mathbb{IR}^n \quad (32)$$

The way equation (31) is working can be explained adding some more comments:

- The contractor-based viability kernel algorithm considers boxes which are twice the dimension of the problem. The boxes the algorithm contracts are $[x_m, x_{m+1}]$.
- K_m is an enclosing lattice: K_m^+.
- For all box $[x]$ inside K_m, we compute elements of $[x]$ that do not satisfy (for sure) the ODE constraints with other boxes $[y]$ of K_m.
- For doing so,

  - We perform the computation for one box $[x]$ with respect to all $[y]$ boxes,
  - do the intersection between all the resulting boxes,
  - add the remaining boxes to $K_m^{\complement}$.

- The intersection process described in the item list just above has to be done for all $[x]$ (first union operator in equation (31)).
- $[x] - Proj_x\left(C_{\Phi\exists u_p}([x, y])\right)$ is the complementary set of $Proj_x\left(C_{\Phi\exists u_p}[.]\right)$ contractor into $[x]$. Using a projection operator and a complementary set procedure, we compute initial conditions of flow $\Phi([x_k], u_p, [u_e])$ which are not (for sure, $\nexists u_p$) reaching $[y]$ boxes and in a sequel reaching K_m.
- We do the union on all $[x] \subset K_m / C_k$ when we compute viability kernels with target; otherwise, $[x] \subset K_m$ has to be considered.
- It has to be noticed that contrary to the forward test and bisect algorithm (Figure 1), the set (lattice) we build is not directly $Viab_\Phi(K, C)$ but its complementary into X ($Viab_\Phi(K, C)^C$).
- Equation (31) translates the fact that Index m is related to the maximum capture time we consider when computing backward reachable sets. It has also to be noticed that for convergence reasons, the dt algorithm time step is nonconstant and has to be smaller when reaching the t_f time limit. In Section (3), more explanations are provided about capture basin computation error estimates.

6.2 Capture Basin with Contractors

In a similar manner, we rewrite the capture basin algorithm applying contractor programming. We define a new ODE contractor $C_{\Phi\exists u_e}$ with $\exists$ quantifier on the uncertainty parameter, i.e., on the u_e control.

$$C_{\Phi\exists u_e}([x_k, x_{k+1}]) : \exists u_e \in V \ \forall u_p \in U \mid [x_{k+1}] = \Phi([x_k], [u_p], u_e) \quad (33)$$

Using the $C_{\Phi\exists u_e}$ contractor defined in equation (33), we compute the states that are not remaining in K_m / C_k. The boxes that are not remaining in K_m / C_k are necessary going in C_k because the boxes we consider are part of the viability kernel. As a consequence, these boxes have to be included into the under construction capture basin $Capt_\Phi(K, C)$. k is the capture basin algorithm iteration index. m is the last index we use for computing $Viab_\Phi(K, C)$ (best computation of $Viab_\Phi(K, C) \simeq K_m$).

$$C_0 = C \quad (34)$$

$$C_{k+} = \left(\bigcup_{[x] \subset K_m / C_k} \ \bigcap_{[y] \subset K_m / C_k} ([x] - Proj_x\left(C_{\Phi\exists u_e}([x, y])\right)) \right) \bigcup C_k \quad (35)$$

$$k = 0 .. \left(\left\lfloor \frac{t_f}{dt} \right\rfloor - 1 \right) \quad (36)$$

7   The Game of Two Cars

7.1   *Kinematics*

The noncooperative perfect information differential game of two cars (Figure 2)
is used to describe ground mobile robot motion. In inertial coordinate frame (cf.
equation 37), a minimum of six state variables is necessary to model coplanar movements involving two vehicles (a pursuer P and an evader E) with constant velocities,
respectively, V_P and V_E. Both players have minimum curvature constraints (R_P and
R_E) and bounded controls (u_p and u_e). u_p and u_e are angular rotation rates acting
on angles θ_P and θ_E describing in which direction velocity vectors $\overrightarrow{V_P}$ and $\overrightarrow{V_E}$ are
pointing with respect to an horizontal reference.

$$
\begin{cases}
\dot{x}_P = V_P \cos\theta_P \\
\dot{y}_P = V_P \sin\theta_P \\
\dot{\theta}_P = \frac{V_P}{R_P} u_p
\end{cases}
\quad \text{and} \quad
\begin{cases}
\dot{x}_E = V_E \cos\theta_E \\
\dot{y}_E = V_E \sin\theta_E \\
\dot{\theta}_E = \frac{V_E}{R_E} u_e
\end{cases}
\tag{37}
$$

$$
u_p \in [-1, 1] \qquad\qquad u_e \in [-1, 1]
$$

For the sake of simplicity, without losing any generality, we assume $R_P = R_E = 1$
in the kinematics equations provided in the paper. However, when performing real
experimentations (see Section 7.3), other minimum curvature parameters have been
considered. In a way to decrease the state vector dimension, we chose E as the origin
of a new relative referential. The horizontal axis is defined by the evader velocity

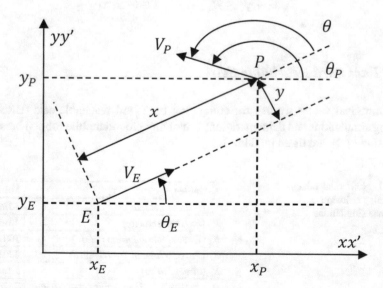

Fig. 2 Game of two cars kinematics

vector, the vertical axis being perpendicular anti-clockwise to the horizontal one. It is usual to choose one player as the origin to decrease the number of state variables. There exist different manners to write these dynamic equations in three dimensions. In the book of Rufus Isaacs, the moving referential is centered on P, and $\overrightarrow{V_P}$ is along the vertical axis. The referential transformation we perform is similar to the one used in [21]. The new state vector in this coordinate frame in translation and rotation with respect to ground references is as follows:

$$X = \begin{pmatrix} x \\ y \\ \theta \end{pmatrix} \tag{38}$$

$$\begin{pmatrix} x \\ y \end{pmatrix} = \begin{pmatrix} \cos\theta_E & \sin\theta_E \\ -\sin\theta_E & \cos\theta_E \end{pmatrix} \begin{pmatrix} x_P - x_E \\ y_P - y_E \end{pmatrix} \tag{39}$$

$$\theta = \theta_P - \theta_E \tag{40}$$

Then, after differentiation (with respect to time) of equations (39) and (40), we obtain evolution equations of dimension 3:

$$\begin{cases} \dot{x} = V_E (u_e\, y - 1) + V_P \cos\theta \\ \dot{y} = V_P \sin\theta - V_E\, u_e\, x \\ \dot{\theta} = V_P\, u_p - V_E\, u_e \end{cases} \tag{41}$$

As in the regular game of two cars, the criterion maximized by the evader and minimized by the pursuer is capture time t. The capture set is defined by:

$$\sqrt{x^2 + y^2} \le R_0 \qquad \theta \in [-\pi, \pi] \tag{42}$$

7.2 Backward Reachable Sets

The numerical values we use for computing backward reachable sets (BRS), for running simulations and demonstrations (involving ground mobile robots) presented in Section (7.3), are listed in Table 1.

Table 1 Numerical values for running collision avoidance simulations

Variables	Values
V_E (Evader velocity)	$30\,m/sec$
V_P (Pursuer velocity)	$30\,m/sec$
R_E (Evader minimum curvature)	$300\,m$
R_P (Pursuer minimum curvature)	$300\,m$
R_0 (target radius)	$50\,m$
$t_f - t_0$ (backward reachable set maximum time horizons we use in simulation)	$5\,sec$

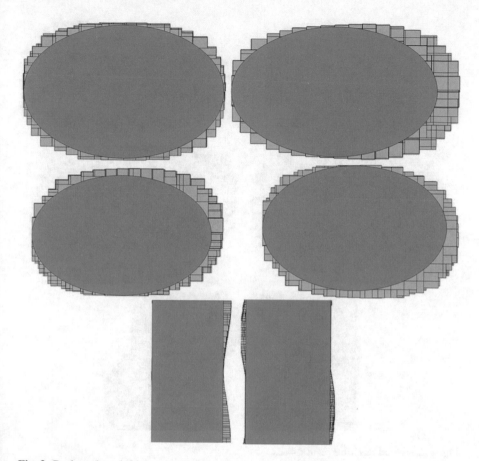

Fig. 3 Backward reachable set at $\tau = 1s$

BRS or level sets (LS) corresponding to $\tau = t_f - t = 1\,s.$, $\tau = 5\,s.$, and $\tau = 10\,s.$ are, respectively, plotted in Figures 3, 4, and 5, t_f being maximum capture time and t being regular time. The BRS figures are 2D cross sections of 3D BRS in the game of two cars rotating coordinate frame. Each BRS figure (corresponding to $\tau = 1\,s$, $5\,s$, and $10\,s$) consists in six drawings. First drawing (top left) is $\theta = 0$. Second drawing (top right) is $\theta = -\pi$, then (middle left) $\theta = -\frac{\pi}{2}$, and (middle right) $\theta = \frac{\pi}{2}$. The two bottom pictures are vertical cross section xz (bottom left) and yz (bottom right) corresponding to $\theta = 0$. On all the drawings, the uniform patterns (circles appearing as ellipsoids because the axes are not orthonormal axes in xy) and cylinders in xz and yz projections are the target set. The surrounding area depicted by collections of boxes are the BRS we compute with interval-based viability algorithms. For having clear view of the axes, it is recommended to refer to the 3D BRS plotting (Figure 6 with the accompanying explanations provided in Section 7.3).

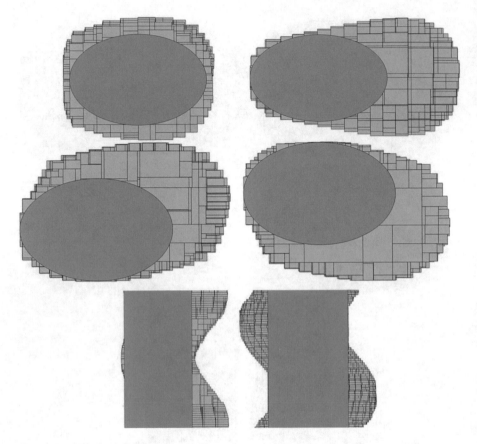

Fig. 4 Backward reachable set at $\tau = 5s$

7.3 Robotic Simulation

The backward reachable sets computed so far have been used for robust collision avoidance when operating two ground mobile robots (Figure 8). A scenario is detailed in Figure 6 (relative trajectories) and Figure 7 (trajectories in inertial coordinate frame) for illustration purpose. Figure 7 shows the target set (cylinder), i.e., the unsafe area at $\tau = 5s$. and the two-player game trajectory. Vertical axis of top Figure 6 is θ in radians. Vertical axis of bottom Figure 6 is a zoom of θ around 180° (axis in degrees). Figure 8 shows the two-player robots with LiDaR (Light Detection and Ranging) sensors for sensing the second player position. Plots of the laser rays are provided. No ray means that an incoming robot has been detected. Based on this information plus its own position and heading, the evader robot (robot at the front of the picture, the robot at the back being the pursuer robot to avoid) reconstructs the game of two cars' state vector and performs robust collision avoidance maneuvers.

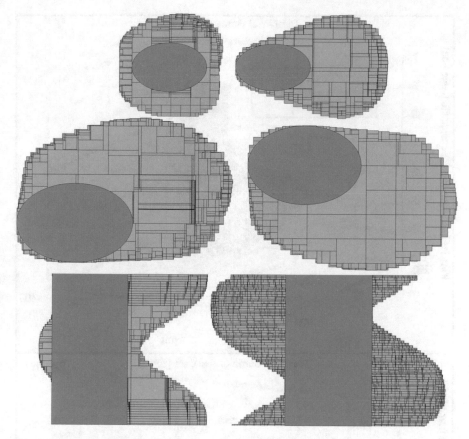

Fig. 5 Backward reachable set at $\tau = 10\,s$

8 Conclusion

After proposing a viability kernel algorithm and a capture basin algorithm based on interval contractors, we compute backward reachable sets (level sets of the value function t^*, i.e., capture time) of the differential game of two cars. Backward reachable sets are differential game capture zones over a limited time horizon. Therefore, we are able to compute safe and unsafe initial conditions with respect to collision in a noncooperative game setting (robust collision avoidance with respect to a moving obstacle performing unknown maneuvers). Then, an illustrative example (simulation and real experimentation involving two ground mobile robots) playing real-time collision avoidance is presented. Safe areas defined by backward reachable sets mean that there exists for an evader robot at least a trajectory to avoid a pursuer robot during the time duration defined by the backward reachable set we consider. The time duration parameter of the backward reachable set we use is related to the sensors' range and to the vehicles' velocities we consider.

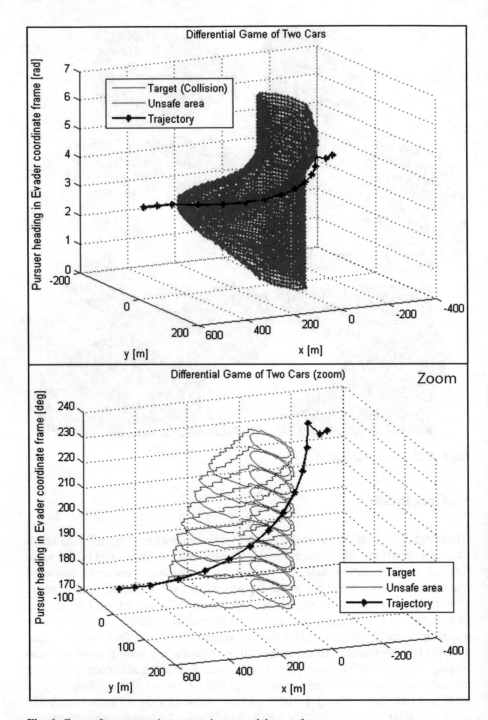

Fig. 6 Game of two cars trajectory turning around the unsafe area

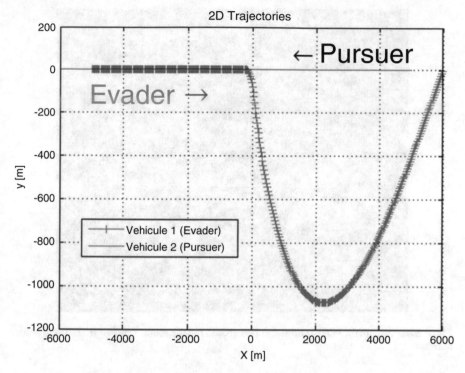

Fig. 7 Simulation trajectories in a robotic experimentation (pursuer in straight line from the right to the left)

Some more work has still to be performed to complete this study, including work on guaranteed integration in interval context which is an essential part of the algorithm, at least from the theoretical point of view. The algorithms developed so far (interval-based viability) can be applied to various problems including collision avoidance between other types of vehicles and for checking the safety of autonomous platforms in other scenarios and applications. Last but not least, the algorithms have to be extended or have to be used in conjunction with other techniques for collision avoidance when facing several (more than two) noncooperative platforms.

Acknowledgements This work was supported by "Agence Nationale de la Recherche" through the ANR ASTRID Maturation funding scheme, VIATIC[2] project. The author would like also to thanks the viability theoreticians who suggested remarks and advices to improve this work.

Fig. 8 Robotic experimentation involving two ground mobile robots (pursuer in the back, evader in front)

References

1. Mark Ainsworth and J. Tinsley Oden. *A Posteriori Error Estimation in Finite Element Analysis*. John Wiley & Sons, Inc., 2000.
2. Julien Alexandre Dit Sandretto and Alexandre Chapoutot. Dynibex: a differential constraint library for studying dynamical systems. In *Conference on Hybrid Systems: Computation and Control (HSCC 2016)*. "http://www.cs.ox.ac.uk/conferences/hscc2016/index.html", "hal-01297273", April 2016, Vienne, Austria. 2016.
3. Julien Alexandre Dit Sandretto and Alexandre Chapoutot. Contraction, propagation and bisection on a validated simulation of ODE. Summer Workshop on Interval Methods. In *Conference on Hybrid Systems: Computation and Control (HSCC 2016)*. hal-01325068, June 2016, Lyon, France. 2016.
4. J.-P. Aubin, A. Bayen, and P. Saint-Pierre. *Viability Theory New Directions, Second Edition*. Number ISBN 978-3-642-16683-9. Springer-Verlag, Berlin Heidelberg, 2011.
5. Jean-Pierre Aubin and Anna Désilles. *Traffic Networks as Information Systems. A viability approach*. Mathematical Engineering. Springer, 1 2016.
6. J.P. Aubin. *Approximation of elliptic boundary-value problems*. Pure and applied mathematics. Wiley-Interscience, 1972.
7. R.E. Bellman. *Dynamic Programming*. Dover Books on Computer Science Series. Dover Publications, 2003.
8. F. Blanchini and S. Miani. *Set-Theoretic Methods in Controls, Second Edition*. Number ISBN 978-0-8176-4606-6. Birkhauser, Boston, 2008.
9. Franco Blanchini. Set Invariance in Control. *Automatica*, 35(11):1747–1767, 1999.
10. L.M. Bujorianu. *Stochastic Reachability Analysis of Hybrid Systems*. Communications and Control Engineering. Springer London, 2012.

11. Manuela L Bujorianu and John Lygeros. Reachability questions in piecewise deterministic Markov processes. In *International Workshop on Hybrid Systems: Computation and Control*, pages 126–140. Springer, 2003.
12. Pierre Cardaliaguet, Marc Quincampoix, and Patrick Saint-Pierre. *Differential Games Through Viability Theory: Old and Recent Results*, pages 3–35. Birkhäuser Boston, Boston, MA, 2007.
13. Ioannis Exarchos, Panagiotis Tsiotras, and Meir Pachter. On the suicidal pedestrian differential game. *Dynamic Games and Applications*, 5(3):297–317, 2015.
14. Rufus Isaacs. Differential Games. 1965.
15. L. Jaulin and E. Walter. Set inversion via interval analysis for nonlinear bounded-error estimation. *Automatica*, 29(4):1053–1064, 1993.
16. Luc Jaulin, Michel Kieffer, Olivier Didrit, and Eric Walter. *Applied Interval Analysis with Examples in Parameter and State Estimation, Robust Control and Robotics*. Springer London Ltd, August 2001.
17. N.N. Krasovskii and A.I. Subbotin. *Game-Theoretical Control Problems*. Springer Verlag, New York, NY, 1988.
18. Sergey S. Kumkov, Stéphane Le Ménec, and Valerii S. Patsko. Level Sets of the Value Function in Differential Games with Two Pursuers and One Evader. Interval Analysis Interpretation. *Mathematics in Computer Science*, 8(3):443–454, 2014.
19. AW Merz. The game of two identical cars. *Journal of Optimization Theory and Applications*, 9(5):324–343, 1972.
20. Ian M Mitchell. A toolbox of level set methods version 1.0. 2004.
21. Ian M Mitchell, Alexandre M Bayen, and Claire J Tomlin. A time-dependent Hamilton-Jacobi formulation of reachable sets for continuous dynamic games. *IEEE Transactions on automatic control*, 50(7):947–957, 2005.
22. Valery S Patsko and Varvara L Turova. Level sets of the value function in differential games with the homicidal chauffeur dynamics. *International Game Theory Review*, 3(01):67–112, 2001.
23. Michel Schmitt and Juliette Mattioli. *Morphologie Mathématique, Mathématiques et Informatiques*. Presses des Mines, 60, Boulevard Saint-Michel, Paris, France, 2013.
24. D. M. Stipanović, I. Hwang, and C. J. Tomlin. Computation of an over-approximation of the backward reachable set using subsystem level set functions. In *European Control Conference (ECC), 2003*, pages 300–305, Sept 2003.

On Linear-Quadratic Gaussian Dynamic Games

Meir Pachter

Abstract Stochastic dynamic games with partial information are considered. Because the LQG hypothesis is common in control theory for its analytic convenience, we exclusively focus on linear-quadratic Gaussian dynamic games (LQGDGs) – we confine our attention to "zero-sum" LQGDGs where the players have partial information. However, LQGDGs with a nonclassical information pattern can be problematic, as Witsenhausen's counterexample illustrates. The state of affairs concerning dynamic games and dynamic team problems with partial information is not satisfactory, and the solution of "zero-sum" LQGDGs where the players have partial information has been the Holy Grail/long-standing goal of the controls and games communities. Delayed commitment strategies as opposed to prior commitment strategies are required. The goal is to obtain a Nash equilibrium in delayed commitment strategies such that the optimal solution is time consistent/subgame perfect. In this article the current literature is discussed, and informational aspects are emphasized. Information patterns which render the game amenable to solution by the method of dynamic programming are of interest, so that the correct solution of LQGDGs with partial information can be obtained in closed-form.

1 Introduction

The complete solution of linear-quadratic Gaussian dynamic games (LQGDGs) has been a long-standing goal of the controls and games communities [1]. That LQGDGs with a nonclassical information pattern can be problematic has been

The views expressed in this article are those of the author and do not reflect the official policy or position of the United States Air Force, Department of Defense, or the US Government.

M. Pachter (✉)
Department of Electrical and Computer Engineering, Air Force Institute of Technology, Wright-Patterson A.F.B., OH 45433, USA
e-mail: meir.pachter@afit.edu

© Springer International Publishing AG 2017
J. Apaloo, B. Viscolani (eds.), *Advances in Dynamic and Mean Field Games*,
Annals of the International Society of Dynamic Games 15,
https://doi.org/10.1007/978-3-319-70619-1_14

amply illustrated in Witsenhausen's seminal paper [2] – see also [3]. Control theorists have traditionally emphasized control theoretic aspects and the backward induction/dynamic programming solution method, which however is not applicable to dynamic games with partial information – one notable exception notwithstanding, being the classes of LQGDGs with partial information addressed in this paper. And game theorists have focused on information economics, that is, the role of information in games but, for the most part, discrete games. The solution of "zero-sum" LQGDGs where the players have partial information has been the Holy Grail/long-standing goal of the controls and games communities. However, the state of affairs concerning dynamic games with partial information is not satisfactory. In this respect, the situation is not much different now than it was in 1971 when Witsenhausen made a similar observation [4]. In this article a careful analysis of dynamic games with partial information is undertaken. We exclusively focus on LQGDGs, which are more readily amenable to analysis. Indeed, (deterministic) linear-quadratic dynamic games (LQDGs) with perfect information stand out as far as applications of the theory of dynamic games are concerned. A canonical instance of an application of the theory of LQDGs can be found in [5] where it has been shown that its solution yields the proportional navigation (PN) guidance law which is universally used in air-to-air missiles. Furthermore, the theory of LQDGs has been successfully applied to the synthesis of H_∞ control laws [6]. The theory of LQDGs with perfect information has therefore received a great deal of attention [6–8]. In these works, the concepts of state, and state feedback, are emphasized, and the solution method entails backward induction, a.k.a. dynamic programming (DP). In this respect, the complete solution of linear-quadratic dynamic games (LQDGs) with perfect information, a.k.a. deterministic LQDGs, was derived in [8, Theorem 2.1] where the Schur complement concept was used to invert a blocked $(m_u + m_v) \times (m_u + m_v)$ matrix which contains four blocks, its two diagonal blocks being a $m_u \times m_u$ matrix and a $m_v \times m_v$ matrix. We further improve on the results of [8] by noting that a matrix with four blocks has *two* Schur complements, say S_B and S_C. This allows one to obtain *explicit* and *symmetric* formulae for the P and E players' optimal state feedback strategies, thus yielding the *complete* solution of the deterministic LQDG. These results are used in this article, and for the sake of completeness, the closed-form solution of the perfect information/deterministic zero-sum LQDG is included herein.

Concerning informational issues in (stochastic) LQGDGs: In the previous work [9] and [10], a static linear-quadratic Gaussian (LQG) team problem was addressed, and a static "zero-sum" LQG game where the players have private information was analyzed in [11]. In this article dynamic "zero-sum" LQG games, that is, LQGDGs, where the players have partial information, are addressed. In this respect, LQGDGs with one-step-delay sharing pattern were addressed in [12] and [13]: the players' past inputs as well as their past observations are shared information. Recently, the more general situation was considered where the past controls, but not the past observations, are shared – the correct solution of LQGDGs with a control-sharing information pattern was obtained in [14]. In the quest for a solution of the LQGDG, it is natural to also consider the "symmetric" information pattern where the past observations, but not the past controls, are shared – a line of attack on this problem is discussed in this paper.

This paper is concerned with the exact solution of LQGDGs with a nonclassical information pattern/partial observations, albeit in the rather restrictive, but analytically convenient, LQG framework. The herein obtained closed-form results could be used to benchmark and validate AI algorithms for POMDPs and Markov games based on Monte Carlo (MC) methods, Q-learning [19], and reinforcement learning (RL) [20, 21], which yield approximate solutions to more general decision problems.

The paper is organized as follows. To make the paper self-contained and for future reference, the complete solution of the baseline/deterministic LQDG game with perfect information is included in Section 2. The careful problem statement of the general LQGDG with partial information is presented in Section 3. The solution of the LQGDG with a control-sharing information pattern which was analyzed in [14] is discussed in Section 4 where the crucial state estimation problem is closely examined. In Section 5 we turn our attention to the general LQGDG and the LQGDG with an observation-sharing information pattern. A novel joint control and state estimation algorithm which is required for the solution of LQGDGs with an observation-sharing information pattern is developed. Concluding remarks are presented in Section 6. The somewhat laborious exposition could perhaps be excused in light of Witsenhausen's observation when discussing decentralized LQG control: "The most confused derivations of the correct results are also among the shortest" [4].

2   Solution of the LQGDG

The complete solution of the perfect information/deterministic LQ zero-sum dynamic game [15] is presented.

The deterministic linear dynamics are

$$x_{k+1} = Ax_k + Bu_k + Cv_k, \quad x_0 \equiv x_0 \in R^n, \quad k = 0, 1, \ldots, N-1 \tag{1}$$

Player P is the minimizer and his control $u_k \in R^{m_u}$. Player E is the maximizer and his control $v_k \in R^{m_v}$. The planning horizon is N.

The cost/payoff functional is quadratic:

$$J(\{u_k\}_{k=0}^{N-1}, \{v_k\}_{k=0}^{N-1}; x_0) = x_N^T Q_F x_N + \sum_{k=0}^{N-1} (x_{k+1}^T Q x_{k+1} + u_k^T R_u u_k - v_k^T R_v v_k) \tag{2}$$

and Q and Q_F are real symmetric matrices. The players' control effort weighting matrices R_u and R_v are typically real symmetric and positive definite. Oftentimes it is stipulated that also the state penalty matrices Q and Q_F be positive definite or, at least, positive semi-definite; these assumptions can be relaxed.

The following holds.

Theorem 1 A necessary and sufficient condition for the existence of a solution to the deterministic zero-sum LQDG (1) and (2) is

$$R_u + B^T P_k B > 0 \tag{3}$$

and

$$R_v > C^T P_k C \tag{4}$$

$\forall \, k = 1, \ldots, N - 1$, where the real, symmetric matrices P_k are the solution of the matrix Riccati difference equation

$$
\begin{aligned}
P_{k+1} = A^T \{ P_k - P_k [BS_B^{-1}(P_k) B^T + BS_B^{-1}(P_k) B^T P_k C (R_v \\
- C^T P_k C)^{-1} C^T + C(R_v - C^T P_k C)^{-1} C^T P_k BS_B^{-1}(P_k) B^T \\
+ C(R_v - C^T P_k C)^{-1} C^T P_k BS_B^{-1}(P_k) B^T P_k C (R_v \\
- C^T P_k C)^{-1} C^T + C(C^T P_k C - R_v)^{-1} C^T] P_k \} A + Q \,, \\
P_0 = Q + Q_F \,, \quad k = 0, \ldots, N - 1
\end{aligned} \tag{5}
$$

In equation (5), the first Schur complement matrix function

$$S_B(P_k) \equiv B^T P_k B + R_u + B^T P_k C (R_v - C^T P_k C)^{-1} C^T P_k B$$

In addition, the problem's parameters must satisfy the conditions

$$R_u + B^T (Q + Q_F) B > 0 \tag{6}$$

and

$$R_v > C^T (Q + Q_F) C \tag{7}$$

The value of the LQDG is

$$V_0(x_0) = x_0^T (P_N - Q) x_0 \tag{8}$$

The players' optimal strategies are the linear state feedback control laws

$$u_k^* = (\gamma_k^{(P)}(x_k))^* = (F_k^{(P)})^* x_k$$

$$v_k^* = (\gamma_k^{(E)}(x_k))^* = (F_k^{(E)})^* x_k$$

where the optimal gains

$$(F_k^{(P)})^* = -S_B^{-1}(P_{N-k-1})B^T[I + P_{N-k-1}C(R_v$$
$$-C^T P_{N-k-1}C)^{-1}C^T]P_{N-k-1}A \cdot x_k, \tag{9}$$

$$(F_k^{(E)})^* = -S_C^{-1}(P_{N-k-1})C^T[I - P_{N-k-1}B(R_u$$
$$+B^T P_{N-k-1}B)^{-1}B^T]P_{N-k-1}A \cdot x_k \tag{10}$$

In equation (10) the second Schur complement matrix function

$$S_C(P_{k+1}) \equiv -\{R_v - C^T(Q + P_{k+1})C + C^T(Q$$
$$+P_{k+1})B[B^T(Q + P_{k+1})B + R_u]^{-1}B^T(Q + P_{k+1})C\}$$

The herein presented solution of the (deterministic) LQDG prominently features in the analysis of the LQGDGs with partial information discussed in the sequel.

3 Linear-Quadratic Gaussian Dynamic Game

The linear-quadratic Gaussian dynamic game (LQGDG) game with two players, P and E, is specified as follows:
  Stochastic dynamics: Linear

$$x_{k+1} = A_k x_k + B_k u_k + C_k v_k + \Gamma_k w_k, \quad x_0 \equiv x_0, \quad k = 0, \ldots, N-1 \tag{11}$$

At decision time k, the controls of P and E are u_k and v_k, respectively.
  The process noise $w_k \sim \mathcal{N}(0, Q_p), \quad k = 0, \ldots, N-1$.
  The planning horizon $N \geq 2$.
  Noise-corrupted measurements: Linear
  The N measurements of player P are
  $\bar{x}_0^{(P)}$ at time $k = 0$ – player P believes that the initial state

$$x_0 \sim \mathcal{N}(\bar{x}_0^{(P)}, P_0^{(P)}), \tag{12}$$

and thereafter he takes the measurements

$$z_{k+1}^{(P)} = H_{k+1}^{(P)} x_{k+1} + v_{k+1}^{(P)}, \quad v_{k+1}^{(P)} \sim \mathcal{N}(0, R_m^{(P)}), \quad k = 0, \ldots, N-2 \tag{13}$$

At decision time k, his measurement record is

$$Z_k^{(P)} = \{\bar{x}_0^{(P)}, z_1^{(P)}, \ldots, z_k^{(P)}\}$$

The N measurements of player E are
$\overline{x}_0^{(E)}$ at time $k = 0$ – player E believes that the initial state

$$x_0 \sim \mathcal{N}(\overline{x}_0^{(E)}, P_0^{(E)}), \tag{14}$$

and thereafter he takes the measurements

$$z_{k+1}^{(E)} = H_{k+1}^{(E)} x_{k+1} + v_{k+1}^{(E)}, \quad v_{k+1}^{(E)} \sim \mathcal{N}(0, R_m^{(E)}), \ k = 0, \ldots, N - 2 \tag{15}$$

At decision time k, his measurement record is

$$Z_k^{(E)} = \{\overline{x}_0^{(E)}, z_1^{(E)}, \ldots, z_k^{(E)}\}$$

Taxonomy

a) If the observation matrices $H^{(P)} = I_n$, $H^{(E)} = I_n$ and there is NO measurement noise, that is, the covariance matrices $R_m^{(P)} = 0$ and $R_m^{(E)} = 0$, the players' "measurement" is $z_k = x_k$. We have a game with *perfect information* – even in the presence of process noise.

b) If the observation matrices $H^{(P)} = I_n$, $H^{(E)} = I_n$ and there is measurement noise, we have a game with *imperfect information*. In other words, the players' measurements are of the form $z_k^{(P)} = x_k + v_k^{(P)}$, $z_k^{(E)} = x_k + v_k^{(E)}$.

c) If the observation matrices H are "short" matrices and there is measurement noise, as in $z_k^{(P)} = H^{(P)} x_k + v_k^{(P)}$, $z_k^{(E)} = H^{(E)} x_k + v_k^{(E)}$, we have a game with *partial information*.

Cost/payoff function: Quadratic
Have noncooperative "zero-sum" game: Player P strives to minimize his cost – he $min_{\{u_k\}_{k=0}^{N-1}} J$, while at the same time, player E $max_{\{v_k\}_{k=0}^{N-1}} J$, where the cost/payoff function

$$J = x_N^T Q_F x_N + \sum_{k=0}^{N-1} [x_{k+1}^T Q_{k+1} x_{k+1} + u_k^T R_c^{(P)} u_k - v_k^T R_c^{(E)} v_k] \tag{16}$$

At each time instant $k = 0, 1, \ldots, N - 1$, players P and E minimize and maximize their expected cost/payoff $E(J \mid \cdot)$, conditional on their so far accumulated private information.

The expectation operator is liberally used in the dynamic game literature, but oftentimes it is not clearly stated with respect to which random variables the expectation is calculated and on which random variables the expectation is conditional. This tends to mask the fact that what appear to be "zero-sum" games are in fact nonzero-sum games. Upon considering "zero-sum" games with partial information, the illusion is then created that a zero-sum game is considered. One then tends to rely on the uniqueness of the saddle point value and the interchangeability of non-unique optimal saddle point strategies in zero-sum games.

This argument is flawed because, as previously discussed, in "zero-sum" games with partial information the P and E players calculate their respective cost and payoff conditional on their private information, as is correctly done in this article; that's why the term zero-sum is in quotation marks. Thus, although high-powered mathematics is oftentimes used, serious conceptual errors make the "results" not applicable. Contrary to statements sometimes encountered in the literature, in "zero-sum" games with partial information, one cannot and one should not look for a saddle point solution: the correct solution concept is a Nash equilibrium, that is, person-by-person satisfactory (PBPS) solution. Our "zero-sum" LQGDG with partial information *and* delayed commitment strategies is *not* zero-sum, and the correct solution concept is a Nash equilibrium or person-by-person satisfactory (PBPS) solution. A *unique* Nash equilibrium must be provided, and the P and E players' separate value functions must be calculated.

In this article a unique Nash equilibrium is provided, and the P and E players' value functions are calculated.

Information pattern

1. Public information

   a) The problem parameters: A_k, B_k, C_k, $H_k^{(P)}$, $H_k^{(E)}$, Q_p, Q_k, Q_F, $R_c^{(P)}$, $R_c^{(E)}$, $R_m^{(P)}$, $R_m^{(E)}$.

   b) The prior information: $P_0^{(P)}$, $P_0^{(E)}$, but not $\bar{x}_0^{(P)}$ and $\bar{x}_0^{(E)}$.

2. Private information

   At decision time $k = 0$, the (prior) information of player P is $\bar{x}_0^{(P)}$.

   At decision time $k = 0$, the (prior) information of player E is $\bar{x}_0^{(E)}$.

   At decision time $1 \le k \le N - 1$, the information of player P are his measurements $\bar{x}_0^{(P)}, z_1^{(P)}, \ldots, z_k^{(P)}$ and ownship control history $U_k \equiv \{u_0, \ldots, u_{k-1}\}$, with $U_0 \equiv \emptyset$.

   At decision time $1 \le k \le N - 1$, the information of player E are his measurements $\bar{x}_0^{(E)}, z_1^{(E)}, \ldots, z_k^{(E)}$ and ownship control history $V_k \equiv \{v_0, \ldots, v_{k-1}\}$ with $V_0 \equiv \emptyset$.

3.1 Sufficient Statistics

To make the dynamic game amenable to analysis, players P and E must be able, based on their respective information states, to independently calculate the following statistics.

First, as in stochastic optimal control:

At decision time $k = 0$, player P has the prior information $x_0 \sim \mathcal{N}(\bar{x}_0^{(P)}, P_0^{(P)})$.

At decision time $k = 0$, player E has the prior information $x_0 \sim \mathcal{N}(\bar{x}_0^{(E)}, P_0^{(E)})$.

The sufficient statistics of player P at decision time $1 \leq k \leq N - 1$: The p.d.f. $f_k^{(P)}(\cdot)$ of the physical state x_k, as calculated by player P using his information state which consists of the public information and private information available to him at time k.

The sufficient statistics of player E at decision time $1 \leq k \leq N - 1$: The p.d.f. $f_k^{(E)}(\cdot)$ of the physical state x_k, as calculated by player E his information state which consists of the public information and private information available to him at time k.

The P and E players' optimal strategies, $(\gamma_k^{(P)}(\cdot))^*$ and $(\gamma_k^{(E)}(\cdot))^*$, are mappings from their respective sufficient statistics into their action spaces R^{m_u} and R^{m_v} – a player's sufficient statistic being determined by his information state, which includes both the public information and his private information up to decision time k, $k = 0, \ldots, N - 1$; and the optimal strategies of P and E must provide a unique Nash equilibrium.

In addition, this sets stochastic dynamic games apart from stochastic optimal control:

Based on his sufficient statistic/information state at decision time $k = 0, 1, \ldots, N - 1$, players P and E must be able to independently calculate the statistics of their opponent's sufficient statistics. This precludes an infinite regress in reciprocal reasoning.

The LQGDG, as specified, allows for

1. No shared past controls information.
2. No shared observations
3. The players' Nash equilibrium strategies must be delayed commitment strategies which are subgame perfect/time consistent.

The solution of LQGDGs with partial information is certainly of great theoretical interest and is important for applications [16]. However, the solution of the LQGDG, as stated, is a tall order – it is the Holy Grail of LQGDGs with partial information. In conclusion, partial information renders dynamic games, including LQGDGs, not amenable to dynamic programming and intractability raises its ugly head.

4 Control-Sharing Information Pattern

To get a better appreciation of the issues germane to the analysis of LQGDGs, a summary of the main result from [14], where the solution of the LQGDG was derived under the assumption that the players' *past* inputs, but not measurements, are shared information, is included herein. In this section the above-listed requirement 1.) is relaxed: The LQGDG (11)–(16) is considered, and it is assumed that the players' information sets are augmented as follows. At decision time k, $k = 1, \ldots, N - 1$, player P is endowed with the additional information regarding the control history $V_k \equiv \{v_0, \ldots, v_{k-1}\}$ of player E. Thus, player P observed the past inputs $v_0, \ldots, v_{k-1}$ of player E. Similarly, at decision time k, $k = 1, \ldots, N-1$, player E is endowed with the additional information regarding the control history $U_k \equiv \{u_0, \ldots, u_{k-1}\}$ of player P. Thus, player E observed the past inputs $u_0, \ldots, u_{k-1}$ of player P.

This information pattern is referred to as the control-sharing information pattern. The dynamics and the measurement equations are linear, and the cost/payoff function is quadratic, but the information pattern is not classical. Strictly speaking, the information pattern is not partially nested because E's measurements, which he used to form his controls, are not known to P and vice versa, P's measurements, which he used to form his controls, are not known to E. However, this is now a moot point because the information pattern is s.t. the control history of player E is known to player P, and vice versa, the control history of player P is known to player E. This, and the fact that player P and player E, each separately, perceive the initial state x_0 to be Gaussian, causes the state estimation problems faced by the players at decision time k to be linear and Gaussian (LG). Hence, at decision time k, the knowledge of the complete control history $u_0, v_0, \ldots, u_{k-1}, v_{k-1}$ and their private measurement records makes it possible for both players to separately apply the linear Kalman filtering algorithm. LQGDGs with a control-sharing information pattern are linear-Gaussian (LG), and consequently at decision time k, each player, based on his private measurement record and the complete input history, separately calculates his estimate $\bar{x}_k$ of the physical state x_k using a *linear* Kalman filter (KF), whereupon the players' optimal strategies $\forall\, k = 0, 1, \ldots, N-1$ are

$$u_k^* = (\gamma_k^{(P)}(\bar{x}_k^{(P)}))^* \tag{17}$$

and

$$v_k^* = (\gamma_k^{(E)}(\bar{x}_k^{(E)}))^* \tag{18}$$

Indeed, the solution of LQGDGs hinges on the ability of the P and E players at decision time k to independently and *correctly* estimate the physical state x_k. That's why in Sections 4.1 and 4.2 in the sequel, we dwell on the state estimation facet in LQGDGs. When considering downrange the LQGDG in its full glory, and, in particular, the case where rather than sharing the past inputs information, the past observations are instead being shared, we'll be confronted with the need to develop a novel Kalman filtering algorithm.

4.1   Decentralized Kalman Filtering

Player P runs the KF

$$(\bar{x}_k^{(P)})^- = A\bar{x}_{k-1}^{(P)} + Bu_{k-1} + Cv_{k-1}, \quad \bar{x}_0^{(P)} \equiv \bar{x}_0^{(P)} \tag{19}$$

$$(P_k^{(P)})^- = AP_{k-1}^{(P)}A^T + \Gamma Q_p \Gamma^T, \quad P_0^{(P)} \equiv P_0^{(P)} \tag{20}$$

$$K_k^{(P)} = (P_k^{(P)})^-(H^{(P)})^T[H^{(P)}(P_k^{(P)})^-(H^{(P)})^T + R_m^{(P)}]^{-1} \tag{21}$$

$$\bar{x}_k^{(P)} = (\bar{x}_k^{(P)})^- + K_k^{(P)}[z_k^{(P)} - H^{(P)}(\bar{x}_k^{(P)})^-] \tag{22}$$

$$P_k^{(P)} = (I - K_k^{(P)}H^{(P)})(P_k^{(P)})^- \tag{23}$$

and so at decision time k player P obtains his estimate $\bar{x}_k^{(P)}$ of the state x_k. Similarly, player E runs the KF

$$(\bar{x}_k^{(E)})^- = A\bar{x}_{k-1}^{(E)} + Bu_{k-1} + Cv_{k-1}, \quad \bar{x}_0^{(E)} \equiv \bar{x}_0^{(E)} \tag{24}$$

$$(P_k^{(E)})^- = AP_{k-1}^{(E)}A^T + \Gamma Q_p \Gamma^T, \quad P_0^{(E)} \equiv P_0^{(E)} \tag{25}$$

$$K_k^{(E)} = (P_k^{(E)})^-(H^{(E)})^T[H^{(E)}(P_k^{(E)})^-(H^{(E)})^T + R_m^{(E)}]^{-1} \tag{26}$$

$$\bar{x}_k^{(E)} = (\bar{x}_k^{(E)})^- + K_k^{(E)}[z_k^{(E)} - H^{(E)}(\bar{x}_k^{(E)})^-] \tag{27}$$

$$P_k^{(E)} = (I - K_k^{(E)}H^{(E)})(P_k^{(E)})^- \tag{28}$$

and so at decision time k, player E obtains his estimate $\bar{x}_k^{(E)}$ of the state x_k.

Let the respective state estimation errors of the Kalman filters of players P and E be

$$e_k^{(P)} \equiv \bar{x}_k^{(P)} - x_k \ and \ e_k^{(E)} \equiv \bar{x}_k^{(E)} - x_k$$

At the decision time instants $k = 1, \ldots, N-1$, the P and E players' respective state estimation errors $e_k^{(P)}$ and $e_k^{(E)}$ are correlated – this is caused by the process dynamics being driven in part by process noise. Thus, concerning decision time $k \geq 1$: Let the correlation matrix

$$\tilde{P}_k^{(E,P)} \equiv E(\, e_k^{(P)}(e_k^{(E)})^T\,) \tag{29}$$

It can be shown that the recursion for the correlation matrix $\tilde{P}_k^{(E,P)}$ is

$$\tilde{P}_{k+1}^{(E,P)} = (I - K_{k+1}^{(P)}H^{(P)})(A\tilde{P}_k^{(P,E)}A^T + \Gamma Q_p \Gamma^T)(I - K_{k+1}^{(E)}H^{(E)})^T, \quad \tilde{P}_0^{(P,E)} = 0,$$

$$k = 0, \ldots, N-1 \tag{30}$$

The KF covariance matrices $P_k^{(P)}, P_k^{(E)}$, and $\tilde{P}_k^{(E,P)}$ are calculated by P and E ahead of time and offline by solving the respective recursions (20), (21), (23); (25), (26), (28); and (30). Also the P and E players' Kalman filter gains $K_{k+1}^{(P)}$ and $K_{k+1}^{(E)}$ are calculated ahead of time and offline.

4.2 Statistics of Sufficient Statistics

In LQGDGs with a control-sharing information pattern the players' sufficient statistic is their state estimate; the latter is the argument of their strategy functions (17) and (18). Hence, in the process of countering E's action at time k, P must compute the *statistics* of E's state estimate $\bar{x}_k^{(E)}$, and, vice versa, while planning his move at time k, E must compute the *statistics* of P's state estimate $\bar{x}_k^{(P)}$. Momentarily assume the point of view of player P: As far as P is concerned, the unknown to him state estimate of player E at time k, $\bar{x}_k^{(E)}$, is a random variable. Similarly, player E will consider the unknown to him state estimate of player P at time k, $\bar{x}_k^{(P)}$, to be a random variable. Hence, in the LQGDG with a control-sharing information pattern, at time k, player P will estimate E's state estimate $\bar{x}_k^{(E)}$ using his calculated ownship state estimate $\bar{x}_k^{(P)}$, and vice versa, player E will estimate P's state estimate $\bar{x}_k^{(P)}$ using his calculated ownship state estimate $\bar{x}_k^{(E)}$. Thus, in the LQGDG with a control-sharing information pattern and with his state estimate $\bar{x}_k^{(P)}$ at time k in hand, player P calculates the statistics of E's state estimate $\bar{x}_k^{(E)}$, conditional on the public and private information available to him at time k. Similarly, having obtained at time k his state estimate $\bar{x}_k^{(E)}$, player E calculates the statistics of the state estimate $\bar{x}_k^{(P)}$ of player P, conditional on the public and private information available to him at time k. Let's start at decision time $k = 0$.

Player P models his measurement/estimate $\bar{x}_0^{(P)}$ of the initial state x_0 as

$$\bar{x}_0^{(P)} = x_0 + e_0^{(P)} , \tag{31}$$

where x_0 is the true physical state and $e_0^{(P)}$ is player P's measurement/estimation error, whose statistics, in view of equation (12), are $e_0^{(P)} \sim \mathcal{N}(0, P_0^{(P)})$. In addition, and according to equation (14), player P models player E's measurement $\bar{x}_0^{(E)}$ of the initial state x_0 as

$$\bar{x}_0^{(E)} = x_0 + e_0^{(E)} , \tag{32}$$

where, as before, x_0 is the true physical state and $e_0^{(E)}$ is player E's measurement/estimation error, whose statistics, which are known to P (see equation (14)), are $e_0^{(E)} \sim \mathcal{N}(0, P_0^{(E)})$. The Gaussian random variables $e_0^{(P)}$ and $e_0^{(E)}$ are *independent* – by hypothesis. From player P's point of view, $\bar{x}_0^{(P)}$ is known, but $\bar{x}_0^{(E)}$ is a random variable. Subtracting equation (31) from equation (32), at time $k = 0$ player P concludes that as far as he is concerned, player E's measurement upon which he will decide, according to equation (18), on his optimal control v_0^*, is the random variable

$$\bar{x}_0^{(E)} = \bar{x}_0^{(P)} + e_0^{(E)} - e_0^{(P)} , \tag{33}$$

In other words, as far as P is concerned, E's estimate $\bar{x}_0^{(E)}$ of the initial state x_0 is the Gaussian random variable

$$\bar{x}_0^{(E)} \sim \mathcal{N}(\bar{x}_0^{(P)}, P_0^{(P)} + P_0^{(E)}) \tag{34}$$

Thus, at decision time $k = 0$, player P has used his measurement/private information $\bar{x}_0^{(P)}$ and the public information $P_0^{(P)}$ and $P_0^{(E)}$ to calculate the statistics of the sufficient statistic $\bar{x}_0^{(E)}$ of player E, which is the argument of E's strategy function $\gamma_0^{(E)}(\cdot)$; the latter, along with P's control u_0, will affect the state x_1 and feature in player's P cost functional. Similarly, as far as player E is concerned, at time $k = 0$ the statistics of the sufficient statistic $\bar{x}_0^{(P)}$ of player P are

$$\bar{x}_0^{(P)} \sim \mathcal{N}(\bar{x}_0^{(E)}, P_0^{(P)} + P_0^{(E)}) \tag{35}$$

Similar to the case where $k = 0$, as far as player P is concerned, the state estimate of player E at decision time $k \geq 1$ is the random variable

$$\bar{x}_k^{(E)} = \bar{x}_k^{(P)} + e_k^{(E)} - e_k^{(P)},$$

that is, at decision time k, player P believes that the state estimate $\bar{x}_k^{(E)}$ of player E is

$$\bar{x}_k^{(E)} \sim \mathcal{N}(\bar{x}_k^{(P)}, P_k^{(E,P)}) \tag{36}$$

where the matrix

$$P_k^{(E,P)} \equiv E((e_k^{(E)} - e_k^{(P)})(e_k^{(E)} - e_k^{(P)})^T)$$
$$= P_k^{(P)} + P_k^{(E)} - E(e_k^{(P)}(e_k^{(E)})^T) - (E(e_k^{(P)}(e_k^{(E)})^T))^T$$

Similarly, as far as he is concerned, player E believes that at decision time $k \geq 1$ the state estimate $\bar{x}_k^{(P)}$ of player P is the random variable

$$\bar{x}_k^{(P)} \sim \mathcal{N}(\bar{x}_k^{(E)}, P_k^{(E,P)}) \tag{37}$$

In summary, at decision time $k = 0, \ldots, N - 1$, player P believes that the statistics of E's estimate $\bar{x}_k^{(E)}$ of the physical state x_k are given by equation (36) and player E believes that the statistics of P's estimate $\bar{x}_k^{(P)}$ of the physical state x_k are given by equation (37) where

$$P_k^{(E,P)} = P_k^{(P)} + P_k^{(E)} - \tilde{P}_k^{(E,P)} - (\tilde{P}_k^{(E,P)})^T$$

Finally, since in LQGDGs with a control-sharing information pattern the sufficient statistic is the players' state estimate, then upon employing the method of dynamic programming and as in partially observable Markov decision processes (POMDPs), at decision time k, player P must project ahead the *estimate* of the physical state x_{k+1} that the Kalman filtering algorithm will provide at time $k+1$. It can be shown that at time k, player P believes that the future state x_{k+1} at time $k+1$ will be the Gaussian random variable

$$\bar{x}_{k+1}^{(P)} = A\bar{x}_k^{(P)} + Bu_k + C\gamma_k^{(E)}(\bar{x}_k^{(P)} + e_k^{(P)} - e_k^{(E)}) + K_{k+1}^{(P)}(H^{(P)}\Gamma w_k + v_{k+1}^{(P)}$$
$$-H^{(P)}Ae_k^{(P)}) \tag{38}$$

Similarly, at decision time k, player E's estimate of the state x_{k+1} at time $k+1$ will be the Gaussian random variable

$$\bar{x}_{k+1}^{(E)} = A\bar{x}_k^{(E)} + B\gamma_k^{(P)}(\bar{x}_k^{(E)} + e_k^{(E)} - e_k^{(P)}) + Cv_k + K_{k+1}^{(E)}(H^{(E)}\Gamma w_k + v_{k+1}^{(E)}$$
$$-H^{(E)}Ae_k^{(E)}) \tag{39}$$

4.3 Value Functions

The value function of player P is

$$V_k^{(P)}(\bar{x}_k^{(P)}) = (\bar{x}_k^{(P)})^T P_{N-k}\bar{x}_k^{(P)} + c_k^{(P)}$$

The intercept in P's value function $\forall k = N - 2, \ldots, 0$ is obtained using the retrograde recursion

$$c_k^{(P)} = c_{k+1}^{(P)} - Trace(((F_k^{(E)})^*)^T(R_c^{(E)} - C^TQC)(F_k^{(E)})^*P_k^{(E,P)})$$
$$+ Trace(A^TQAP_k^{(P)}) + Trace(\Gamma^TQ\Gamma Q_p)$$
$$+ 2\,Trace(A^TQC(F_k^{(E)})^*((\tilde{P}_k^{(E,P)})^T - P_k^{(P)}))$$
$$+ Trace(((F_k^{(E)})^*)^TC^TP_{k+1}C(F_k^{(E)})^*P_k^{(E,P)})$$
$$+ Trace(\Gamma^T(H^{(P)})^T(K_{k+1}^{(P)})^TP_{k+1}K_{k+1}^{(P)}H^{(P)}\Gamma Q_p)$$
$$+ Trace((K_{k+1}^{(P)})^TP_{k+1}K_{k+1}^{(P)}R_m^{(P)})$$
$$+ Trace(A^T(H^{(P)})^T(K_{k+1}^{(P)})^TP_{k+1}K_{k+1}^{(P)}H^{(P)}AP_k^{(P)})$$
$$- Trace(((F_k^{(E)})^*)^TC^TP_{k+1}K_{k+1}^{(P)}A\tilde{P}_k^{(E,P)})$$
$$+ Trace(((F_k^{(E)})^*)^TC^TP_{k+1}K_{k+1}^{(P)}AP_e^{(P)}) \tag{40}$$

and we recall:

$$P_k^{(E,P)} = P_k^{(P)} + P_k^{(E)} - \tilde{P}_k^{(E,P)} - (\tilde{P}_k^{(E,P)})^T$$

The retrograde recursion (40) is "initialized" with $c_{N-1}^{(P)}(P_{N-1}^{(P)}, P_{N-1}^{(E)}, \tilde{P}_{N-1}^{(P,E)})$

$$
\begin{aligned}
&= Trace(A^T Q_F A P_{N-1}^{(P)}) + Trace(\Gamma^T Q_F \Gamma Q_p) \\
&\quad + 2\, Trace((P_{N-1}^{(P)} - \tilde{P}_{N-1}^{(P,E)})A^T Q_F C(F_{N-1}^{(E)})^*) \\
&\quad - Trace(((F_{N-1}^{(E)})^*)^T (R_v - C^T Q_F C)(F_{N-1}^{(E)})^* P_{N-1}^{(E,P)})
\end{aligned}
\tag{41}
$$

The value function of player E is

$$V_k^{(E)}(\bar{x}_k^{(E)}) = (\bar{x}_k^{(E)})^T P_{N-k} \bar{x}_k^{(E)} + c_k^{(E)}$$

The intercept in E's value function $\forall \ k = N - 2, \ldots, 0$ is obtained using the retrograde recursion

$$
\begin{aligned}
c_k^{(E)} &= c_{k+1}^{(E)} + Trace(((F_k^{(P)})^*)^T (R_c^{(P)} + B^T QB)(F_k^{(P)})^* P_k^{(E,P)}) \\
&\quad + Trace(A^T QA P_k^{(E)}) + Trace(\Gamma^T Q\Gamma Q_p) \\
&\quad + 2\, Trace((A^T QB(F_k^{(P)})^*(\tilde{P}_k^{(E,P)} - P_k^{(E)}) \\
&\quad + Trace(((F_k^{(P)})^*)^T B^T P_{k+1} B(F_k^{(P)})^* P_k^{(E,P)}) \\
&\quad + Trace(\Gamma^T (H^{(E)})^T (K_{k+1}^{(E)})^T P_{k+1} K_{k+1}^{(E)} H^{(E)} \Gamma Q_p) \\
&\quad + Trace((K_{k+1}^{(E)})^T P_{k+1} K_{k+1}^{(E)} R_m^{(E)}) \\
&\quad + Trace(A^T (H^{(E)})^T (K_{k+1}^{(E)})^T P_{k+1} K_{k+1}^{(E)} H^{(E)} A P_k^{(E)}) \\
&\quad - Trace(((F_k^{(P)})^*)^T B^T P_{k+1} K_{k+1}^{(E)} A(\tilde{P}_k^{(E,P)}))^T \\
&\quad + Trace(((F_k^{(P)})^*)^T B^T P_{k+1} K_{k+1}^{(E)} A P_e^{(E)})
\end{aligned}
\tag{42}
$$

The retrograde recursion (42) is "initialized" with $c_{N-1}^{(E)}(P_{N-1}^{(E)}, P_{N-1}^{(P)}, \tilde{P}_{N-1}^{(E,P)})$

$$
\begin{aligned}
&= Trace(A^T Q_F A P_{N-1}^{(E)}) + Trace(\Gamma^T Q_F \Gamma Q_p) \\
&\quad + Trace(((F_{N-1}^{(P)})^*)^T (R_u + B^T Q_F B)(F_{N-1}^{(P)})^* P_{N-1}^{(E,P)}) \\
&\quad + 2\, Trace(((F_{N-1}^{(P)})^*)^T B^T Q_F A(P_{N-1}^{(E)} - \tilde{P}_{N-1}^{(E,P)}))
\end{aligned}
\tag{43}
$$

The scalar recursions (40) and (42) amount to straight summations.

4.4 Summary

A careful analysis in [14] revealed that although this is a game with partial information, the control-sharing information pattern renders the game amenable to solution by the method of DP. The solution of the LQGDG with a control-sharing information pattern is similar in structure to the solution of the LQG optimal control problem in so far as the principle of certainty equivalence/decomposition holds. The *correct* closed-form solution of LQGDGs with a control-sharing information pattern is summarized below.

Theorem 2 Consider the LQGDG (11), (16) with the information pattern:

- The P and E players' prior information is $x_0 \sim \mathcal{N}(\overline{x}_0^{(P)}, P_0^{(P)})$ and $x_0 \sim \mathcal{N}(\overline{x}_0^{(E)}, P_0^{(E)})$, respectively. The covariances $P_0^{(P)}$ and $P_0^{(E)}$ are finite.
- Players P and E have *Partial Information*: At time $1 \leq k \leq N - 1$ the respective measurements P and E are (13) and (15).
- The prior information $\overline{x}_0^{(P)}$ and $\overline{x}_0^{(E)}$, and the measurements (13) and (15), that is, the respective data records $Z_k^{(P)}$ and $Z_k^{(E)}$, are the P and E players' *private* information. It is *not* shared among the P and E players.
- The P and E players have complete recall of ownship control histories U_k and V_k, respectively.
- The players observed their opponent's *past* moves: At decision time $1 \leq k \leq N - 1$, the control *history* $V_k = \{v_0, \ldots, v_{k-1}\}$ of player E is known to player P, and, similarly, player E knows the control *history* $U_k = \{u_0, \ldots, u_{k-1}\}$ of player P.

The players obtain their respective private state estimates $\overline{x}_k^{(P)}$ and $\overline{x}_k^{(E)}$ by running two *separate* Kalman filters in parallel: Player P initialized his KF with his prior information (12) and uses his measurements $z_k^{(P)}$. Similarly, E runs his KF initialized with his prior information (14) and using his measurements $z_k^{(E)}$. Specifically:

- Player P runs his KF (19)–(23) which is driven by his private measurements (13) and his initial state information (12) + P and E's control *histories* U_k, V_k.
- Player E runs his KF (24)–(28) which is driven by his private measurements (15) and his initial state information (14) + E and P's control *histories* U_k, V_k.

A Nash equilibrium in *delayed commitment* strategies for the "zero-sum" LQGDG with a control-sharing information pattern is established. The optimal strategies are linear in the players' sufficient statistics – the linear strategies from the deterministic LQDG are reused as follows:

$$(\gamma_k^{(P)}(\overline{x}_k^{(P)}))^* = (F_k^{(P)})^* \cdot \overline{x}_k^{(P)} \tag{44}$$

$$(\gamma_k^{(E)}(\overline{x}_k^{(E)}))^* = (F_k^{(E)})^* \cdot \overline{x}_k^{(E)} \tag{45}$$

where the respective P and E players' optimal gains are (9) and (10), respectively.

The respective affine quadratic value functions of players P and E are

$$V_k^{(P)}(\bar{x}_k^{(P)}, P_k^{(P)}; P_k^{(E)}, \tilde{P}_k^{(E,P)}) = (\bar{x}_k^{(P)})^T P_k \bar{x}_k^{(P)} + c_k^{(P)},$$

$$V_k^{(E)}(\bar{x}_k^{(E)}, P_k^{(E)}; P_k^{(P)}, \tilde{P}_k^{(E,P)}) = (\bar{x}_k^{(E)})^T P_k \bar{x}_k^{(E)} + c_k^{(E)}$$

The matrices P_k are the solution of the Riccati equation (5) for the deterministic LQDG.

The "intercepts" $c_k^{(P)}$ and $c_k^{(E)}$ are obtained by solving the respective scalar recursions (40) (using equations (10), (21), (41)) and (42) (using equations (9), (26), (43)).

The KF calculated covariance matrices $P_k^{(P)}$, $P_k^{(E)}$ and the matrices $\tilde{P}_k^{(E,P)}$ which are given by the solution of the Lyapunov equation (30) exclusively feature in the intercepts $c_k^{(P)}$ and $c_k^{(E)}$. This requires each player to solve a Kalman filtering Riccati equation + both players solve the deterministic LQDG Riccati equation, in total:

Three *decoupled* Riccati equations

The deterministic control Riccati equation for P_k, the KF Riccati equation (20), (21), (23) of player P, the KF Riccati equation (25), (26), (28) of player E, and the Lyapunov equation (30) are solved ahead of time and offline.

Once the three Riccati equations and the Lyapunov equation have been solved, the recursions (40), (41) and (42), (43) for the "intercepts" $c_k^{(P)}$ and $c_k^{(E)}$ are also solved offline.

The P and E players' optimal Nash strategies are *Delayed Commitment* strategies. The solution of the "zero-sum" LQGDG is *Time Consistent/Subgame Perfect*.

5 The Opponent's Past Inputs

Based on their respective information states, the players must be able to obtain their sufficient statistics, that is, they must be able to independently compute the statistics of the current state of the control system. From the discussion so far, it is evident that a major stumbling block in the treatment of LQGDGs is the non-nested information pattern brought about by partial information: The non-nested information pattern stands in the way of a player's characterization of his opponent's past inputs, and, consequently, the calculation of the statistics, a.k.a. expectation, of the current state of the control system. None of this is an issue in the one-step-delay sharing pattern [12, 13] or the control-sharing information pattern.

Thus, it stands to reason to require that, based on his current information state, player P must be able to jointly estimate both the current state and the past inputs of player E. The same applies to player E who must be able to obtain his own estimate of the current state, and to do this he must be able to jointly estimate both the current state and the past inputs of player P.

5.1 State Estimation in LQGDGs

Although the players are exclusively interested in the current state estimate, the lack of information on the adversary's past inputs necessitates that they be able to jointly estimate both the current state and their opponent's past inputs. To this end the novel KF for joint state and control estimation from reference [17] will be adapted to the analysis of the LQGDG with partial information. We shall require

Assumption 1 The matrices $H^{(P)}C$ and $H^{(E)}B$ are square and invertible.

Assumption 1 is not outlandish. It is standard in control design using dynamic inversion. As an aside, the flight control system of the F-35 JSF is designed using dynamic inversion. In direct adaptive control [18], an even stronger assumption is invoked: it is required that the product of the control's system observation and control matrices is a square and positive definite matrix, namely, the high frequency gain $CB > 0$.

The following holds.

Theorem 3 Consider the LQGDG with partial information (11)–(16). The P and E players have full recall of ownship control history, U_k and V_k, respectively. However, at time $k \geq 1$ player P has no information about the past inputs $v_0, \ldots, v_{k-1}$ of player E, and in the same vein, player E has no information about the past inputs $u_0, \ldots, u_{k-1}$ of player P.

Hypothesis: Assumption 1 holds.

The min variance/ML (optimal) state estimate of player P is

$$\bar{x}_{k+1}^{(P)} = [I_n - C(H^{(P)}C)^{-1}H^{(P)}](A\bar{x}_k^{(P)} + Bu_k) + C(H^{(P)}C)^{-1}z_{k+1}^{(P)},$$
$$\bar{x}_0^{(P)} \equiv \bar{x}_0^{(P)} \tag{46}$$

and $\forall\, k = 0, \ldots, N - 1$ the covariance of his state estimation error is given by the linear recursion

$$P_{k+1}^{(P)} = [I_n - C(H^{(P)}C)^{-1}H^{(P)}](AP_k^{(P)}A^T + \Gamma Q_p \Gamma^T)[I_n$$
$$-(H^{(P)})^T(C^T(H^{(P)})^T)^{-1}C^T]$$
$$+C(H^{(P)}C)^{-1}R^{(P)}(C^T(H^{(P)})^T)^{-1}C^T, \quad P_0^{(P)} \equiv P_0^{(P)} \tag{47}$$

The min variance/ML (optimal) state estimate of player E is

$$\bar{x}_{k+1}^{(E)} = [I_n - B(H^{(E)}B)^{-1}H^{(E)}](A\bar{x}_k^{(E)} + Cv_k) + B(H^{(E)}B)^{-1}z_{k+1}^{(E)},$$
$$\bar{x}_0^{(E)} \equiv \bar{x}_0^{(E)} \tag{48}$$

and $\forall\, k = 0, \ldots, N - 1$ the covariance of his state estimation error is given by the linear recursion

$$P_{k+1}^{(E)} = [I_n - B(H^{(E)}B)^{-1}H^{(E)}](AP_k^{(E)}A^T + \Gamma Q_p \Gamma^T)[I_n$$
$$- (H^{(E)})^T(B^T(H^{(E)})^T)^{-1}B^T]$$
$$+ B(H^{(E)}B)^{-1}R^{(E)}(B^T(H^{(E)})^T)^{-1}B^T, \quad P_0^{(E)} \equiv P_0^{(E)} \tag{49}$$

As before, let the state estimation errors of the filters of P and E be $e_k^{(P)} \equiv \bar{x}_k^{(P)} - x_k$, $e_k^{(E)} \equiv \bar{x}_k^{(E)} - x_k$. The error equation of P's filter is

$$e_{k+1}^{(P)} = [I_n - C(H^{(P)}C)^{-1}H^{(P)}](Ae_k^{(P)} - \Gamma w_k) + C(H^{(P)}C)^{-1}\zeta_{k+1}^{(P)}, \ k = 0, \ldots, N-1$$

and the error equation of E's filter is

$$e_{k+1}^{(E)} = [I_n - B(H^{(E)}B)^{-1}H^{(E)}](Ae_k^{(E)} - \Gamma w_k)$$
$$+ B(H^{(E)}B)^{-1}\zeta_{k+1}^{(E)}, \ k = 0, \ldots, N-1$$

Process noise causes the P and E filters' state estimation errors to be correlated. Recall $\tilde{P}_k^{(E,P)} \equiv E(\, e_k^{(P)}(e_k^{(E)})^T\,)$, whereupon

$$\tilde{P}_{k+1}^{(E,P)} = [I_n - C(H^{(P)}C)^{-1}H^{(P)}](A\tilde{P}_k^{(E,P)}A^T + \Gamma Q_p\Gamma^T)[I_n - B(H^{(E)}B)^{-1}H^{(E)}]^T,$$
$$\tilde{P}_0^{(E,P)} = 0$$

When solving the LQGDG using the method of dynamic programming and similar to partially observable Markov decision processes (POMDPs), at decision time $0 \leq k \leq N-2$, each player must project ahead his estimate of the physical state x_{k+1}: In this respect, the projected ahead state estimate of player P will be the random variable

$$\bar{x}_{k+1}^{(P)} = A\bar{x}_k^{(P)} + Bu_k + C\gamma_k^{(E)}(\hat{x}_k^{(P)} + e_k^{E} - e_k^{P})$$
$$- C(H^{(P)}C)^{-1}[H^{(P)}(Ae_k^{(P)} - \Gamma w_k) - \zeta_{k+1}^{(P)}]$$

and the projected ahead state estimate of player E will be the random variable

$$\bar{x}_{k+1}^{(E)} = A\bar{x}_k^{(E)} + B\gamma_k^{(P)}(\hat{x}_k^{(E)} + e_k^{P} - e_k^{E}) + Cv_k$$
$$- B(H^{(E)}B)^{-1}[H^{(E)}(Ae_k^{(E)} - \Gamma w_k) - \zeta_{k+1}^{(E)}]$$

Remark The players' projected state estimates are as in the control-sharing scenario [14], except the Kalman gains $K_{k+1}^{(P)}$ and $K_{k+1}^{(E)}$ therein are replaced by the $C(H^{(P)}C)^{-1}$ and $B(H^{(E)}B)^{-1}$ gains, respectively.

It would seem that, in principle at least, one has the tools necessary to obtain a result for LQGDGs which is analogous to Theorem 2 – this, courtesy of Assumption 1.

5.2 Shared Observations

The shared observation information pattern where the players' *past*, but not current, measurements are shared, is interesting. The following holds.

Corollary 4 Consider the special case of shared observations, that is, $\bar{x}_0^{(P)} = \bar{x}_0^{(E)} = \bar{x}_0$, $P_0^{(P)} = P_0^{(E)} = P_0$ and, of course, $H^{(P)} = H^{(E)} = H$, $R_m^{(P)} = R_m^{(E)} = R_m$. The min variance/ML (optimal) state estimate of player P is

$$\bar{x}_{k+1}^{(P)} = [I_n - C(HC)^{-1}H](A\bar{x}_k^{(P)} + Bu_k) + C(HC)^{-1}z_{k+1},$$
$$\bar{x}_0^{(P)} \equiv \bar{x}_0 \tag{50}$$

and $\forall\, k = 0, \ldots, N-1$ the covariance of his state estimation error is

$$P_{k+1}^{(P)} = [I_n - C(HC)^{-1}H](AP_k^{(P)}A^T + \Gamma Q_p \Gamma^T)[I_n$$
$$-H^T(C^T H^T)^{-1}C^T]$$
$$+C(HC)^{-1}R(C^T(H)^T)^{-1}C^T, \ P_0^{(P)} \equiv P_0 \tag{51}$$

The min variance/ML (optimal) state estimate of player E is

$$\bar{x}_{k+1}^{(E)} = [I_n - B(HB)^{-1}H](A\bar{x}_k^{(E)} + Cv_k) + B(HB)^{-1}z_{k+1}^{(E)},$$
$$\bar{x}_0^{(E)} \equiv \bar{x}_0 \tag{52}$$

and $\forall\, k = 0, \ldots, N-1$ the covariance of his state estimation error is

$$P_{k+1}^{(E)} = [I_n - B(HB)^{-1}H](AP_k^{(E)}A^T + \Gamma Q_p \Gamma^T)[I_n$$
$$-H^T(B^T H^T)^{-1}B^T]$$
$$+B(HB)^{-1}R(B^T(H)^T)^{-1}B^T, \ P_0^{(E)} \equiv P_0 \tag{53}$$

The error equation of the state estimation filter of player P is

$$e_{k+1}^{(P)} = [I_n - C(HC)^{-1}H](Ae_k^{(P)} - \Gamma w_k) + C(HC)^{-1}\zeta_{k+1}^{(P)}, \ k = 0, \ldots, N-1$$

and the error equation of the state estimation filter of player E is

$$e_{k+1}^{(E)} = [I_n - B(HB)^{-1}H](Ae_k^{(E)} - \Gamma w_k) + B(HB)^{-1}\zeta_{k+1}^{(E)}, \ k = 0, \ldots, N-1$$

The P and E filters' state estimation errors are correlated:

$$\tilde{P}_{k+1}^{(E,P)} = [I_n - C(HC)^{-1}H](A\tilde{P}_k^{(E,P)}A^T + \Gamma Q_p\Gamma^T)[I_n$$

$$-B(HB)^{-1}H]^T, \ \tilde{P}_0^{(E,P)} = P_0 \tag{54}$$

Finally, when solving the LQGDG using the method of dynamic programming, at decision time $0 \le k \le N-2$, each player must project ahead his estimate of the physical state x_{k+1}: The projected ahead state estimate of P is the random variable

$$\bar{x}_{k+1}^{(P)} = A\bar{x}_k^{(P)} + Bu_k + C\gamma_k^{(E)}(\bar{x}_k^{(P)} + e_k^{(E)} - e_k^{(P)})$$

$$-C(HC)^{-1}[H(Ae_k^{(P)} - \Gamma w_k) - \zeta_{k+1}] \tag{55}$$

and the projected ahead state estimate of E is the random variable

$$\bar{x}_{k+1}^{(E)} = A\bar{x}_k^{(E)} + B\gamma_k^{(P)}(\hat{x}_k^{(E)} + e_k^{(P)} - e_k^{(E)}) + Cv_k$$

$$-B(HB)^{-1}[H(Ae_k^{(E)} - \Gamma w_k) - \zeta_{k+1}] \tag{56}$$

5.3 Discussion

In order to propagate a dynamic game with partial information forward in time and therefore be able to decompose the optimization problem using the method of dynamic programming (DP) and render the problem on hand tractable, the players must be able to independently calculate their respective sufficient statistics using their information states; the latter comprises a player's public and private information at decision time k. The question of what constitutes a player's sufficient statistic is thus paramount. In a game against nature, that is, in a stochastic optimal control problem with partial information a.k.a., in a POMDP, the decision-maker's (DM's) sufficient statistic is the p.d.f. of the physical state. This allows POMDPs to be solved using the method of DP, where the state is now lifted – it is the p.d.f. of the physical state. The solution of POMDPs is challenging. Also in dynamic games with partial information, the argument of a players' strategy functions is the p.d.f. of the physical state, as perceived by the player; it is a function of the player's information state and can be referred to as the player's belief. Thus, the players' sufficient statistics are the arguments of their respective strategy functions $\gamma_k^{(P)}(\cdot)$ and $\gamma_k^{(E)}(\cdot)$. But in dynamic games with partial information, this is not the end of the story. Because an intelligent opponent, and not randomness, is at work, the sufficient statistic of a player must be s.t. he can also calculate the statistics of the sufficient statistics of his opponent: At decision time k, the player must be able to calculate the statistics of the parameters of his opponent's – derived p.d.f. of the physical state x_k. This is a tall order.

Theorem 3 and Corollary 4 specify linear KFs which at decision time k allow the players in LQGDGs with partial information to independently obtain estimates of the physical state x_k using linear regression, provided Assumption 1 holds. Assumption 1 is critical – it rendered the players' state estimation problem linear-Gaussian (LG). At decision time k, the novel KFs provide the P and E players with their sufficient statistics, their independently calculated respective minimum variance/ML state estimates $\bar{x}_k^{(P)}$ and $\bar{x}_k^{(E)}$. Furthermore, the herein derived error dynamics of their respective filters, and having obtained the ownship state estimate, allows a player to also calculate the statistics of his opponent's minimum variance/ML state estimate, namely, the statistics of the opponent's sufficient statistics. The availability of sufficient statistics makes the LQGDG with partial information amenable to solution using the method DP. Consequently, similar to LQG optimal control, it appears that LQGDGs with partial information are amenable to solution by the method of stochastic DP. It would thus appear that Assumption 1 is all that is needed to render the LQGDG with partial information tractable. This is certainly so in the case of shared past observations. Corollary 4 is the key to the solution of LQGDGs with an observation-sharing information pattern.

6 Conclusion

In this paper conceptual issues germane to the solution of linear-quadratic Gaussian dynamic games (LQGDGs) with partial information are addressed. In addition to LQGDGs with a one-step-delay sharing pattern and LQGDGs with a control-sharing information pattern, and to complete the picture, LQGDGs with an observation-sharing information pattern are also addressed. The herein derived novel KF algorithm holds the key to an attack on LQGDGs with an observation-sharing information pattern, to be further addressed in future work. The exact solution of LQGDGs with a nonclassical information pattern/partial observations could be used to benchmark and validate AI algorithms for POMDPs based on Monte Carlo (MC) methods and reinforcement learning (RL) which yield approximate solutions to more general decision problems.

References

1. W. W. Willman: "Formal Solutions for a Class of Pursuit-Evasion Games", IEEE Trans. on AC, Vol. 14 (1969), pp. 504–509.
2. Hans S. Witsenhausen: "A Counterexample in Stochastic Optimum Control", SIAM Journal of Control, Vol. 6, 1968, pp. 131–147.
3. Pachter, M. and K. Pham: "Informational Issues in Decentralized Control", in *Dynamics of Information Systems: Computational and Mathematical Challenges*, C. Vogiatzis, J. Walteros and P. Pardalos, Eds., Springer 2014.

4. Hans S. Witsenhausen: "Separation of Estimation and Control for Discrete Time Systems", Proceedings of the IEEE, Vol. 59, No. 11, November 1971, pp. 1557–1566.
5. Y. C. Ho, A. E. Bryson, S. Baron: "Differential Games and Optimal Pursuit-Evasion Strategies", IEEE Trans. on AC, Vol. 10, No. 4, October 1965, pp. 385–389.
6. Basar, T., P. Bernhard: "H^∞ - Optimal Control and Related Minimax Design Problems: a dynamic game approach", Birkhhauser, Boston, 2008.
7. Engwerda, J.: "LQ Dynamic Optimization and Differential Games", Wiley, 2005.
8. Pachter, M. and K. Pham: "Discrete-Time Linear-Quadratic Dynamic Games", Journal of Optimization Theory and Applications, Vol. 146 No 1, July 2010, pp 151–179.
9. M. Pachter: "Static Linear-Quadratic Gaussian Games", in *Advances in DynamicGames*, V. Krivan and G. Zaccour, Eds., pp. 85–105, Birkhauser 2013.
10. Radner, R.: "Static Teams and Stochastic Games", The Annals of Mathematical Statistics, Vol. 33, No. 3, Sept. 1962, pp. 857–867.
11. Ben-Zion Kurtaran and Raphael Sivan: "Linear-Quadratic-Gaussian Control with One-Step-Delay Sharing Pattern", IEEE Trans. on AC, October 1974, pp. 571–574.
12. M. Pachter: "Static Linear-Quadratic Gaussian Games" in *Advances in Dynamic Games*, V. Krivan and G. Zaccour Eds., pp. 85–105, Birkhauser 2013.
13. Ben-Zion Kurtaran: "A Concise Derivation of the LQG One-Step-Delay Sharing Problem Solution", IEEE Trans. on AC, December 1975, pp.108–110.
14. M. Pachter: "LQG Dynamic Games With a Control-Sharing Information Pattern", Dynamic Games And Applications, Vol. 7, No 2, pp. 289–322, DOI 10.1007/s13235-016-0182-6, 2017.
15. Pachter, M. and K. Pham: "Discrete-Time Linear-Quadratic Dynamic Games", Journal of Optimization Theory and Applications, Vol. 146 No 1, July 2010, pp 151–179.
16. A. W. Merz: "Noisy Satellite Pursuit-Evasion Guidance", AIAA J. Guidance, Control and Dynamics, Vol. 12, No. 6, pp.901–905, 1989.
17. M. Pachter: "Kalman Filtering When the Large Bandwidth Control is Not Known", IEEE Trans. on AES, Vol. 48, No. 1, pp.542–551, January 2012.
18. I. Barkana, Marcelo C. M. Teixeira, L. Hsu: "Mitigation of Symmetry Condition in Positive Realness for Adaptive Control", Automatica, Vol. 42, Issue 9, pp.1611–1616, September 2006.
19. M. L. Littman: "Markov Games as a Framework for Multi-Agent Reinforcement Learning", Proceedings of the Eleventh International Conference on Machine Learning, pp. 157–163, San Francisco, CA, 1994.
20. M. L. Littman: "Value Function Reinforcement Learning in Markov Games" Journal of Cognitive Systems Research 2 (2001), pp. 55–66.
21. S. M. LaValle: Planning Algorithms, Cambridge University Press, 2006.

Part IV
Computational Methods for Dynamic Games

Viability Approach to Aircraft Control in Wind Shear Conditions

Nikolai Botkin, Johannes Diepolder, Varvara Turova, Matthias Bittner,
and Florian Holzapfel

Abstract This paper addresses the analysis of aircraft control capabilities in the
presence of wind shears. The cruise flight phase (flying at the established level
with practically constant configuration and speed) is considered. The study utilizes
a point-mass aircraft model describing both vertical and lateral motions. As a
particular case, a reduced model of lateral motion is derived from the full one.
State variables of the models are constrained according to aircraft safety conditions,
and differential games where a guiding system, the first player, works against
wind disturbances, the second player, are considered. Viability theory is used
to find the leadership kernel, the maximal subset of the state constraint where
the aircraft trajectories can remain arbitrary long if the first player utilizes an
appropriate feedback control, and the second player generates any admissible
disturbances. The computations are based on a theoretical background resulting in a
grid method developed by the authors. The corresponding software is implemented
on a multiprocessor computer system.

1 Introduction

There is permanent interest in designing aircraft guidance schemes ensuring safe
aircraft performance in the presence of wind disturbances. In papers [6, 10, 15–19,
21], optimal control theory, robust control techniques, and differential game theory
have been used to design appropriate controls. Both the case of known wind velocity
field and the case of unknown wind disturbances have been considered.

N. Botkin (✉) • V. Turova
Department of Mathematics, Technische Universität München, Boltzmannstr. 3, 85748 Garching
near Munich, Germany
e-mail: botkin@ma.tum.de; turova@ma.tum.de

J. Diepolder • M. Bittner • F. Holzapfel
Institute of Flight System Dynamics, Technische Universität München, Boltzmannstr. 15, 85748
Garching near Munich, Germany
e-mail: johannes.diepolder@tum.de; m.bittner@tum.de; florian.holzapfel@tum.de

© Springer International Publishing AG 2017
J. Apaloo, B. Viscolani (eds.), *Advances in Dynamic and Mean Field Games*,
Annals of the International Society of Dynamic Games 15,
https://doi.org/10.1007/978-3-319-70619-1_15

While the papers quoted above deal with obtaining control strategies for survival enhancing, it is reasonable to choose an appropriate flight domain (AFD), defined by constraints imposed on the state variables, and to find controls keeping the system arbitrary long there (see [3, 23, 24]). Such an approach is concerned with viability theory (see [1, 2] for basic concepts and methods). In recent times, there has been an essential progress in the numerical approximation of viability and discriminating/leadership kernels for control and conflict control problems. A comprehensive overview of corresponding abstract algorithms and their numerical implementations can be found in [20]. A geometrical analysis and game-based interpretation of discriminating and leadership kernels as well as formal algorithms for approximate computing them are given in [8] and [9].

We propose numerical grid-based algorithms for computing leadership and discriminating kernels. For our purposes, the leadership kernel of the AFD is especially interesting because all trajectories can be kept in the leadership kernel using a pure feedback control, which does not require any information about current wind gusts.

The dynamics of the aircraft is considered as a differential game (see, e.g., [14]) where the first player is associated with control inputs, whereas the second player forms the worst wind disturbances. It is assumed that the first player can measure the current state, whereas the second player measures both the current state and control of the first player ("future" values are not available). In other words, the second player uses the so-called feedback counterstrategies (see [14]).

The paper is organized as follows:

Section 2 presents a point-mass model describing a generic modern regional jet transport aircraft. The model is written in the kinematic reference system, and therefore, it does not contain the time derivatives of the wind components.

In Section 3, conflict control problems are formulated, state constraints corresponding to the cruise phase are imposed, and computed leadership kernels are presented.

Section 4 outlines the concept of differential games and discriminating/leadership kernels. A new theory substantiating the computation of these kernels through time limits of value functions of differential games with state constraints is developed. A grid algorithm implementing this theoretical basis is sketched, and a method of designing optimal feedback strategies and counterstrategies is given.

2   Model Equations

This section introduces a point-mass model describing the dynamics of a modern generic regional jet transport aircraft. To describe the influence of the relevant forces on the aircraft dynamics, the following coordinate systems (COS) (see [13] for the exhaustive treatment) with their origin either in the center of gravity of the aircraft (CG) or at a fixed reference point on the Earth's surface (O) are considered (see Table 1).

Table 1 Aircraft coordinate systems.

COS	Index	x, y, z-axis	Origin
Local	N	Parallel to the Earth's surface	O
Earth (NED)	O	x_O – positive in the direction of north; y_O – positive in the direction of east; z_O – positive toward the center of the Earth	CG
Kinematic	K	x_K, in the direction of $\mathbf{V}_K$; z_K, perpendicular to the x_K axis, in the plane of symmetry of the aircraft, positive below the aircraft; y_K, perpendicular to the x_K, z_K plane, positive to the right	CG
Aerodynamic	A	x_A, in direction of $\mathbf{V}_A$; z_A, perpendicular to the x_A axis, in the plane of symmetry of the aircraft, positive below the aircraft; y_A, perpendicular to the x_A, z_A plane, positive to the right	CG
Body-fixed	B	x_B, positive out the nose in the symmetry plane of the aircraft; z_B, perpendicular to the x_B axis, in the plane of symmetry of the aircraft, positive below the aircraft; y_B, perpendicular to the x_B, z_B plane, positive to the right	CG

Here, $\mathbf{V}_K$ and $\mathbf{V}_A$ are kinematic and aerodynamic aircraft velocities, respectively. The angles which define the relationship between the coordinate systems are:

the kinematic flight-path bank angle μ_K,
the kinematic angle of attack α_K,
the kinematic angle of sideslip β_K,
the kinematic path inclination angle γ_K,
the kinematic azimuthal path angle χ_K,
the aerodynamic angle of attack α_A,
the aerodynamic angle of sideslip β_A.

Matrices defining transformations $O \rightarrow K$ (Earth to kinematic), $A \rightarrow B$ (aerodynamic to body), and $B \rightarrow K$ (body to kinematic) look as follows:

$$M_{KO} = \begin{bmatrix} \cos(\chi_K)\cos(\gamma_K) & \sin(\chi_K)\cos(\gamma_K) & -\sin(\gamma_K) \\ -\sin(\chi_K) & \cos(\chi_K) & 0 \\ \cos(\chi_K)\sin(\gamma_K) & \sin(\chi_K)\sin(\gamma_K) & \cos(\gamma_K) \end{bmatrix},$$

$$M_{BA} = \begin{bmatrix} \cos(\alpha_A)\cos(\beta_A) & -\cos(\alpha_A)\sin(\beta_A) & -\sin(\alpha_A) \\ \sin(\beta_A) & \cos(\beta_A) & 0 \\ \sin(\alpha_A)\cos(\beta_A) & -\sin(\alpha_A)\sin(\beta_A) & \cos(\alpha_A) \end{bmatrix},$$

$$M_{KB} = \begin{bmatrix} a_{ij} \end{bmatrix}, \ i, j = 1, 2, 3,$$

where

$$a_{11} = \cos(\alpha_K)\cos(\beta_K), \quad a_{12} = -\cos(\alpha_K)\sin(\beta_K)\cos(\mu_K) + \sin(\alpha_K)\sin(\mu_K),$$

$$a_{13} = -\cos(\alpha_K)\sin(\beta_K)\sin(\mu_K) - \sin(\alpha_K)\cos(\mu_K),$$

$$a_{21} = \sin(\beta_K), \quad a_{22} = \cos(\beta_K)\cos(\mu_K), \quad a_{23} = \cos(\beta_K)\sin(\mu_K),$$

$$a_{31} = \sin(\alpha_K)\cos(\beta_K), \quad a_{32} = -\sin(\alpha_K)\sin(\beta_K)\cos(\mu_K) - \sin(\mu_K)\cos(\alpha_K),$$

$$a_{33} = -\sin(\alpha_K)\sin(\beta_K)\sin(\mu_K) + \cos(\alpha_K)\cos(\mu_K).$$

The translation dynamics are derived in the kinematic coordinate system (K), and the position propagation is given in the local coordinate system (N). The model looks as follows:

$$\dot{x} = V_K \cos(\gamma_K) \cos(\chi_K) \tag{1}$$

$$\dot{y} = V_K \cos(\gamma_K) \sin(\chi_K) \tag{2}$$

$$\dot{z} = -V_K \sin(\gamma_K) \tag{3}$$

$$\dot{V}_K = \frac{X_T}{m} \tag{4}$$

$$\dot{\chi}_K = \frac{Y_T}{mV_K \cos(\gamma_K)} \tag{5}$$

$$\dot{\gamma}_K = \frac{-Z_T}{mV_K} \tag{6}$$

In the above equations, X_T, Y_T, and Z_T denote the components of the total force, $\mathbf{F}_T$, in the kinematic coordinate system (K), and m is the aircraft mass. Here

$$\mathbf{F}_T = \mathbf{F}_A + \mathbf{F}_P + \mathbf{F}_G,$$

where the right-hand side contains aerodynamic, propulsion, and gravitation forces.

Aerodynamic Forces. They are defined as follows:

$$\mathbf{F}_A = M_{KB}M_{BA} \begin{bmatrix} C_D \\ C_Q \\ C_L \end{bmatrix} \frac{1}{2}\rho V_A^2 S,$$

where C_D, C_Q, and C_L are the drag, sideforce, and lift coefficients, respectively, $\rho = \rho(h)$ is the air density depending on the altitude, V_A the aerodynamic velocity, and S is the wing area.

The coefficients C_D and C_L are functions of α_A, β_A, and the Mach number M. They are taken in the polynomial form:

$$C_D = \sum c_{ijk}^{D} \alpha_A^i \beta_A^j M^k, \quad C_L = \sum c_{ijk}^{L} \alpha_A^i \beta_A^j M^k, \tag{7}$$

where the constants c_{ijk}^{D} and c_{ijk}^{L}, $i \in \overline{0,3}$, $j \in \overline{0,3}$, $k \in \overline{0,2}$, and $i + j + k \leq 3$ are found from least square fitting to experimental data. The sideforce coefficient is approximated as a linear function of β_A:

$$C_Q = c^Q \cdot \beta_A.$$

The absolute value of the aerodynamic velocity V_A can be derived using its relation to the kinematic velocity $\mathbf{V}_K$ considered in the Earth frame (O) and the wind velocities W_x, W_y, and W_z in the x_O, y_O, and z_O directions, respectively. Therefore,

$$V_A = \left\| (\mathbf{V}_K)^O - \begin{bmatrix} W_x \\ W_y \\ W_z \end{bmatrix} \right\|, \tag{8}$$

and finally, using the inverse of M_{KO}, this implies the formula

$$V_A^2 = \left(V_K \cos(\chi_K) \cos(\gamma_K) - W_x \right)^2$$
$$+ \left(V_K \sin(\chi_K) + W_y \right)^2$$
$$+ \left(V_K \cos(\chi_K) \sin(\gamma_K) + W_z \right)^2.$$

Moreover, the following expressions hold:

$$\alpha_A = \arctan\left(\frac{w_A}{u_A} \right), \quad \beta_A = \arcsin\left(\frac{v_A}{V_A} \right), \quad \beta_K = \arcsin\left(\frac{v_K}{V_K} \right),$$

where u_A, v_A, and w_A are x_A, y_A, and z_A components of the aerodynamic velocity, respectively, and v_K is y_K-component of the kinematic velocity.

The Mach number M is defined as the ratio between the absolute value of the aerodynamic velocity V_A and the speed of sound c:

$$M = \frac{V_A}{c}, \quad c = \sqrt{\kappa R T(h)},$$

with c depending on the adiabatic index for air, κ; the gas constant for ideal gases, R; and the temperature of air, $T(h)$, at the altitude h.

From the formulas of the International Standard Atmosphere ISA (DIN ISO 2533) for the troposphere layer ($h = -2 \ldots 11$ km, relative to sea level), the temperature of air, $T(h)$, and air density, $\rho(h)$, are approximated as follows:

$$T(h) = T_s \cdot \left[1 - \frac{n-1}{n} \frac{g}{R \cdot T_s} \cdot H_G\right],$$

$$\rho(h) = \rho_s \cdot \left[1 - \frac{n-1}{n} \frac{g}{R \cdot T_s} \cdot H_G\right]^{\frac{1}{n-1}},$$

$$H_G = \frac{r_E \cdot h}{r_E + h}.$$

In the formulas above, n is the polytropic exponent, g the gravitational constant, r_E the Earth's radius, and T_s and ρ_s are the reference temperature and density of air, respectively.

Note that, in all following state constraints, the altitude h is defined relative to the altitude $h_0 = 5000$ m.

Propulsion Forces. For modeling thrust forces, a two-engine setup is considered:

$$\mathbf{F}_P = M_{KB}(\mathbf{F}_{Pnet,left} + \mathbf{F}_{Pnet,right}),$$

$$\mathbf{F}_{Pnet,i} = \begin{bmatrix} f_V(\delta_T, \alpha_A, \beta_A, M) \\ f_\chi(\delta_T, \alpha_A, \beta_A, M) \\ f_\gamma(\delta_T, \alpha_A, \beta_A, M) \end{bmatrix}, \quad i = \text{left, right},$$

where δ_T is the thrust command and the functions $f_V, f_\chi,$ and f_γ are approximated similar to that in formula (7).

Gravitation Force. Obviously,

$$\mathbf{F}_G = M_{KO} \begin{bmatrix} 0 \\ 0 \\ g \end{bmatrix}.$$

Equations of Lateral Motion. In the case of lateral dynamics, it is necessary to ensure that the inclination angle remains equal to zero, and therefore, the altitude is kept constant. This can be achieved by choosing an appropriate value of the angle of attack α_A. The required value of α_A is being determined from the equation

$$\dot{\gamma}_K = \frac{-Z_T(\alpha_A, \text{other variables})}{mV_K} = 0. \tag{9}$$

The equations of lateral motion read:

$$\dot{y} = V_K \sin(\chi_K), \tag{10}$$

$$\dot{V}_K = \frac{X_T}{m}, \tag{11}$$

$$\dot{\chi}_K = \frac{Y_T}{mV_K}. \tag{12}$$

3 Problem Statement and Simulation Results

Problem 1 First, consider the model of lateral motion. The model consists of equations (10)–(12) and the altitude controller (9). Thus, the state vector has three state variables: y, V_K, and χ_K. The kinematic flight-path bank angle, μ_K, and the thrust command, δ_T, are considered as controls, whereas W_x and W_y are regarded as disturbances; see formula (8). The following constraints are imposed on the controls and disturbances:

$$|\mu_K| \le 5 \deg, \quad \delta_T \in [0.3, 1], \quad |W_x| \le 8 \, \text{m/s}, \quad |W_y| \le 8 \, \text{m/s}. \tag{13}$$

Therefore, instantaneous changes in the angle of attack and thrust command are permitted. The following state constraints are imposed:

$$|y| \le 150 \, \text{m}, \quad V_K \in [100, 170] \, \text{m/s}, \quad |\chi_K| \le 20 \deg. \tag{14}$$

Figure 1 shows the leadership kernel in Problem 1. Additionally, an optimal trajectory emanating from an initial point lying in the leadership kernel near to its boundary is shown. The trajectory is computed when the first player (control) uses its optimal feedback strategy, whereas the second player (disturbance) utilizes its optimal feedback counterstrategy. The simulated flight time is equal to 30 min. The trajectory goes to its attraction point and stays there.

In the numerical construction, the box $[-160, 160] \times [90, 180] \times [-25, 25]$ of the space (y, V_K, χ_K) was divided in $320 \times 100 \times 50$ grid cells, and the grid method (35) described in Section 4.1 is applied. The sequence of time steps, $\{\delta_\ell\}$, was chosen either as $\delta_\ell = 0.005/\ln(3 + \ell^2)$ or $\delta_\ell \equiv 0.005$, and the computations were performed until $|\overline{\Psi}^h_{\ell+1} - \overline{\Psi}^h_{\ell}| \le \epsilon = 10^{-6}$ for all grid nodes. About 30000 time steps were carried out. In both cases of choosing $\{\delta_\ell\}$, the results were identical. Moreover, the discriminating kernel was computed with the grid schema (36). It is only a bit larger than the leadership kernel, which shows that the additional information about current values of the wind disturbances does not provide too much improvement of the control quality.

It should be noted that the constraints imposed on the disturbances (see (13)) are near to critical. That is, the leadership kernel is empty if the bound on the disturbances is larger than 8 m/s.

Problem 2 Consider now both lateral and vertical motions. The model consists of equations (2)–(6). Thus, the state vector has five state variables: y, z, V_K, χ_K, and

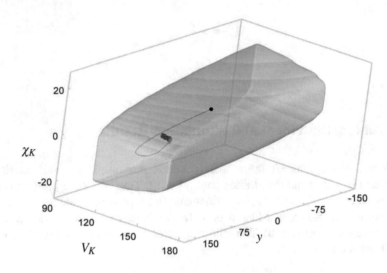

Fig. 1 The leadership kernel in Problem 1. The black dot shows the start point of the trajectory.

γ_K. The kinematic flight-path bank angle, μ_K; the kinematic angle of attack, α_K; and the thrust command, δ_T, are considered as controls, whereas W_x, W_y, and W_z are regarded as disturbances; see formula (8). The following constraints are imposed on the controls and disturbances:

$$|\mu_K| \leq 5\,\text{deg}, \quad |\alpha_k| \leq 20\,\text{deg}, \quad \delta_T \in [0.3, 1],$$
$$|W_x| \leq 2\,\text{m/s}, \quad |W_y| \leq 2\,\text{m/s}, \quad |W_z| \leq 2\,\text{m/s}. \tag{15}$$

The following state constraints are imposed:

$$|y| \leq 200\,\text{m}, \quad |z| \leq 200\,\text{m}, \quad V_K \in [100, 170]\,\text{m/s}, \quad |\chi_K| \leq 20\,\text{deg}, \quad |\gamma_K| \leq 20\,\text{deg}. \tag{16}$$

Figure 2 shows the cross section of the leadership kernel if the last two state variables are fixed: $\chi_K = \gamma_k = 0$.

In the numerical construction, the box $[-220, 220] \times [-220, 220] \times [90, 180] \times [-25, 25] \times [-25, 25]$ of the space $(y, z, V_K, \chi_K, \gamma_k)$ was divided in $100 \times 100 \times 50 \times 12 \times 12$ grid cells, and the grid method (35) described in Section 4.1 is applied. The sequence of time steps, $\{\delta_\ell\}$, was chosen as in Problem 1.

The computation was performed on the SuperMUC system at the Leibnitz Supercomputing Centre of the Bavarian Academy of Sciences and Humanities. The problem was parallelized between 200 computers with 16 cores per computer and 2 threads per core. About 30000 time steps were done. The runtime was about 10 min for Problem 1 and about 1 hour for Problem 2.

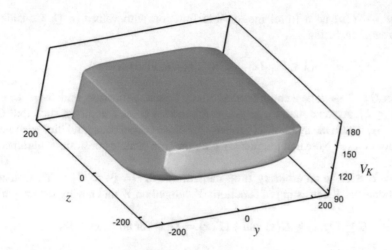

Fig. 2 Cross section ($\chi_K = 0$ and $\gamma_K = 0$) of the five-dimensional leadership kernel in Problem 2.

4 Differential Games and Discriminating/Leadership Kernels

Let us shortly outline a method for computing discriminating and leadership kernels. The description will be given in the framework of general nonlinear differential games.

Consider a conflict control system with the autonomous dynamics

$$\dot{x} = f(x, u, v), \; x \in R^n, \; u \in P \subset R^p, \; v \in Q \subset R^q. \tag{17}$$

Here, x is the state vector; u and v are control parameters of the first and second players, respectively; and P and Q are compacts of the corresponding dimensions. In the following, it is assumed that all functions of x are defined on the whole R^n and have global regularity properties. Thus, the right-hand side f is supposed to be globally bounded, continuous in (x, u, v), and Lipschitzian in x.

The following relation is called saddle point condition:

$$H^{mima}(x, \ell) := \min_{u \in P} \max_{v \in Q} \langle \ell, f(x, u, v) \rangle =$$
$$\max_{v \in Q} \min_{u \in P} \langle \ell, f(x, u, v) \rangle =: H^{mami}(x, \ell), \; \ell \in R^n, \; x \in R^n. \tag{18}$$

Note that this condition does not hold for Problems 1 and 2.

Bearing in mind the conflict control system (17), consider for any $v \in Q$ the differential inclusion

$$\dot{x} \in F_v(x) = \overline{co}\{f : f(x, u, v), u \in P\}. \tag{19}$$

Let $u \rightarrow v(u)$ be a Borel measurable function with values in Q. Consider the differential inclusion

$$\dot{x} \in F_{v(\cdot)}(x) = \overline{\mathrm{co}}\{f : f(x, u, v(u)), u \in P\}. \tag{20}$$

Let $G : R^n \rightarrow R$ be a continuous globally bounded function, and $G_\lambda = \{x \in R^n : G(x) \leq \lambda\}$. Assume that, for any $\lambda \in R^n$, the set G_λ is a compact such that $G_\lambda = \mathrm{cl}\,\mathrm{int}\,G_\lambda$, where the symbol "cl" means the closure operation, and "int" denotes the interior of a set. Note that the sets G_λ play the role of state constraints, parametrized by λ.

Let $T \in R$ be an arbitrary time instant, and $W = (-\infty, T] \times R^n$. Consider a functional set **V** consisting of functions V defined on W and having the properties

$$C \geq V(t, x) \geq G(x) \text{ and } V(T, x) = G(x) \text{ for all } (t, x) \in W, \tag{21}$$

where C is a large constant.

The following definitions describe stability properties of functions V.

Definition 1 (maximin u-stability property [14] for functions) A function $V \in \mathbf{V}$ is said to be u-stable in maximin sense on the interval $(-\infty, T]$ if for any initial position $(t_*, x_*) \in W$, for any time instant $t^* \in [t_*, T]$, and for any $v \in Q$, there exists a solution $x(\cdot)$ to the differential inclusion (19) with the initial state $x(t_*) = x_*$ such that $V(t^*, x(t^*)) \leq V(t_*, x_*)$.

Definition 2 (maximin u-stability property [14] for sets) A set $D \subset W$ is said to be u-stable in maximin sense on the interval $(-\infty, T]$ if for any initial position $(t_*, x_*) \in D$, for any time instant $t^* \in [t_*, T]$, and for any $v \in Q$, there exists a solution $x(\cdot)$ to the differential inclusion (19) with the initial state $x(t_*) = x_*$ such that $(t^*, x(t^*)) \in D$.

Definition 3 (minimax u-stability property [14] for functions) A function $V \in \mathbf{V}$ is said to be u-stable in minimax sense on the interval $(-\infty, T]$ if for any initial position $(t_*, x_*) \in W$, for any time instant $t^* \in [t_*, T]$, and for any Borel measurable function $u \rightarrow v(u)$ with values in Q, there exists a solution $x(\cdot)$ to the differential inclusion (20) with the initial state $x(t_*) = x_*$ such that $V(t^*, x(t^*)) \leq V(t_*, x_*)$.

Definition 4 (minimax u-stability property [14] for sets) A set $D \subset W$ is said to be u-stable in minimax sense on the interval $(-\infty, T]$ if for any initial position $(t_*, x_*) \in D$, for any time instant $t^* \in [t_*, T]$, and for any Borel measurable function $u \rightarrow v(u)$ with values in Q, there exists a solution $x(\cdot)$ to the differential inclusion (20) with the initial state $x(t_*) = x_*$ such that $(t^*, x(t^*)) \in D$.

Below, properties of u-stable (both in maximin and minimax sense) functions are given. Note that the superscripts pointing out to minimax or maximin u-stability will be omitted:

1. The lower semicontinuous regularization of an u-stable function is u-stable function.
2. The infimum of any family of u-stable functions is u-stable function.
3. There exists a unique minimal, lower semicontinuous, u-stable function if the set **V** contains at least one u-stable function.
4. If V is a minimal u-stable function, then $V(t_1, x) \geq V(t_2, x)$ whenever $t_1 \leq t_2 \leq T$, and, obviously, $V(T, x) = G(x)$ for all $x \in R^n$.
5. Remembering that there are two types of u-stability, the following holds for minimal u-stable functions, $V^{mima} \geq V^{mami}$, where the upper indexes point out to minimax and maximin u-stability, respectively.
6. If the saddle point condition (18) holds, then $V^{mima} = V^{mami}$.
7. There exists $t_0(C)$ (see inequality (21)) such that the restriction of the minimal u-stable function to the interval $[t_0(C), T]$ is the value function (minimax or maximin depending on the kind of u-stability) of the differential game (17) with the payoff functional $\gamma(x(\cdot)) = \max_{\tau \in [t, T]} G(x(\tau))$.

Proof (Property 1) Consider the case of minimax u-stability. Let t_*, t^*, x_*, and $v(\cdot)$ be chosen, and $\delta := t^* - t_*$. For each $(\tau, y) \in W$, there exists a solution $X(\cdot; \tau, y, v(\cdot))$ of the differential inclusion (20), with the initial state (τ, y) and the function $v(\cdot)$, such that $V(\tau + \delta, X(\tau + \delta; \tau, y, v(\cdot))) \leq V(\tau, y)$, and therefore

$$\inf_{X(\cdot)} V(\tau + \delta, X(\tau + \delta; \tau, y, v(\cdot))) \leq V(\tau, y),$$

where "inf" is taken over all solutions of (20) with the initial state (τ, y) and the function $v(\cdot)$.

The last inequality implies the relation

$$\inf_{\substack{(\tau, y) \in \\ B_\epsilon(t_*, x_*)}} \inf_{X(\cdot)} V(\tau + \delta, X(\tau + \delta; \tau, y, v(\cdot))) \leq \inf_{\substack{(\tau, y) \in \\ B_\epsilon(t_*, x_*)}} V(\tau, y),$$

where $B_\epsilon(t_*, x_*) \subset W$ is the ball of radius ϵ with the center at (t_*, x_*).

Let $(\tau^\epsilon, y^\epsilon) \in B_\epsilon(t_*, x_*)$ and $X^\epsilon(\cdot)$ are chosen such that

$$V(\tau^\epsilon + \delta, X^\epsilon(\tau^\epsilon + \delta; \tau^\epsilon, y^\epsilon, v(\cdot))) - \epsilon \leq \inf_{\substack{(\tau, y) \in \\ B_\epsilon(t_*, x_*)}} \inf_{X(\cdot)} V(\tau + \delta, X(\tau + \delta; \tau, y, v(\cdot))),$$

which implies, up to a subsequence, the inequality

$$\lim_{\epsilon \to 0} V(\tau^\epsilon + \delta, X^\epsilon(\tau^\epsilon + \delta; \tau^\epsilon, y^\epsilon, v(\cdot))) \leq \lim_{\epsilon \to 0} \inf_{\substack{(\tau, y) \in \\ B_\epsilon(t_*, x_*)}} V(\tau, y) = \overline{V}(t_*, x_*), \quad (22)$$

where $\overline{V}$ is the lower semicontinuous regularization of V.

Using properties of the differential inclusion (20), it is easy to prove, cf. [7], that there is a solution $X(\cdot; t_*, x_*, v(\cdot))$ of (20) such that, up to a subsequence,

$$\lim_{\epsilon \to 0} X^\epsilon(\tau^\epsilon + \delta; \tau^\epsilon, y^\epsilon, v(\cdot)) = X(t^*; t_*, x_*, v(\cdot)) := x^*.$$

Therefore, $\tau^\epsilon + \delta \to t^*$ and $X^\epsilon(\ldots) \to x^*$, which implies the inequality

$$\lim_{\epsilon \to 0} \inf_{\substack{(\tau, y) \in \\ B_\epsilon(t^*, x^*)}} V(\tau, y) \leq \lim_{\epsilon \to 0} V(\tau^\epsilon + \delta, X^\epsilon(\tau^\epsilon + \delta; \tau^\epsilon, y^\epsilon, v(\cdot))). \tag{23}$$

Inequalities (22) and (23) imply the estimate

$$\overline{V}(t^*, x^*) \leq \overline{V}(t_*, x_*),$$

which proves Property 1.   ∎

Property 2 immediately follows from the definition of u-stability.

Property 3 is the consequence of the previous two properties.

Proof (Property 4) Let $V \in \mathbf{V}$ be the minimal u-stable lower semicontinuous function. Consider the function

$$V^{\mathrm{inf}}(t, x) = \inf_{\tau \leq t} V(\tau, x)$$

and prove its u-stability. To do that, choose t_*, t^*, x_*, and $v(\cdot)$. Let τ^ϵ be such that

$$V(\tau^\epsilon, x_*) \leq V^{\mathrm{inf}}(t_*, x_*) + \epsilon. \tag{24}$$

Due to u-stability of V, there exists a solution $X(\cdot; \tau^\epsilon, x_*, v(\cdot))$ of (20) such that

$$V(\tau^\epsilon + \delta, X(\tau^\epsilon + \delta; \tau^\epsilon, x_*, v(\cdot))) \leq V(\tau^\epsilon, x_*). \tag{25}$$

Taking into account that the inclusion (20) is autonomous yields the equality

$$X(\tau^\epsilon + \delta; \tau^\epsilon, x_*, v(\cdot)) = X(t_* + \delta; t_*, x_*, v(\cdot)).$$

Therefore,

$$\inf_{\tau \leq t_* + \delta} V(\tau, X(t_* + \delta; t_*, x_*, v(\cdot))) \leq V(\tau^\epsilon + \delta, X(\tau^\epsilon + \delta; \tau^\epsilon, x_*, v(\cdot))). \tag{26}$$

Inequalities (24), (25), and (26) yield the relation

$$V^{\mathrm{inf}}(t^*, X(t^*; t_*, x_*, v(\cdot))) \leq V^{\mathrm{inf}}(t_*, x_*) + \epsilon.$$

Since ϵ is arbitrary, this proves u-stability of V^{inf}.

Moreover, it is easy to prove that the lower semicontinuous regularization of V^{inf} inherits the monotonicity property of V^{inf}. On the other hand, the regularized function must coincide with V, which proves Property 4.   ∎

Properties 5 and 6 are true due to the following climes: the minimax u-stability property implies the maximin one, and they are equivalent if the saddle point condition holds.

Proof (Property 7) The choice of $t_0(C)$ is a technical thing ensuring that the value function does not reach the value C if considering on the interval $[t_0(C), T]$. Obviously, $t_0(C) \to -\infty$ as $C \to \infty$. Denote $t_0 = t_0(C)$ for brevity and prove the following assertion:

Let $V \in \mathbf{V}$ be the minimal u-stable lower semicontinuous function and $\hat{V}$ the minimal lower semicontinuous u-stable on the interval $[t_0, T]$ function such that $\hat{V} \geq G$ and $\hat{V}(T, x) = G(x)$, $x \in R^n$. Then $V = \hat{V}$ on $[t_0, T]$.

Indeed, $\hat{V} \leq V$ on $[t_0, T]$. Therefore, the function, $V^* = V$, $t \in (-\infty, t_0)$ and $V^* = \hat{V}$, $t \in [t_0, T]$, lies in $\mathbf{V}$ and being lower semicontinuous and u-stable on the interval $(-\infty, T]$ due to Property 4. On the other hand, $V^* \leq V$ on $(-\infty, T]$, and therefore $V^* = V$. This proves the assertion.

Further, the restriction of V on the interval $[t_0, T]$ will be denoted by the same symbol.

Consider an arbitrary point (t_*, x_*), $t_* \in [t_0, T]$, $x_* \in R^n$, such that $\lambda_* := V(t_*, x_*) > G(x_*)$. The above assertion shows that the set $V_{\lambda_*} = \{(t, x) : V(t, x) \leq \lambda_*\}$ is the maximal u-stable subset of the state constraint G_{λ_*}. Consider the set $V^c_{\lambda_*} = \mathrm{cl}\,(G_{\lambda_*} \setminus V_{\lambda_*})$. Note that $(t_*, x_*) \in V^c_{\lambda_*}$ and $(t_*, x_*) \in \mathrm{int}\,(G_{\lambda_*})$. Then, according to [14], the v-stability condition holds at (t_*, x_*), and therefore, the following condition holds at (t_*, x_*):

For any function $\varphi \in \mathbb{C}^1$ such that $V - \varphi$ attains a local maximum at (t_*, x_*), the inequality is true:

$$\frac{\partial \varphi}{\partial t}(t_*, x_*) + H\left(x_*, \frac{\partial \varphi}{\partial x}(t_*, x_*)\right) \geq 0, \tag{27}$$

where H is one of the Hamiltonians defined in (18), depending on the kind of u-stability.

On the other hand, u-stability of V implies the following condition:
For any function $\varphi \in \mathbb{C}^1$ such that $V - \varphi$ attains a local minimum at (t_*, x_*), the inequality is true:

$$\frac{\partial \varphi}{\partial t}(t_*, x_*) + H\left(x_*, \frac{\partial \varphi}{\partial x}(t_*, x_*)\right) \leq 0. \tag{28}$$

Thus, according to [4], the function V is the value function of the differential game (17) with the payoff functional appearing in Property 7. ∎

The next proposition gives rise to the definition of discriminating and leadership kernels in terms of a limiting function. The proposition deals with both types of u-stability.

Proposition 1 *Let V be a minimal lower semicontinuous u-stable function. Then there exists the limiting function*

$$\Psi(x) = \lim_{t \to -\infty} V(t, x). \tag{29}$$

This function is lower semicontinuous, u-stable, and satisfies the relation $\Psi(x) \geq G(x)$, $x \in R^n$.

Proof (Proposition 1) According to Property 4, the function V is monotone increasing as $t \to -\infty$. Moreover, V is bounded, and therefore, Ψ is correctly defined. In addition, due to Property 4, the representation $\Psi(x) = \sup_{t \in (-\infty, T]} V(t, x)$, $x \in R^n$ holds, and, therefore, Ψ is lower semicontinuous. Let t^*, t_*, x_*, and a function $v(\cdot)$ are chosen. Choose t^ϵ such that

$$V(t^\epsilon, x_*) \leq \lim_{t \to -\infty} V(t, x_*) + \epsilon \tag{30}$$

Obviously, $t^\epsilon \to -\infty$ as $\epsilon \to 0$.

The property of u-stability implies

$$V(t^\epsilon + \delta, X(t^\epsilon + \delta, t^\epsilon, x_*, v(\cdot)) \leq V(t^\epsilon, x_*),$$

where $X(t^\epsilon + \delta, t^\epsilon, x_*, v(\cdot))$ is a solution of (20). Since the differential inclusion (20) is autonomous, we have

$$X(t^\epsilon + \delta, t^\epsilon, x_*, v(\cdot)) = X(t^*, t_*, x_*, v(\cdot)).$$

Therefore,

$$V(t^\epsilon + \delta, X(t^*, t_*, x_*, v(\cdot)) \leq V(t^\epsilon, x_*). \tag{31}$$

Using (30) and (31), imply the relation

$$V(t^\epsilon + \delta, X(t^*, t_*, x_*, v(\cdot)) \leq \lim_{t \to -\infty} V(t, x_*) + \epsilon = \Psi(x_*) + \epsilon$$

Letting ϵ tend to zero yields the inequality

$$\Psi(X(t^*, t_*, x_*, v(\cdot))) = \lim_{\epsilon \to 0} V(t^\epsilon + \delta, X(t^*, t_*, x_*, v(\cdot)) \leq \Psi(x_*).$$

This proves Proposition 1.   ∎

Recalling that there are two variants of the function Ψ, Ψ^{mima} corresponds to the case of minimax u-stability, and Ψ^{mami} is related to the case of maximin u-stability, the following definition is relevant:

Definition 5 (leadership and discriminating kernels) The set $\Psi_\lambda^{mima} = \{x \in R^n : \Psi^{mima}(x) \leq \lambda\}$ is called leadership kernel of the state constraint $G_\lambda =$

$\{x \in R^n : G(x) \le \lambda\}$ and denoted by $lead(G_\lambda)$. Analogously, the set $\Psi_\lambda^{mami} = \{x \in R^n : \Psi^{mami}(x) \le \lambda\}$ is called discriminating kernel of the state constraint G_λ and denoted by $discr(G_\lambda)$. Therefore, we have a family of leadership and discriminating kernels corresponding to the family of state constraints. Obviously, $lead(G_\lambda) \subset discr(G_\lambda) \subset G_\lambda$, and the sets $lead(G_\lambda)$ and $discr(G_\lambda)$ may be empty for small values of λ.

Remark 1 Note that the above definition is consistent with results of paper [9] pointing out to the fact that the epigraph of a minimal lower semicontinuous u-stable function is the discriminating kernel of some suitably extended conflict control system.

The following propositions show the appropriateness of the above-given definitions. Actually, this proposition is closely related to Krasovskii's and Subbotin's alternative theorems for lower and upper differential games (see [14]).

Proposition 2 *Assume that $G_\lambda \ne \emptyset$. Let $x_* \in lead(G_\lambda)$ and $\bar{t} > 0$ be an arbitrary time instant. Then there exists a pure feedback strategy $U(t, x)$ of the first player such that all trajectories generated by U and started at $t = 0$ from x_* remain in the set $lead(G_\lambda)$ for all $t \in [0, \bar{t}]$. If $x^* \notin lead(G_\lambda)$, then there exists a feedback counter strategy $V^c(t, x, u)$ and a time instant t_f such that all trajectories generated by V^c and started at $t = 0$ from x^* violate the state constraint G_λ for $t > t_f$.*

Proposition 3 *The same proposition as proposition 2 is true if considering $discr(G_\lambda)$, feedback counterstrategy $U^c(t, x, v)$, and feedback pure strategy $V(t, x)$.*

Proposition 4 *If the saddle point condition (18) holds, then $lead(G_\lambda) = discr(G_\lambda)$, and the same proposition as proposition 2 is true if considering feedback pure strategies $U(t, x)$ and $V(t, x)$.*

The proof of these propositions is based on results of the monograph [14]. Consider, for example, Proposition 2. Note that the set $[0, \bar{t}] \times lead(G_\lambda)$ is u-stable in the minimax sense, and therefore, the first part of Proposition 2 is exactly a result of [14] (Chapter 10).

If now $x^* \notin lead(G_\lambda)$, then there exists a time instant t^* such that $(t^*, x^*) \notin \{(t, x) : W(t, x) \le \lambda, \ t \in [t^*, T]\} := D_\lambda$, where $W \in \mathbf{V}$ is the lower semicontinuous minimal u-stable in the minimax sense function. Obviously, the set D_λ is the maximal u-stable in the minimax sense subset of $[t^*, T] \times G_\lambda$. Therefore, according to [14], there exists a feedback counterstrategy V^c such that all trajectories generated by V^c and started from x^* at $t = t^*$ remain outside of a closed neighborhood of D_λ. Since $D_\lambda|_{t=T} = G_\lambda$ (see Property 4), all trajectories violate the state constraint G_λ at $t = T$. Since the system (17) is autonomous, we can set $t^* = 0$ and $t_f = T - t^*$, which proves the second part of Proposition 2. ∎

4.1 Grid Method for Computing Leadership and Discriminating Kernels

The numerical method utilizes the idea of computing leadership and discriminating kernels on the base of relation (29), where V is the value function of the differential game (17) with the payoff functional described in Property 7.

A grid approximation of such a value function can be computed using an algorithm described in [4] and [5]; see also [11] and [12].

Let $\delta > 0$ be a time step, and $h := (h_1, \ldots, h_n)$ space discretization steps. Set $|h| := \max\{h_1, \ldots, h_n\}$. Consider the following operators defined on grid functions:

$$\Pi^*(\phi; \delta, h)(x) = \phi(x) + \delta \min_{u \in P} \max_{v \in Q} \sum_{i=1}^{n} (p_i^{\text{right}} f_i^+ + p_i^{\text{left}} f_i^-), \tag{32}$$

$$\Pi_*(\phi; \delta, h)(x) = \phi(x) + \delta \max_{v \in Q} \min_{u \in P} \sum_{i=1}^{n} (p_i^{\text{right}} f_i^+ + p_i^{\text{left}} f_i^-), \tag{33}$$

where f_i are the components of $f(x, u, v)$ and

$$a^+ = \max\{a, 0\}, \quad a^- = \min\{a, 0\},$$

$$p_i^{\text{right}} = [\phi(x_1, \ldots, x_i + h_i, \ldots, x_n) - \phi(x_1, \ldots, x_i, \ldots, x_n)]/h_i, \tag{34}$$

$$p_i^{\text{left}} = [\phi(x_1, \ldots, x_i, \ldots, x_n) - \phi(x_1, \ldots, x_i - h_i, \ldots, x_n)]/h_i.$$

Let $\{\delta_\ell\}$ be a sequence of positive reals such that $\delta_\ell \to 0$ and $\sum_{\ell=0}^{\infty} \delta_\ell = \infty$. Consider the following grid schemes:

$$\overline{\Psi}_{\ell+1}^h = \max\left\{\Pi^*(\overline{\Psi}_\ell^h; \delta_\ell, h), G^h\right\}, \quad \overline{\Psi}_0^h = G^h, \quad \ell = 0, 1, \ldots, \tag{35}$$

$$\underline{\Psi}_{\ell+1}^h = \max\left\{\Pi_*(\underline{\Psi}_\ell^h; \delta_\ell, h), G^h\right\}, \quad \underline{\Psi}_0^h = G^h, \quad \ell = 0, 1, \ldots, \tag{36}$$

where G^h is the restriction of G to the grid.

It can be proven that $\overline{\Psi}_\ell^h$ and $\underline{\Psi}_\ell^h$ monotonically converge point-wise to grid functions $\overline{\Psi}^h$ and $\underline{\Psi}^h$, respectively. These functions define approximations of leadership and discriminating kernels, respectively, given by Definition 5. Note that the relation $\delta_\ell/|h| \leq (\sqrt{n}M)^{-1}$ should hold for all ℓ, where M denotes the bound of the right-hand side of (17). Moreover, the relation $\delta_\ell/|h_\ell| \leq (\sqrt{n}M)^{-1}$ should hold if the grid fineness $|h_\ell|$ goes to zero too.

Remark 2 If the saddle point condition (18) holds, then

$$\lim_{\ell \to \infty} \overline{\Psi}_{\ell}^{h\ell} = \lim_{\ell \to \infty} \underline{\Psi}_{\ell}^{h\ell} = \Psi.$$

The proof follows from the fact that the both operators (32) and (33) satisfy the same consistency condition (see [4]) corresponding to the Hamiltonian

$$H(x, p) := \max_{v \in Q} \min_{u \in P} \langle p, f(x, u, v) \rangle = \min_{u \in P} \max_{v \in Q} \langle p, f(x, u, v) \rangle.$$

Remark 3 Numerical simulations show that the grid functions $\overline{\Psi}_{\ell}^{h}$ and $\underline{\Psi}_{\ell}^{h}$ practically coincide for all ℓ, if the saddle point condition (18) is true.

4.2 Control Design

This section outlines one of possible methods of control design. Consider the case of Proposition 2 where the first player uses pure feedback strategies, whereas the second player utilizes counter feedback controls. Consider the grid scheme (35) assuming that ℓ is sufficiently large so that the desired approximation is reached, i.e., $\left| \overline{\Psi}_{\ell+1}^{h} - \overline{\Psi}_{\ell}^{h} \right|_{L_\infty} \le \epsilon$.

Let x be a current state of the game. The control u and the disturbance $v(u)$ can be found as solutions of the following problem:

$$u, v \to \min_{u \in P} \max_{v \in Q} \mathcal{L}^{h} \left[\overline{\Psi}_{\ell}^{h} \right] (x + \delta f(x, u, v)).$$

Here, $\mathcal{L}^{h}$ is an interpolation operator (see, e.g., [5]) defined on the corresponding grid functions, and δ is a parameter which should be larger than the time step size of the control scheme to provide some stabilization. Note that the function $\overline{\Psi}_{\ell}^{h}$ can be stored in a sparse form (see, e.g., [25] and [22]), which may reduce the storage space to several megabytes. The disadvantage of this method is its slower performance.

5 Conclusion

The current investigation shows that methods of viability theory can by applied to realistic models of aircraft to evaluate its potential control ability in the presence of wind disturbances. The new feature of this approach is the use of leadership and discriminating kernels arising from differential games in the case where the saddle point condition does not hold. In this case, it is reasonable to restrict the control to use pure feedback strategies against feedback counterstrategies of the disturbance, which demands computing leadership kernels. Strategies (counterstrategies) of the players can be computed and stored in a sparse form, which allows us to integrate them into flight simulators and realistic control systems.

Acknowledgements This work was supported by the DFG grant TU427/2-1 and HO4190/8-1. Computer resources for this project have been provided by the Gauss Centre for Supercomputing/Leibniz Supercomputing Centre under grant: pr74lu.

References

1. Aubin, J.-P.: Viability Theory. Birkhäuser, Boston (1991)
2. Aubin, J.-P., Bayen, A. M., Saint-Pierre, P.: Viability Theory: New Directions. Springer-Verlag, Berlin/Heidelberg (2011)
3. Bayen, A. M., Mitchell, I. M., Osihi, M. K., Tomlin, C. J.: Aircraft autolander safety analysis through optimal control-based reach set computation. Journal of Guidance, Control, and Dynamics **30** (1), 68–77 (2007)
4. Botkin, N. D., Hoffmann, K.-H., Mayer, N., Turova, V. L.: Approximation schemes for solving disturbed control problems with non-terminal time and state constraints. Analysis **31**, 355–379 (2011)
5. Botkin, N. D., Hoffmann, K.-H., Turova, V. L.: Stable numerical schemes for solving Hamilton–Jacobi–Bellman–Isaacs equations. SIAM J. Sci. Comput. **33** (2), 992–1007 (2011)
6. Botkin, N. D., Turova, V. L.: Dynamic programming approach to aircraft control in a windshear. In: Křivan, V., Zaccour, G. (eds.) Advances in Dynamic Games: Theory, Applications, and Numerical Methods. Annals of the International Society of Dynamic Games, vol. 13, pp. 53–69. Birkhäuser, Boston (2013)
7. Botkin, N. D., Turova, V. L.: Numerical construction of viable sets for autonomous conflict control systems. Mathematics **2**, 68–82 (2014)
8. Cardaliaguet, P.: A differential game with two players and one target, SIAM J. Control Optim. **34**, 1441–1460 (1996)
9. Cardaliaguet, P., Quincampoix, M., Saint-Pierre, P.: Set valued numerical analysis for optimal control and differential games. In: Bardi, M., Raghavan, T.E.S., Parthasarathy, T. (eds.) Stochastic and Differential Games: Theory and Numerical Methods. Annals of the International Society of Dynamic Games, vol. 4, pp. 177–274. Birkhäuser, Boston (1999)
10. Chen, Y. H., Pandey, S.: Robust control strategy for take-off performance in a windshear. Optim. Contr. Appl. Met. **10** (1), 65–79 (1989)
11. Falcone, M.: Numerical methods for differential games via PDEs. International Game Theory Review **8**(2), 231–272 (2006)
12. Cristiani, E., Falcone, M.: Fully-discrete schemes for value function of pursuit- evasion games with state constraints. In: Bernhard, P., Gaitsgory, V., Pourtallier, O. (eds.) Advances in Dynamic Games and Their Applications, Annals of the International Society of Dynamic Games, vol. 10, pp. 177–206. Birkhäuser, Boston (2009)
13. Fisch, F.: Development of a framework for the solution of high-fidelity trajectory optimization problems and bilevel optimal control problems. Dissertation, Technische Universität München, Institute of Flight System Dynamics (2011)
14. Krasovskii, N. N., Subbotin, A. I.: Game-Theoretical Control Problems. Springer, New York (1988)
15. Leitmann, G., Pandey, S.: Aircraft control under conditions of windshear. In: Leondes, C. T. (ed.) Control and Dynamic Systems, vol. 34, part 1, pp. 1–79. Academic Press, New York (1990)

16. Leitmann, G., Pandey, S.: Aircraft control for flight in an uncertain environment: Takeoff in windshear. J. Optimiz. Theory App. **70** (1), 25–55 (1991)
17. Leitmann, G., Pandey, S., Ryan, E.: Adaptive control of aircraft in windshear. Int. J. Robust Nonlin. **3**, 133–153 (1993)
18. Miele, A., Wang, T., Melvin, W. W.: Optimal take-off trajectories in the presence of windshear. J. Optimiz. Theory App. **49**, 1–45 (1986)
19. Miele, A., Wang, T., Melvin, W. W.: Guidance strategies for near-optimum take-off performance in windshear. J. Optimiz. Theory App. **50** (1), 1–47 (1986)
20. Mitchell, I. M.: A summary of recent progress on efficient parametric approximations of viability and discriminating kernels. EPiC Ser. Comput. Sci. **37**, 23–31 (2015)
21. Patsko, V. S., Botkin, N. D., Kein, V. M., Turova, V. L., Zarkh, M. A.: Control of an aircraft landing in windshear. J. Optimiz. Theory App. **83** (2), 237–267 (1994)
22. Pflüger, D.: Spatially adaptive sparse grids for higher-dimensional problems. Dissertation, Verlag Dr. Hut, München (2010)
23. Seube, N., Moitie, R., Leitmann, G.: Aircraft take-off in windshear: A viability approach. Set-Valued Analysis **8**, 163–180 (2000)
24. Seube, N., Moitie, R., Leitmann, G.: Viability analysis of an aircraft flight domain for take-off in a windshear. Mathematical and Computer Modelling **36** (6), 633–641 (2002)
25. Zenger, C.: Sparse Grids. In: Hackbusch, W. (ed.) Parallel Algorithms for Partial Differential Equations. Notes on Numerical Fluid Mechanics, vol. 31, pp. 241–251. Vieweg, Braunschweig/Wiesbaden (1991)

Modeling Autoregulation of Cerebral Blood Flow Using Viability Approach

Varvara Turova, Nikolai Botkin, Ana Alves-Pinto, Tobias Blumenstein, Esther Rieger-Fackeldey, and Renée Lampe

Abstract A model of autoregulation of cerebral blood flow is under consideration. The flow is described using a blood vessel network, and blood is considered as a micropolar fluid. Sudden changes in partial pressure of oxygen and carbon dioxide in the arterial blood are considered as disturbances, and medicines dilating or restricting blood vessels are referred as control. A viability approach is applied as follows. An appropriate safety domain (ASD) is chosen in the space of state variables, and the discriminating kernel, the largest subset of ASD, where the state vector can be kept, is computed. Feedback controls forcing the state vector to remain in the discriminating kernel are constructed.

1 Introduction

The paper is devoted to the application of differential game theory to a medical problem related to cerebral blood circulation in preterm infants.

The most common congenital disorder in children is cerebral palsy that occurs as a result of brain bleeding mostly arising in preterm infants with low gestational age (less than 32 weeks) and low birth weight (less than 1500 g) [1]. The cause of such a bleeding is very often the so-called germinal matrix (the site of origin for neuronal and glial cells), a cluster of unstable blood vessels, which is being reduced

V. Turova (✉) • N. Botkin
Department of Mathematics, Technische Universität München, Boltzmannstr. 3, 85748 Garching, near Münich, Germany
e-mail: v.turova@tum.de; botkin@ma.tum.de

A. Alves-Pinto • T. Blumenstein • R. Lampe
Orthopedic Department of the Clinic 'rechts der Isar' of the Technical University of Munich, Germany
e-mail: ana.alves-pinto@tum.de; t.blumenstein@tum.de; renee.lampe@tum.de

E. Rieger-Fackeldey
Frauenklinik und Poliklinik of the Clinic 'rechts der Isar' of the Technical University of Munich, Ismaningerstr. 22, 81675 Munich, Germany
e-mail: Esther.Riegerfackeldey@lrz.tu-muenchen.de

© Springer International Publishing AG 2017
J. Apaloo, B. Viscolani (eds.), *Advances in Dynamic and Mean Field Games*,
Annals of the International Society of Dynamic Games 15,
https://doi.org/10.1007/978-3-319-70619-1_16

in term newborns. Preterm infants have impaired cerebral autoregulation, which results in pressure-passive cerebral blood flow (*CBF*) [25, 32] so that variations in blood pressure and blood gases can lead to disturbances of *CBF*. Since unstable blood vessels of the germinal matrix are very vulnerable to such disturbances, they can be easily damaged.

Cerebrovascular system of the mature brain is the object of mathematical study since the recent 20 years. The simplest models are based on Kirchhoff's law and Hagen-Poiseuille's flow of Newtonian fluids. Such models do not require a great amount of data. Information on the approximate number of vessels, their size, length, and reactivity is already available for every vascular level (compartment); see, e.g., [30]. Using these models and experimental data (see, e.g., [4] for results of anatomic analysis of blood vessels in the germinal matrix), a preliminary estimation of pressures in vessels of the germinal matrix and the dependency of these pressures on the CO_2 partial pressure can be obtained. To analyze the influence of different physiological and pathological stimuli such as alterations in cerebral perfusion pressure, changes in circulating oxygen, glucose, carbon dioxide, etc., the model of cerebrovascular system should be coupled to a model of cerebral autoregulation mechanism. Several mathematical models of cerebral autoregulation in the mature brain are known. A mathematical model by Ursino and Lodi [31] simulates interactions between intracranial pressure, cerebral blood volume, and a feedback autoregulator. Better understanding of which factors increase the risk of cerebral ischemia is provided due to electric circuit analog models (lumped parameter models) developed in [2, 29]. Accounting for the peripheral resistance of the cerebrovascular tree is often implemented by introducing blocks of porous media into the model (see [17, 21]). Also phenomenological models (e.g., from [7]) give useful information on brain blood flow.

However, it should be noticed that, in spite of significant progress in understanding of cerebral regulation mechanisms, there is a lack of knowledge on normal and disturbed cerebral circulation in preterm newborns. The role of blood pressure, hypotension, and impaired autoregulation is far from resolved. Therefore, the development of adequate mathematical models for cerebral blood circulation taking into account peculiarities of immature brain is a challenging task. Furthermore, in the majority of existing mathematical models of cerebral circulation, blood is treated as a Newtonian fluid. More appropriate would be to model blood as a micropolar fluid that is characterized by the presence of rotational degrees of freedom in their particles. In this paper, a mathematical model of cerebral autoregulation accounting for realistic data collected from preterm infants is presented. The model is treated with a viability approach [3] in a differential game framework. As state variables, the compliance (ability of vessels to distend with increasing pressure), the partial CO_2, and O_2 pressures in the arterial blood are considered. The control is interpreted as the intake of a drag affecting the vascular volume. The disturbances being unpredictable change in partial CO_2 and O_2 pressures. With this approach, the discriminating kernel (see [3, 8, 14, 15], and [11]) can be computed. From the point view of differential games (see [22, 23]), such a kernel is the largest subset of the state constraint in which a pure feedback control strategy (because the saddle-point condition (18) holds) can keep trajectories of the system arbitrary long regardless of any "admissible" disturbance input.

In recent time, many methods for approximate computing viability and discriminating kernels for control and conflict control problems have been developed. A detailed outline of abstract algorithms and their numerical implementations can be found in [28].

We propose another, simply implementable and well-scalable, grid algorithm which can be interpreted as approximate computing the time limit of viscosity solutions of an appropriate Hamilton-Jacobi equation. The methods of [9] and [10] for computing value functions of differential games are adopted for these purposes (see [5, 6, 16, 19, 27], and [20] for other grid algorithms for solving differential games).

2 Cerebral Vascular Model

In this section, a hierarchical cerebrovascular model proposed in [30] is outlined. The derivation of expressions defining the CBF is repeated here for completeness.

Let $M = 19$ be the number of arterial and venous levels (see Figure 1). Denote by p_a and p_v the mean arterial and venous pressures, respectively. Let R_i, m_i, r_i^0, c_i, and l_i be the resistance, the number, the reference radius, the CO_2 reactivity, and the length of vessels of the ith level, respectively. The radius, r_i, of vessels of the ith level depends on the partial CO_2 pressure, pCO_2, according to the formula

$$r_i = r_i(pCO_2) = r_i^0 \cdot \lambda \cdot (1 + c_i \cdot pCO_2), \tag{1}$$

where λ is a modifier related to the vascular volume change. In contrast to [30], the unshifted value of pCO_2 is used in formula (1). All results remain however true if pCO_2 is replaced by its shifted value, $pCO_2 - pCO_2^*$, where pCO_2^* is a reference level. In this case, simulation results are very similar to that presented in Section 6.

By analogy with electrical circuits (see also [24]), taking into account that all vessels in each compartment are connected in parallel, the following formula is true:

$$CBF = (p_a - p_v) \left(\sum_{i=1}^{19} \frac{R_i}{m_i} \right)^{-1}, \tag{2}$$

where R_i is the hydraulic resistance of each vessel of the ith compartment and m_i is the number of vessels there.

According to the Poiseuille's law, the resistance of vessels is given by the formula $R_i = 8\mu l_i / \pi r_i^4$, where μ is the dynamic viscosity of blood.

Finally, the following formula is true:

$$CBF = (p_a - p_v) \left(\sum_{i=1}^{M} \frac{8\mu l_i}{\pi m_i r_i^4} \right)^{-1}, \tag{3}$$

where r_i is computed by formula (1).

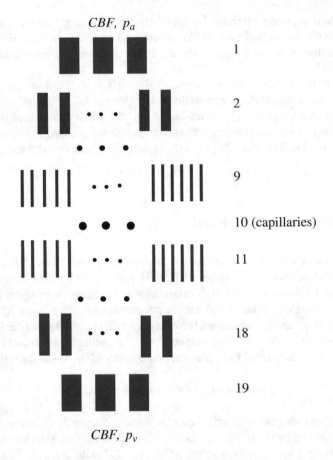

Fig. 1 Cerebrovascular bed considered in [30]. The arterial levels are shown in red; the venous ones are depicted in blue. The blood inflow on the top is equal to the blood outflow on the bottom. The arterial and venous pressures p_a and p_v are applied on the top and bottom, respectively

Remark 1 (Driving pressure) The value $p_a - p_v$ plays the role of driving pressure in the above formulas. Since the venous pressure, p_v, is difficult to measure, it is usually approximated by the intracranial pressure, p_{ic}, which is more easily measurable. Therefore, in the following, p_v will be replaced by p_{ic} which is assumed to be constant.

3 Modeling of Blood as a Micropolar Fluid

The flow of complex fluids, such as blood, in fine vessels is much different from the flow of water. Blood resembles a micropolar fluid due to the presence of rigid, randomly oriented particles suspended in a viscous medium. Such fluids are described using additional rotational degrees of freedom at each point. In practical simulations, the compressibility of such fluids is usually neglected. We adopt the mathematical model of steady-state flow of micropolar fluid through a pipe with circular cross section; see [18]. It should be noted that the last paper gives the complete description of such a flow in terms of modified Bessel functions, which is time-consuming if numerous calls of the corresponding computer code are required. Therefore, we arrive at the idea to derive simplified expressions for the flow velocity profile and hydraulic resistance using power series expansion of solutions. To do that, it is necessary to repeat some calculations presented in [18].

3.1 Micropolar Field Equations for Incompressible Viscous Fluid

We begin with a general steady-state linear model of micropolar fluid; see [18, 26]:

$$-(v + v_r)\Delta\vec{v} + \nabla p = 2v_r \, \text{curl}\,\vec{\omega}$$

$$-(c_a + c_d)\Delta\vec{\omega} - (c_0 + c_d - c_a)\nabla \text{div}\,\vec{\omega} + 4v_r\vec{\omega} = 2v_r \, \text{curl}\,\vec{v} \qquad (4)$$

$$\text{div}\,\vec{v} = 0.$$

Here, $\vec{v}$ is the velocity field, $\vec{\omega}$ is the micro-rotation field, p is the hydrostatical pressure, v is the classical viscosity coefficient, v_r is the vortex viscosity coefficient, and c_a, c_d, and c_0 are the spin gradient viscosity coefficients.

The first and second equations represent the conservation of the momentum and the angular momentum, respectively. The third one (continuity equation) accounts for the incompressibility of the fluid.

Let us rewrite equations (4) in cylindrical polar coordinates (see Figure 2A). Following conventional assumptions when considering flow in a pipe with circular cross section (see [18]), we neglect the radial and angular components of the velocity as well as the radial and vertical components of the micro-rotation field and assume that the vertical component of the velocity and the angular component of micro-rotation depend on the vessel radius only. Therefore,

$$v_r = v_\theta = 0, \quad v_z = u(r), \quad \omega_r = 0, \quad \omega_\theta = \omega(r), \quad \omega_z = 0.$$

Using this, we arrive at the following system of ordinary differential equations (primes denote the derivatives with respect to r):

$$(v + v_r)(u'' + \frac{1}{r}u') + 2v_r\frac{(r\omega)'}{r} = \frac{dp}{dz}, \qquad (5)$$

$$(c_a + c_d)(\omega'' + \frac{1}{r}\omega' - \frac{1}{r^2}\omega) - 2v_r u' - 4v_r\omega = 0. \qquad (6)$$

Fig. 2 A, cylindrical polar coordinates. B, form of the function F

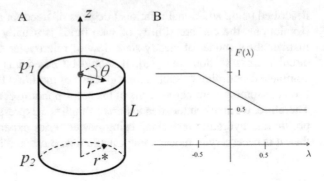

Observe that the continuity equation, the third one of (4), is satisfied identically, and note that the following boundary conditions (cf. [18]) should be added to equations (5) and (6):

$$\omega(0) = 0, \quad u'(0) = 0, \quad u(r^*) = 0, \quad u'(r^*) = -(2+s)\omega, \qquad (7)$$

where r^* is the radius of the pipe and s, $0 \le s < \infty$, is a parameter which can be interpreted as a measure of suspension concentration on the boundary. The value $s = 0$ corresponds to the very dilute suspension, and $s = \infty$ means an extremely high concentration.

The solution to the problem (5)–(7) is given by paper [18] in the form of modified Bessel functions. As it was mentioned above, this form is not suitable for numerical implementation. In the next subsection, an approximate solution based on a power series expansion will be described.

3.2 Computation of Solutions Using Power Series Expansion

First, note that $dp/dz = const$ because the left-hand side of (5) does not depend on z. This implies that

$$dp/dz = (p_1 - p_2)/L := A. \qquad (8)$$

Multiplying equation (5) by r and integrating the result allow us to express ω through u as follows:

$$\omega = -\frac{\nu + \nu_r}{2\nu_r} u' + \frac{Ar}{4\nu_r}.$$

Substituting ω into equation (6) and denoting $w := u'$ yield the equation

$$r^2 w'' + r w' - a r^2 w - w + b r^3 = 0, \tag{9}$$

where $a = \dfrac{4\nu\nu_r}{(\nu + \nu_r)(c_a + c_d)}$, $b = \dfrac{2\nu_r A}{(\nu + \nu_r)(c_a + c_d)}$.

The boundary conditions assume now the form:

$$w(0) = 0, \quad w(r^*) = \frac{A r^*}{2[\nu + (1 - 2/(2 + s))\nu_r]} =: d. \tag{10}$$

Note that the ODE (9) with the boundary conditions (10) describes the velocity profile of the pipe micropolar fluid flow.

Searching a solution to (9)–(10) in the form of a power series in r,

$$w = a_0 + a_1 r + a_2 r^2 + a_3 r^3 + a_4 r^4 + a_5 r^5,$$

and collecting terms in powers of r result in the following system of linear algebraic equations:

$$a_0 = 0,$$
$$3a_2 - a \cdot a_0 = 0,$$
$$8a_3 - a \cdot a_1 = -b,$$
$$15a_4 - a \cdot a_2 = 0,$$
$$24a_5 - a \cdot a_3 = 0,$$
$$a_0 + a_1 r^* + a_2 r^{*2} + a_3 r^{*3} + a_4 r^{*4} + a_5 r^{*5} = d.$$

This system is uniquely solvable, which yields a unique solution $w(r)$. This solution obtained using Maple software for symbolic computations is relatively simple but exceedingly lengthy, and, therefore, we do not show it here.

Then, recalling the notation $w := u'$ and accounting for the boundary conditions for u, we have

$$u(r) = \int_0^r w(\eta)d\eta - \int_0^{r^*} w(\eta)d\eta.$$

Note that the constant A (see (8)) enters the expression for w through the coefficients b and d linearly.

The flow through the pipe is expressed as

$$Q(A) = -2\pi \int_0^{r^*} u(r) r \, dr.$$

It is easy to see that $Q(A)$ is a homogeneous linear function of A, and therefore,

$$Q(A) = Q_1 \cdot A = Q_1 \frac{p_1 - p_2}{L},$$

where Q_1 depends only on r^*.

The hydraulic resistance is given by the formula

$$R^{mp}(r^*, L) = \frac{p_1 - p_2}{Q(A)} = \frac{L}{Q_1}.$$

Thus, coming back to formula (2), we have in the case of micropolar fluid:

$$CBF = (p_a - p_{ic}) \left(\sum_{i=1}^{19} \frac{R^{mp}(r_i, l_i)}{m_i} \right)^{-1}, \tag{11}$$

where, as before, $r_i = r_i^0 \cdot \lambda \cdot (1 + c_i \cdot pCO_2)$.

Finally, we have formula (3) for computing the CBF in the case of Newtonian fluids and expression (11) for obtaining the CBF in the case of micropolar fluids. Note that the last expression is implemented using symbolic computations with Maple software. The resulting C code is rather large but fast. Therefore, in both cases, we have quickly computable functions defining the cerebral blood flow. Denote each of these functions by the same symbol q to have

$$CBF = q(p_a, \lambda, pCO_2).$$

In the following, both functions q will be used in the right-hand side of a control system describing autoregulation dynamics. Numerical experiments related to both Newtonian and micropolar fluids will be conducted.

4   Autoregulation Model

In the previous sections, functions defining the CBF in the case of Newtonian and micropolar fluids are derived. Both functions are denoted by $q(p_a, \lambda, pCO_2)$, which should not lead to a confusion. Using this notation, the following system describing the deviation of the CBF from a prescribed value q_0 is stated:

$$\frac{dC_a}{dt} = \frac{1}{\tau} \left[F\left(\frac{q(p_a, \lambda, pCO_2) - q_0}{q_0} \right) - C_a \right] + u,$$

$$\frac{d}{dt} pCO_2 = k_1(pCO_2 - v_1), \tag{12}$$

$$\frac{d}{dt} pO_2 = k_2(pO_2 - v_2).$$

Here, C_a is the compliance (ability of vessels to distend with increasing pressure), pO_2 the partial oxygen pressure, and q_0 a nominal value of CBF required by tissue metabolism. The form of the function F (see Figure 2B) is adopted from the model of cerebral hemodynamics by Ursino and Lodi [31]. The variable u is a control parameter influencing the rate of the compliance change, which can be interpreted as the effect of some medical exposure. The variables v_1 and v_2 are disturbances affecting the partial CO_2 and O_2 pressures, respectively.

It is worth to mention that the complexity of the dynamics (12) refers to the function q. The computation of this function requires a sequence of calculations described in Sections 2 and 3.

Two additional dependencies are included into the model:

$$\lambda = \alpha \sqrt{V_a}, \quad V_a = C_a(p_a - p_{ic}). \tag{13}$$

Here, V_a is the vascular volume, and p_{ic} is the intracranial pressure, which is supposed to be constant. The second dependence of (13) is indeed the definition of the compliance C_a; see [31]. The first dependence can be derived from the relation $V_a \sim CBF^{1/2}$; see [24] and formula (3) which shows that $CBF \sim \lambda^4$.

For the arterial pressure, the following quasi-linear regression model is used:

$$p_a = \alpha_0 + \alpha_1 \cdot pCO_2 + \alpha_2 \cdot pO_2 + \alpha_{12} \cdot pCO_2 \cdot pO_2. \tag{14}$$

The coefficients α_0, α_1, α_2, and α_{12} are fitted using the experimental values of pCO_2 and pO_2 collected from preterm infants from the Newborns Intensive Station of the Perinatal Medicine Department of the clinic "Rechts der Isar" of the Technical University of Munich.

The following constraints on control, disturbances, and state variables are imposed:

$$|u| \le \gamma, \quad \underline{v}_1 \le v_1 \le \overline{v}_1, \quad \underline{v}_2 \le v_2 \le \overline{v}_2,$$

$$\underline{q} \le q(p_a, \lambda, pCO_2) \le \overline{q}, \tag{15}$$

$$\underline{C_a} \le C_a \le \overline{C}_a, \quad \underline{pCO_2} \le pCO_2 \le \overline{pCO_2}, \quad \underline{pO_2} \le pO_2 \le \overline{pO_2}.$$

Note that the second inequality of (15) ensures the maintenance of CBF in the autoregulation range, which is the main objective of medical treatment.

Numerical values of model parameters are given in Table 1. The coefficients α_0, α_1, α_2, and α_{12} in (14) are computed based on the experimental data for preterm infants with gestational age of 24 weeks.

The problem (12)–(15) is considered as a differential game in which the first player uses the control parameter u to keep the system states within the state constraints (safety domain), whereas the second player strives to violate the state constraints by using the controls v_1 and v_2. Our aim is to construct the so-called discriminating kernel, i.e., the largest set of initial states, lying inside the state

Table 1 Numerical values of model parameters (P, parameter, V, value)

P	V	P	V	P	V	P	V
μ	3.e-3 $\frac{\text{Ns}}{\text{m}^2}$	v_r	2.32e-4 $\frac{\text{Ns}}{\text{m}^2}$	α_1	0.179	$\underline{v}_1$	30 mmHg
$\underline{C}_a$	0.2 $\frac{\text{ml}}{\text{mmHg}}$	v	2.9e-3 $\frac{\text{Ns}}{\text{m}^2}$	α_2	0.071	$\overline{v}_1$	80 mmHg
$\overline{C}_a$	1.8 $\frac{\text{ml}}{\text{mmHg}}$	$c_a + c_d$	1.e-6 Ns	α_{12}	-0.003	$\underline{v}_2$	25 mmHg
$\underline{pCO_2}$	29 mmHg	s	18	k_1	0.1 min^{-1}	$\overline{v}_2$	85 mmHg
$\overline{pCO_2}$	81 mmHg	τ	0.5 min	k_2	0.1 min^{-1}	$\underline{q}$	10 $\frac{\text{ml}}{\text{min}}$
$\underline{pO_2}$	24 mmHg	α	0.1	γ	2 $\frac{\text{ml}}{\text{mmHg min}}$	$\overline{q}$	15 $\frac{\text{ml}}{\text{min}}$
$\overline{pO_2}$	86 mmHg	α_0	27.307	p_{ic}	5 mmHg	q_0	12.4 $\frac{\text{ml}}{\text{min}}$

constraints, from which viable trajectories emanate. Moreover, feedback controls that produce trajectories remaining in the discriminating kernel for all possible admissible disturbances have to be constructed.

In the next section, the theoretical background of our approach to solving the problem will be outlined.

5 Discriminating Kernel

In this section, we recall the definition of discriminating kernels, and outline a numerical scheme for their computation. More details can be found in [3, 8]), [11, 14, 15], and [13].

Consider the following differential game:

$$\dot{x} = f(x, u, v), \tag{16}$$

where $x := (x_1, \ldots, x_n) \in \mathbb{R}^n$ is the state vector and u and v are the control parameters of the first and second player, respectively, restricted as

$$u \in P \subset \mathbb{R}^p, \quad v \in Q \subset \mathbb{R}^q. \tag{17}$$

Here, P and Q are given compacts. The function f is supposed to be Lipschitz continuous in x and continuous in u and v. Additionally, it satisfies a growth condition that provides the continuability of solutions to any time interval. The following saddle point (Isaacs) condition is assumed to hold:

$$\min_{u \in P} \max_{v \in Q} \langle \ell, f(x, u, v) \rangle = \max_{v \in Q} \min_{u \in P} \langle \ell, f(x, u, v) \rangle, \quad \ell \in R^n, \ x \in R^n. \tag{18}$$

The state vector satisfies the following state constraint

$$G = \{x \in \mathbb{R}^n, \ g(x) \leq 0\}, \tag{19}$$

where g is a Lipschitz continuous function.

Consider for any $v \in Q$ the differential inclusion

$$\dot{x} \in \mathscr{F}_v(x) = \overline{co}\{f(x, P, v)\}. \tag{20}$$

Definition 1 (Viability Property) A set $K \subset \mathbb{R}^n$ is said to be viable if $\forall x_* \in K$, $\forall v \in Q$ there exists a solution $x(\cdot)$ of (20), such that $x(0) = x_*, x(t) \in K, \forall t \geq 0$.

Definition 2 (Discriminating Kernel) For a given compact set $G \subset \mathbb{R}^n$ denote by $discr(G)$ the largest subset of G with the viability property. This subset is called the discriminating kernel of G.

5.1 Grid Method for Finding Discriminating Kernels

Below, the numerical procedure for finding discriminating kernels of the differential game (16)–(19) is briefly described (see [13] and [11] for details).

Consider a family of state constraints

$$G_\lambda = \{x \in \mathbb{R}^n, \ g(x) \leq \lambda\}.$$

It is required to construct a function V, such that

$$discr(G_\lambda) = \{x \in \mathbb{R}^n, \ V(x) \leq \lambda\}.$$

Let $\delta > 0$ be a time step and $h := (h_1, \ldots, h_n)$ space sampling with $|h| = \max_{1 \leq i \leq n} h_i$. For any continuous function $\mathscr{V} : R^n \to R$, introduce the following upwind operator:

$$\Pi(\mathscr{V}; \delta, h)(x) = \mathscr{V}(x) + \delta \min_{u \in P} \max_{v \in Q} \sum_{i=1}^{n} (p_i^{right} f_i^+ + p_i^{left} f_i^-), \tag{21}$$

where f_i is the ith component of f and

$$a^+ = \max\{a, 0\}, \quad a^- = \min\{a, 0\},$$

$$p_i^{right} = [\mathscr{V}(x_1, \ldots, x_i + h_i, \ldots, x_n) - \mathscr{V}(x_1, \ldots, x_i, \ldots, x_n)]/h_i, \tag{22}$$

$$p_i^{left} = [\mathscr{V}(x_1, \ldots, x_i, \ldots, x_n) - \mathscr{V}(x_1, \ldots, x_i - h_i, \ldots, x_n)]/h_i.$$

Note that the operator Π can be applied to grid functions considered on the grid defined by the space sampling h to return grid functions of the same art.

Denote $\mathcal{V}^h(x_{i_1},\ldots,x_{i_n}) = \mathcal{V}(i_1h_1,\ldots,i_nh_n)$ and $g^h(x_{i_1},\ldots,x_{i_n}) = g(i_1h_1,\ldots,i_nh_n)$ being the restrictions of $\mathcal{V}$ and g to the grid. Let a sequence $\{\delta_\ell\}$ be chosen, such that $\delta_\ell \to 0$ and $\sum_{\ell=0}^{\infty}\delta_\ell = \infty$.

The numerical scheme reads as follows:

$$\mathcal{V}_{\ell+1}^h = \max\{\Pi(\mathcal{V}_\ell^h;\delta_\ell,h),g^h\}, \quad \mathcal{V}_0^h = g^h, \quad \ell = 0,1,\ldots.$$

Here, $\mathcal{V}_\ell^h$ is the current grid approximation function.

Proposition 1 (see [12] for the proof) *Let $|\mathcal{F}_v(x)| \leq B$ for all $x \in \mathbb{R}^n$ and $v \in Q$. If $\delta_\ell/|h| \leq 1/(B\sqrt{n})$ for all ℓ, then $\mathcal{V}_\ell^h \nearrow V^h$ as $\ell \to \infty$.*

The estimate $|\mathcal{V}^h - V| \leq C\sqrt{|h|}$ is expected (cf. [10, 11]), and therefore $\mathcal{V}_\ell^h$ approximates V if ℓ is large and $|h|$ is small. The stopping criterion in the computation is $|\mathcal{V}_{\ell+1}^h - \mathcal{V}_\ell^h|_{L_\infty} \leq \varepsilon$.

5.2   Control Design

The control ensuring that the trajectories of system (12)–(15) remain in the discriminating kernel is designed in the following way. Let ℓ be sufficiently large so that the desired approximation is reached, i.e., $\left|\mathcal{V}_{\ell+1}^h - \mathcal{V}_\ell^h\right|_{L_\infty} \leq \varepsilon = 10^{-6}$.

Let x be a current state of the game. The control $u(x)$ can be taken as any minimizer in the following expression:

$$\min_{u\in P}\max_{v\in Q} \mathcal{L}^h\left[\mathcal{V}_\ell^h\right](x + \tau f(x,u,v)), \tag{23}$$

which means that $u(x)$ strives to shift the state of the control system to decrease the function $\mathcal{L}^h\left[\mathcal{V}_\ell^h\right]$. Here, $\mathcal{L}^h$ is an interpolation operator (see, e.g., [9]) defined on the above-considered grid functions, and τ is a parameter which should be larger than the time step size of the control scheme to provide some stabilization. Note that the control design based on interpolation of grid approximations of the value function is discussed in [20].

In addition to the approximate optimal feedback control found from (23), the following heuristic feedback control will be used in the simulations:

$$u = -2[q(p_a, \alpha\sqrt{V_a}, pCO_2) - q_0]. \tag{24}$$

Here, the control is proportional to the deviation of the *CBF* from its nominal value, which can be interpreted as some therapy strategy stabilizing the *CBF* around its nominal value.

Below, we will see that the control found from (23) is not acceptable because it produces chattering regimes, whereas the control (24) yields smooth time realizations and also works good against all admissible disturbances.

Remark 2 It should be noted that the saddle point condition (18) holds for the differential game (12), which implies the following observations:

- The operations min and max can be swapped in the operator (21), see [13].
- The operations min and max can also be swapped in formula (23) to obtain a feedback counter control $u(x, v)$. Nevertheless, this additional advantage of the first player cannot improve his result; see [23], Section 10.5.2, pp. 470–471.

6 Simulation Results

In Figure 3, two computed discriminating kernels are shown. The left (resp. right) column corresponds to the variant where blood is modeled as Newtonian (resp., micropolar) fluid. The upper row shows discriminating kernels in the space of original state variables. In the middle row, the compliance is replaced by the vascular volume and by the CBF in the lower row. For all representations, an essential difference between the discriminating kernels is observed depending on the blood model. Therefore, accounting for micropolar properties of blood is important.

Concerning the simulation of trajectories, responses of the control system to some open-loop step-shaped disturbances v_1 and v_2 shown in Figure 4a and 4b are tested. For all simulations, blood is modeled as micropolar fluid. Figure 4c presents the time realization of u constructed using the expression (23). The horizontal axis measures time in minutes. In Figure 5, the corresponding trajectories (the compliance, the partial CO_2 and O_2 pressures) and the time realization of CBF are plotted. One can see that the state constraints are kept, but, as usually in differential games, the performances of feedback controls show chattering regimes, which are impossible to implement.

Consider now the heuristic feedback control given by formula (24). The same step functions v_1 and v_2 are used as disturbances. The corresponding trajectories are shown in Figure 6a-d. One can see that the trajectories substantially lie within the state constraints, and the control strategy is implementable (see Figure 7a). Note that this control maintains an approximately constant CBF (cf. Figure 6d).

To prove that the heuristic control can properly work against all admissible disturbances, we put this control into the model and compute the discriminating kernel for the resulting system. In this case, the grid algorithm contains only maximizations over the disturbances. If such a discriminating kernel is nonempty, the heuristic control is classified as acceptable. Figure 7b shows the resulting, nonempty, discriminating kernel, and therefore the heuristic control can successfully work against any admissible disturbances.

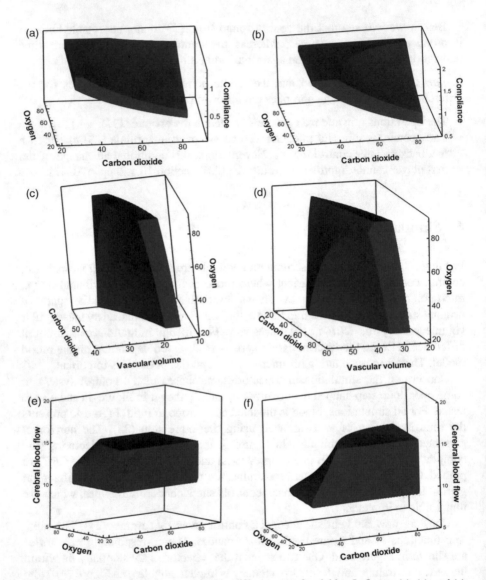

Fig. 3 Discriminating kernels represented by different sets of variables. In figures (a), (c), and (e), blood is modeled as Newtonian fluid; in figures (b), (d), and (f), blood is modeled as micropolar fluid

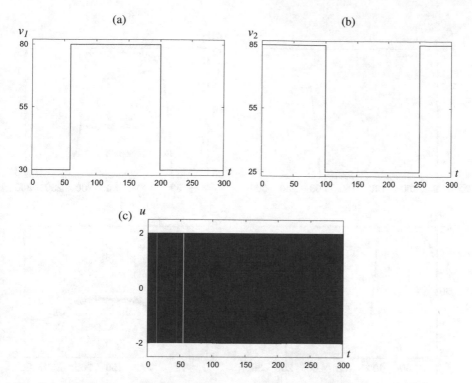

Fig. 4 Approximate optimal feedback control found from (23) plays against step-shaped disturbances: (a) disturbance input v_1, (b) disturbance input v_2, (c) time realization of the approximate optimal feedback control

The computation of the discriminating sets was performed on a Linux SMP computer with 8xQuad-Core AMD Opteron processors (Model 8384, 2.7 GHz) and shared 64 GB memory. The programming language C with OpenMP (Open Multi-Processing) support was used. The efficiency of the parallelization is up to 80%. The box $[0.1, 2] \times [20, 90] \times [20, 90]$ of the state space (C_a, pCO_2, pO_2) was divided in $100 \times 140 \times 120$ grid cells.

The time-step width δ was equal to 0.0005, and the accuracy ε was equal to 10^{-6}. The computation time for the discriminating kernels was about 15 min.

7 Conclusion

A mathematical model of cerebral circulation based on experimental data is proposed. The model is handled using the viability approach, which assumes finding the sets where trajectories of the control system can be confined for infinitely long time. The model can be used to evaluate different therapy strategies: A therapy

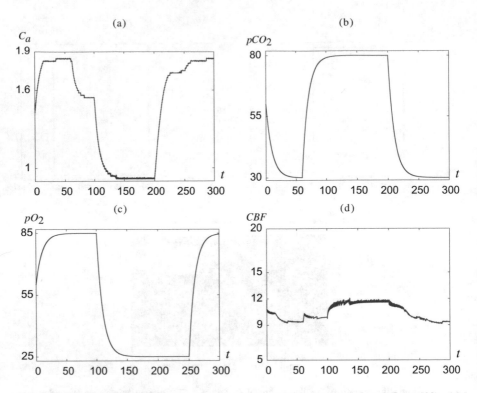

Fig. 5 Trajectories corresponding to the approximate optimal control found from (23): (a) compliance, (b) CO_2 partial pressure, (c) O_2 partial pressure, (d) CBF

strategy, formulated as feedback control, can be plugged into the model to compute the discriminating kernel. If the last is nonempty, the strategy may be considered as acceptable because it ensures the admissible range for the CBF. It should be noted that the model has to be enhanced in many instances by accounting more regulation mechanisms and experimental data. The next step toward the improvement of the model concerns the replacement of the simplified cerebrovascular bed by more realistic networks accounting for the germinal matrix.

Authors' contribution. Renée Lampe, Nikolai Botkin, and Varvara Turova equally contributed to this work.

Acknowledgements The authors are thankful to Christine Klindt-Schuster (clinic "Rechts der Isar," Technical University of Munich) for her help with collecting the experimental data. Financial supports of the Klaus Tschira Foundation, Würth Foundation, and Buhl-Strohmaier-Foundation are gratefully acknowledged. Computer resources have been provided by the Gauss Centre for Supercomputing/Leibniz Supercomputing Centre under grant: pr74lu.

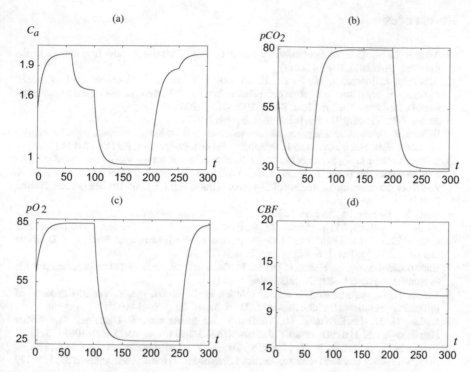

Fig. 6 Trajectories corresponding to the heuristic feedback control (24): (a) compliance, (b) CO_2 partial pressure, (c) O_2 partial pressure, (d) CBF

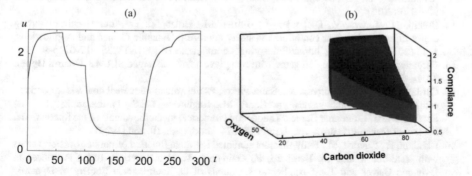

Fig. 7 Simulation results for (a) time realization of the heuristic feedback control (24), (b) the discriminating kernel corresponding to the heuristic feedback control

References

1. Adcock, L. M.: Clinical manifestations and diagnosis of intraventricular hemorrhage in the newborn. UpToDate, Dec 14 (2015)
2. Alastruey, J., Moore, S. M., Parker, K. H., David, T., Peiró, J., Scherwin, S. J.: Reduced modeling of blood flow in cerebral circulation: Coupling 1-D, 0-D and cerebral autoregulation models. Int J Num Meth in Fluids 56(8): 1061–1067 (2008)
3. Aubin, J.-P.: Viability Theory. Birkhäuser, Boston (1991)
4. Ballabh, P., Braun, A., Nedergaard, M.: Anatomic analysis of blood vessels in germinal matrix, cerebral cortex, and white matter in developing infants. Pediatr Res. 56(1), 117–124 (2004)
5. Bardi, M., Soravia, P., Falcone, M.: Fully discrete schemes for the value function of pursuit-evasion games. In: Başar, T., Hauri, A. (eds) Advances in Dynamic Games and Applications. Annals of the International Society of Dynamic Games, vol. 1, pp. 89–105. Birkhäuser, Boston (1994)
6. Bardi, M., Falcone, M., Soravia, P.: Numerical methods for pursuit-evasion games via viscosity solutions. In: Bardi, M., Raghavan, T.E.S., Parthasarathy, T. (eds) Stochastic and Differential Games: Theory and Numerical Methods. Annals of the International Society of Dynamic Games, vol. 4, pp. 105–176. Birkhäuser, Boston (1999)
7. Battisti-Charbonney, A., Fisher, J., Duffin, J.: The cerebrovascular response to carbon dioxide in humans. J. Physiol. 589(12): 3039–3049 (2011)
8. Botkin, N. D.: Asymptotic behavior of solution in differential games. Viability domains of differential inclusions. Russian Acad. Sci. Dokl. Math. 46(1), 8–11 (1993)
9. Botkin, N. D., Hoffmann, K.-H., Turova, V. L.: Stable numerical schemes for solving Hamilton–Jacobi–Bellman–Isaacs equations. SIAM J. Sci. Comp. 33(2), 992–1007 (2011)
10. Botkin, N. D., Hoffmann, K.-H., Mayer, N., Turova, V. L.: Approximation schemes for solving disturbed control problems with non-terminal time and state constraints. Analysis 31, 355–379 (2011)
11. Botkin, N. D., Turova, V. L.: Numerical construction of viable sets for autonomous conflict control systems. Mathematics 2, 68–82 (2014)
12. Botkin, N. D., Turova, V. L.: Examples of computed viability kernels. Proceedings of the Institute of Mathematics and Mechanics (Trudy Instituta Matematiki i Mekhaniki) 21(2), 306–319 (2015)
13. Botkin, N., Turova, V., Diepolder, J., Bittner, M., Holzapfel, F.: Aircraft control during cruise flight in windshear conditions: viability approach. Dynamic Games and Applications, 1–15 (2017) Open access: https://link.springer.com/article/10.1007/s13235-017-0215-9
14. Cardaliaguet, P.: A differential game with two players and one target. SIAM J. Control Optim. 34, 1441–1460 (1996)
15. Cardaliaguet, P., Quincampoix, M., Saint-Pierre, P.: Set valued numerical analysis for optimal control and differential games. In: Bardi, M., Raghavan, T.E.S., Parthasarathy, T. (eds) Stochastic and Differential Games: Theory and Numerical methods. Annals of the International Society of Dynamic Games, vol. 4, pp. 177–274. Birkhäuser, Boston (1999)
16. Cristiani, E., Falcone, M.: Fully-discrete schemes for value function of pursuit-evasion games with state constraints. In: Bernhard, P., Gaitsgory, V., Pourtallier, O. (eds.) Advances in Dynamic Games and Their Applications, Annals of the International Society of Dynamic Games, vol. 10, pp. 177–206. Birkhäuser, Boston (2009)
17. David, T., Brown, M., Ferrandez, A.: Auto-regulation and blood flow in cerebral circulation. Int. J. Num. Meth. Fluids 43, 701–713 (2003)
18. Erdoğan, M. E.: Polar effects in apparent viscosity of a suspension. Rheol. Acta 9, 434–438 (1970)
19. Falcone, M.: Numerical methods for differential games via PDEs. International Game Theory Review 8(2), 231–272 (2006)
20. Falcone, M., Ferretti, R.: Semi-Lagrangian Approximation Schemes for Linear and Hamilton-Jacobi Equations. Society for Industrial and Applied Mathematics (2013)

21. Ferrandez, A., David, T., Brown, M. D.: Numerical models of auto-regulation and blood flow in the cerebral circulation. Comput. Methods Biomech. Biomed. Engin. **5**(1): 7–19 (2002)
22. Krasovskii, N. N., Subbotin, A. I.: Positional Differential Games. Nauka, Moscow (1974) (in Russian)
23. Krasovskii, N. N., Subbotin, A. I.: Game-Theoretical Control Problems. Springer, New York (1988)
24. Lampe, R., Botkin, N., Turova, V., Blumenstein, T., Alves-Pinto, A.: Mathematical modelling of cerebral blood circulation and cerebral autoregulation: Towards preventing intracranial hemorrhages in preterm newborns. Computational and Mathematical Methods in Medicine, 1–9 (2014)
25. Lou, H. C., Lassen, N. A., Friis-Hansen, B.: Impaired autoregulation of cerebral blood flow in the distressed newborn infant. J. Pediatr. **94**, 118–121 (1979)
26. Lukaszewicz, G: Micropolar Fluids: Theory and Applications. Birkhäuser, Boston (1999)
27. Mitchell, I. M.: Application of level set methods to control and reachability problems and hybrid systems. PhD Thesis, Stanford University (2002)
28. Mitchell, I. M.: A summary of recent progress on efficient parametric approximations of viability and discriminating kernels. EPiC Ser. Comput. Sci. **37**, 23–31 (2015)
29. Olufsen, M. S., Nadim, A., Lipsitz, L. A. Dynamics of cerebral blood flow regulation explained using a lumped parameter model. Am J Physiol Regul Integr Comp Physiol **282**(2), R611–R622 (2002)
30. Piechnik, S. K., Chiarelli, P. A., Jezzard, P.: Modelling vascular reactivity to investigate the basis of the relationship between cerebral blood volume and flow under CO_2 manipulation. NeuroImage **39**, 107–118 (2008)
31. Ursino, M., Lodi, C. A.: A simple mathematical model of the interaction between intracranial pressure and cerebral hemodynamics. J. Appl. Physiol. **82**(4), 1256–1269 (1997)
32. van de Bor, M., Walther, F. J.: Cerebral blood flow velocity regulation in preterm infants. Biol. Neonate **59**, 329–335 (1991)

Printed in the United States
By Bookmasters